Data-Driven iOS Apps for iPad™ and iPhone®

with FileMaker Pro®, Bento® by FileMaker and FileMaker Go

Jesse Feiler

800 East 96th Street, Indianapolis, Indiana 46240 USA

Data-Driven iOS Apps for iPad and iPhone with FileMaker Pro, Bento by FileMaker, and FileMaker Go

ISBN-13: 978-0-7897-4786-0

ISBN-10: 0-7897-4786-3

Library of Congress Cataloging-in-Publication data is on file.

Printed in the United States of America

First Printing: April 2010

Trademarks

Warning and Disclaimer

Bulk Sales

Que Publishing offers excellent discounts on this book when ordered in quantity for bulk purchases or special sales. For more information, please contact

U.S. Corporate and Government Sales

1-800-382-3419

corpsales@pearsontechgroup.com

For sales outside of the U.S., please contact

International Sales

international@pearson.com

Editor in Chief
Greg Wiegand

Acquisitions Editor
Loretta Yates

Development Editor
The Wordsmithery, LLC

Managing Editor
Sandra Schroeder

Project Editor
Seth Kerney

Copy Editor
Sarah Kearns

Indexer
Ken Johnson

Proofreader
Dan Knott

Technical Editor
Charles Edwards

Publishing Coordinator
Cindy Teeters

Book Designer
Anne Jones

Compositor
Bronkella Publishing, Inc.

Contents at a Glance

Contents

About the Author

Jesse Feiler is a developer, web designer, trainer, and author. He has worked with mobile devices starting with Apple's Newton and continuing with the iOS products (iPhone, iPod touch, and iPad).

His books include *Using FileMaker Bento* (Sams/Pearson), *Sams Teach Yourself Drupal in 24 Hours* (Sams/Pearson), *Get Rich with Apps! Your Guide to Reaching More Customers and Making Money NOW* (McGraw-Hill), *Database-Driven Web Sites* (Harcourt), *How to Do Everything with Web 2.0 Mashups* (McGraw-Hill), *iWork '09 For Dummies* (Wiley), *The Bento Book* (Sams/Pearson), and *FileMaker Pro in Depth* (Sams/Pearson).

He has developed software on a range of platforms ranging from Mac OS X, iOS, Windows, and mainframe computers. His clients represent a variety of organizations, including nonprofits and governments, production management, publishing, and banking.

He is the author of MinutesMachine, the meeting management software for iPad. There are more details at champlainarts.com.

A native of Washington, DC, he has lived in New York City and currently lives in Plattsburgh, NY.

He can be reached at northcountryconsulting.com.

Acknowledgments

Thanks go most of all to the people at Apple and FileMaker, who have built an incredibly elegant and powerful platform for databases and mobile devices. If you go back to the very roots of what is now Mac OS X, you will find the key concepts that make the technologies in this book possible. For example, networking data in what is now called "the cloud" was part of the very first versions of the operating system. The tremendous work at Apple (and, to be fair, at other places) to let mobile devices discover one another has been integral to the mobile world we have today. With that focus, developers, engineers, and end users were able to make the most of mobile technologies as they have matured.

In all seriousness, there's another thank you. There have been ups and downs in the world of technology, and the subset of it that is Apple. To all the people who stayed with it and tried to get out of the occasional bad times through innovation and yet more creativity, thanks for helping to make today's environment possible.

At FileMaker, Kevin Mallon, Senior Manager of Public Relations, once again provided valuable and much-welcomed help and support, along with Terry Barwegen, Ryan Griggs, and Rick Kalman. The engineers and designers who have developed FileMaker and Bento products have helped millions (literally) of people around the world organize and use their data in powerful and innovative ways, and they also deserve thanks from us all. Delfina Daves and the staff at FileMaker Developer Relations help share the knowledge through FileMaker Business Alliance (http://www.filemaker.com/fba/) and FileMaker Technical Network (http://www.filemaker.com/technet/). And in the Bento Template Exchange (http://solutions.filemaker.com/database-templates/), more than 600 templates have been contributed to the Bento community; as of this writing, there have been more than 650,000 downloads of these freely shared shared database tools.

At Pearson, Loretta Yates, acquisitions editor, has taken a concept and moved it from an idea through the adventures along the way to printed books. She is always a pleasure to work with. Seth Kerney, the project editor, has helped to make the complex project involving four products on multiple platforms run smoothly. Thanks are also due to Anne Jones for the cover and interior layout (along with Tricia Bronkella), as well as copy editor Sarah Kearns.

As always, Carole Jelen at Waterside Productions has provided help and guidance in bringing this book to fruition.

We Want to Hear from You!

As the reader of this book, *you* are our most important critic and commentator. We value your opinion and want to know what we're doing right, what we could do better, what areas you'd like to see us publish in, and any other words of wisdom you're willing to pass our way.

As an associate publisher for Que Publishing, I welcome your comments. You can email or write me directly to let me know what you did or didn't like about this book—as well as what we can do to make our books better.

Please note that I cannot help you with technical problems related to the topic of this book. We do have a User Services group, however, where I will forward specific technical questions related to the book.

When you write, please be sure to include this book's title and author as well as your name, email address, and phone number. I will carefully review your comments and share them with the author and editors who worked on the book.

Email: feedback@quepublishing.com

Mail: Greg Wiegand
Editor in Chief
Que Publishing
800 East 96th Street
Indianapolis, IN 46240 USA

Reader Services

Visit our website and register this book at quepublishing.com/register for convenient access to any updates, downloads, or errata that might be available for this book.

Introduction

Welcome to FileMaker on the Move

The rise of mobile devices such as iPhone and iPad during the first decade of the twenty-first century has been as dramatic an event in the technology world as the introductions of personal computers and the World Wide Web. In fact, some (including the author) would argue that mobile computing is a much bigger innovation because it has quickly and dramatically changed the ways in which people use technology. Personal computers were smaller and more flexible than mainframe computers, and the Web is a fast and powerful way to present and consume organized data, but what you can do with a FileMaker database in your hand wherever you go is not just smaller, more flexible, or faster; it is different.

The significance of the mobile revolution was immediately apparent to some people and organizations (yes, this refers to the people at Apple among others), but it was not quite so apparent to many others. When the first ads for iPad were shown, one commentator pointed out that people using iPads were sitting with their feet up on sofas or footstools. They were not using iPads while sitting at a desk in an office, and the commentator reasoned that meant that iPads were not for business use.

Some people objected to that line of reasoning. Perhaps it is desks and offices that are no longer relevant to business. And within the first year, iPads became Apple's long-desired entree to the world of corporate enterprises.

Through its wholly owned FileMaker subsidiary, Apple is uniquely positioned to merge the worlds of databases and mobile computing. This book introduces you to the FileMaker products that are designed for mobile devices; it also shows you how to tweak existing FileMaker databases and solutions to make them easier to use on mobile devices.

Who Should Read This Book

This book is for anyone who wants to develop or use mobile data on iPhone, iPod touch, or iPad using any of FileMaker's tools: FileMaker Pro, FileMaker Server, FileMaker Go, and Bento by FileMaker. If you are not familiar with FileMaker or Bento, you might want to refer to the author's book *FileMaker Pro in Depth* or the extensive documentation that comes with those products. The focus here is just on mobile data and the issues you need to understand to use it.

If you are not certain that you want to put your data on a mobile device, this book might help you make up your mind. Although there are a few decisions you have to make carefully, the benefits of having your data with you (literally in your hand or your pocket) outweigh a little bit of work in getting things set up.

Downloading the Example Files

Example files can be downloaded from the author's website at northcountryconsulting.com or from the publisher's site at quepublishing.com.

How This Book Is Organized

There are four parts to this book. You can focus on whichever one addresses an immediate problem, or you can get a good overview by reading straight through.

Part I: Data to Go

This part introduces the basic issues of the book and shows you principles and techniques that apply to all the products discussed:

- **Chapter 1, "Making Data Mobile"**—This chapter introduces you to the products described in this book. You also take a close look at the interfaces in order to see how they differ from desktop and laptop computers to mobile devices. These observations help you design your own mobile layouts later on in the book.
- **Chapter 2, "Introducing the FileMaker Architecture"**—FileMaker Go lets you access FileMaker databases from mobile devices, so you need to know what those databases look like and what they can do. You also use this knowledge in Part IV when you use web publishing to put FileMaker databases on the web and from there onto mobile devices.
- **Chapter 3, "Managing Data on the Move"**—When your data is moving around, how do you manage it? This chapter shows you the principles and techniques involved in sharing, copying, and synchronizing data so that wherever it is, it is correct and timely.
- **Chapter 4, "Working with Mobile Devices"**—A lot of what you have learned about interface design does not apply to mobile devices (after all, there are no menus and no resizable windows). This chapter introduces you to the issues you need to think about as you prepare data and layouts for the mobile world.

- **Chapter 5, "Preparing FileMaker for Mobile Use"**—Here are specific issues and techniques to think about and use as you create and modify FileMaker databases for mobile use and for the web.
- **Chapter 6: "Introducing FileMaker Server"**—FileMaker Server enables you to share FileMaker databases among large groups of people. It supports access from FileMaker Pro and FileMaker Go, and also provides tools for publishing databases on the web. When you publish a database on the web, that means it is accessible to any web browser; that, in turn, means that you have a way to put your databases on mobile devices that do not run FileMaker Go but have a browser built into them.

Part II: FileMaker Go

FileMaker Go runs on iPhone, iPod touch, and iPad (there are separate versions for iPhone/iPod touch and for iPad). It lets you access FileMaker databases that are shared over a network or that are located on the mobile devices. Here are the details:

- **Chapter 7, "Introducing FileMaker Go"**—This is a hands-on guide to what FileMaker Go can do and how you can use it. It includes the new features for FileMaker Go 1.2 such as signature capture.
- **Chapter 8, "Optimizing FileMaker Databases for FileMaker Go"**—There are a few features of FileMaker that work differently in FileMaker Go. This chapter explores the differences and shows you how to work around issues that might arise. In addition, you see how to take advantage of some of the differences to do things that you could not do on FileMaker.
- **Chapter 9, "Designing a FileMaker Go Solution"**—This chapter provides a step-by-step guide to building a FileMaker Go solution for iPhone and a variation for iPad.
- **Chapter 10: "Using Printing and Charting with FileMaker Go"**—FileMaker has long had sophisticated printing features that let you move and resize fields during the print process so as to close up blank space and do other ad hoc formatting. With FileMaker Pro 11, charting functionality was added to FileMaker. As a result, FileMaker Pro and FileMaker Go can not only store and manipulate data, but they can also present it attractively and effectively on screens of various sizes, in print, and in graphical formats. This chapter shows you how to make the most of these features for FileMaker Go.

Part III: Bento by FileMaker

Bento is the personal database from FileMaker. It is designed for individuals and small workgroups. It is simple to deploy and use, which makes it perfect for many mobile projects:

- **Chapter 11, "Using Bento and Bento Libraries"**—This chapter provides a detailed look at Bento on Mac OS X to get you up to speed if you have never used it before. If you are an experienced Bento user, you will probably find new tips and techniques to make you even more productive.
- **Chapter 12, " Using Bento Records, Fields, Forms, and Tables"**—Inside your Bento libraries, records and fields store your data just as they do with FileMaker. Instead of FileMaker layouts, Bento gives you forms and tables for the presentation and manipulation of data. To make life even easier, it automatically creates these for you, although you can customize them if you like.

- **Chapter 13, "Working with Location and Media Fields"**—Bento takes advantage of the location services on mobile devices so that you can automatically store geographic data in your Bento libraries and manipulate it. Those features as well as the features for media fields that incorporate the camera are the topic of this chapter.
- **Chapter 14, "Importing and Exporting Bento and FileMaker Data"**—This chapter helps you move your data back and forth between FileMaker and Bento. These tools also let you move data back and forth between Bento and other application programs.

Part IV: FileMaker Web Publishing: Instant Web Publishing (IWP) and Custom Web Publishing (CWP)

Because mobile devices have web browsers, moving your FileMaker database to the web enables it to be used on mobile devices. Here are some special tips and issues to consider:

- **Chapter 15, "Deploying FileMaker/IWP with FileMaker Server Advanced"**—Here are the settings required to use IWP to share your database. The point of IWP is to develop a web-based solution that looks as much as possible like your desktop version. On a mobile device you might want to re-think the look and feel, particularly if iPhone is your target.
- **Chapter 16, "Deploying FileMaker/CWP with FileMaker Server"**—FileMaker supports a PHP interface, so if you want to write code for your FileMaker-driven website, here is how you do it. The chapter also explores the built-in iPhone template to make mobile development even easier.

Appendix

Many people find it convenient to pair a wireless keyboard with a mobile device when data entry is involved. This section gives you the details on how to do that.

- **Appendix A, "Connecting a Wireless Keyboard to a Mobile Device"**—If you are doing a lot of typing, a wireless keyboard can be a valuable accessory to an iPad or an iPhone. This appendix describes how to connect and use one. Even more important, it includes how to switch a wireless keyboard between a desktop or laptop Mac and your mobile device.

IN THIS CHAPTER

Making Data Mobile

How many weeks worth of music do you have on your iPod? How many thousands—scores of thousands—of pages of books do you have on your iPad? And do you have enough video on your iPhone to get you across an ocean...and back?

When information is digitized, it now can be stored in unimaginable quantities on small and even tiny devices. Also, thanks to that digitization, high-speed communications, and powerful compression and decompression algorithms, that information can speed across continents in the blink of an eye or two.

That means that the way in which we deal with data—particularly large amounts of data—has changed. That goes for all forms of data, including books, music, and even the most basic types of data, such as spreadsheets and databases. Not very many years ago, people marveled at the possibility that they could carry around the digitized contents of an encyclopedia in a small portable device. Today, no one thinks twice about carrying around the equivalent of a public library in a combination of memory chips and always-on Internet access.

Not only do we carry data around with us on mobile devices, but we also collect data as we move around. Adding photos to your databases from the albums on your computer has been possible for a long time, but now you can take a photo and insert it into a database without leaving your FileMaker software. You can capture your current location and store it in your database without leaving Bento. And in addition to its many other features, FileMaker Go now has signature capture built in. You can take a photo of a job site, type in an estimate of the work to be done, and have a client sign off on it right then and there. Send the estimate from your mobile device using email or use FileMaker's sophisticated print formatting tools to print right from your mobile device over a local area network.

This book helps you to understand the issues involved in using mobile data and shows you how to get started with (and, in fact, do much more than just start) putting your data on mobile devices using FileMaker's suite of tools (FileMaker Go, Bento, and FileMaker Server) for mobile devices and specifically for Apple's iOS devices (iPhone, iPod touch, and iPad).

Introducing the FileMaker Products for Mobile Computing

FileMaker is a wholly owned subsidiary of Apple. Its products are centered around FileMaker. As you see in Chapter 2, "Introducing the FileMaker Architecture," FileMaker combines a desktop and a powerful yet easy-to-use development environment so that it's not hard to create interfaces that make the databases easy to use for specific purposes. Networking is built into the product line from top to bottom. The next sections provide a summary of the FileMaker products.

FileMaker

On the desktop, two products enable you to access and create databases, as well as to develop the interfaces other people (and you) can use:

- **FileMaker Pro** is the product most commonly used to access and create databases as well as to develop the interfaces.
- **FileMaker Pro Advanced** includes all the FileMaker Pro functionality but adds some development tools for advanced developers.

Both of these products allow up to nine other users to access shared FileMaker databases over a network using their own copies of FileMaker Pro or FileMaker Pro Advanced. For larger deployments, there are two other products, as follows:

- **FileMaker Server** can support up to 250 users of FileMaker Pro or FileMaker Pro Advanced. It provides tools for the processes needed to support networked databases—even for just a handful of users. (Automated backups and scheduling of scripts are just two examples of these advanced features.) Custom Web Publishing such as the PHP Site Assistant is included. Learn more about FileMaker Server in Chapter 16, "Deploying FileMaker/CWP with FileMaker Server."
- **FileMaker Server Advanced** adds Instant Web Publishing (discussed in Chapter 15, "Deploying FileMaker/IWP with FileMaker Server Advanced"), as well as the ability to integrate FileMaker databases with external ODBC/JDBC databases. In addition, there is no hard and fast limit to the number of FileMaker Pro users that can be supported. The limit depends on your hardware configuration.

USING FILEMAKER RUNTIME VERSIONS

You can use FileMaker Pro Advanced to save a *runtime* version of your database. The runtime version includes code from FileMaker that enables people to run it on a Mac or a Windows computer without needing a copy of FileMaker Pro. Runtime versions cannot be networked; they are strictly single-user versions. Also, runtime versions allow manipulation of data but not of the database structure itself.

FileMaker Go

If you have a FileMaker database running on either a desktop version or one of the server versions, you can access it from a mobile device with FileMaker Go. There are actually two FileMaker Go apps: one for iPhone and one for iPad. They let you access FileMaker databases, but you cannot create or modify the database structure or the layouts with FileMaker Go. You can certainly modify the data contained in the database, but you cannot, for example, add new fields to it.

This means that somewhere along the line, the database you access with FileMaker Go has to have been created (and possibly modified) with a copy of FileMaker Pro or FileMaker Pro Advanced.

Bento

Bento is a personal database that runs on Mac OS X. You can share your Bento data with other Bento users on your local area network (up to five in all). You can access Bento from a pair of apps: one for iPhone and one for iPad.

What Mobile Data Means

People expect access to data no matter where they are, and for the most part, they can have that access. Yes, there are many areas where Internet access remains problematic, but many of those areas are gaining access (and high-speed access at that) rapidly. In addition, the vast majority of the world's information is not accessible in a digitized and portable manner, despite the efforts of individuals and groups to digitize long out-of-print books and to move public and private records to digitized data stores. Although the end is nowhere in sight, more and more of the mound of undigitized information is making its way into digitized databases.

The process of bringing high-speed communications to homes and offices has been lengthy, and it serves as a model for the process of bringing digitized information to entire populations. When it comes to high-speed Internet, high-capacity backbones of the Internet have been created using fiber optic cables. This is an expensive process, but it has been proceeding rapidly over the last years.

As a result, people expect their iPhones to provide them with phone and Internet connectivity almost everywhere (yes, there are problems, but they are being addressed rapidly). Users can read books and play movies and music on iPhones, iPods, iPod touches, and now on iPads.

What is not keeping pace is access to interactive databases. The issues involved in moving the movie *Gone with the Wind* to your iPod in a few seconds were solved a long time ago; you can get the latest headline news and interviews over the phone network or a Wi-Fi network on your iPad. But what about the sales and orders for your business? What about the reservations for dinner at your restaurant? What about the enrollments in that baking class you're interested in; is there still room for you to register and learn how to make Tarte aux Pommes a la Internet?

Interactive access to data and databases is the next frontier, the information industry's equivalent of the communication industry's challenge of "the last mile" in providing high-speed Internet. This interactive access can be the back-and-forth of data retrieval and updates as orders are placed or class enrollments are processed; interactive access can also consist of the opening of sophisticated databases to mobile devices so that all kinds of searches and data analyses can be conducted even if no data is changed in the process.

People expect to be able to access information and databases from their mobile devices in exactly the same way that they've been able to access it from desktop computers and, before them, from mainframe computers. They want to be able to view and update the data as they see fit, no matter where they are.

But there's a paradox here. People also want to be able to access information and databases from their mobile devices in exactly the same way that they've been able to access photos, movies, email, the web, games, and video conferencing on their iPhone, iPod touch, or iPad.

That's the topic of this chapter. It explores the issues involved in making data mobile and provides you with some strategies for implementing your choices regarding those issues.

For many people, and certainly for this book, providing access to data on mobile devices means providing that access on iPad and iPhone. (iPod touch implements almost all the iPhone features with the exception of telephony; it is not specifically called out in this book, but references to the non-telephone features of iPhone are intended to include iPod touch.)

If you have used an iPad or iPhone for even a brief period of time, you probably have discovered how easy it is to use. This is a tribute to designers and engineers at Apple who have been working on usability and interfaces for three decades, as well as to those who have been thinking about mobile devices since the days of the Newton at Apple in 1987 and the Knowledge Navigator concept video made by Apple at the same time (you can find it at http://www.digibarn.com/collections/movies/knowledge-navigator.html).

Now that you're thinking about putting your data on a mobile device, you must start looking at the devices very differently. Look past the ease-of-use to understand the interface because you will need to create your own database interface. The first step is to look at the traditional user interface more closely and watch yourself as you use it.

Introducing the Reservations Example

You need a mobile interface to explore, and as you start to develop your own interfaces and put your own data on mobile devices, you'll need a project that is simple but useful. The Reservations example is available for download from the author's website (http://northcountryconsulting.com), as well as from the publisher's site (http://quepublishing.com) as described in the Introduction.

Reservations was chosen as an example of one type of database solution that fits well on both desktop and mobile computers. It implements a scenario that can apply to reservations for a restaurant, as well as for events such as performances, classes, or even hikes, rallies, races, and walking tours. What you need from a reservations system in each of these cases is the ability to enter events, a few simple features to be used when people make a reservation, and finally a few features to be used when people check in for the meal, performance, race, or other event.

This is a basic type of database application, and you could implement it easily in FileMaker Pro on a desktop or laptop running Windows or Mac OS X. You could also implement it in Bento on Mac OS X or on an iOS device. However, this scenario envisions a slightly more complex—and for many cases more practical—scenario that uses a networked database running under FileMaker Server or using FileMaker Pro to share the database with up to nine other computers. That means that one or more people can respond to reservation requests that are made via email, phone, or walk-up transactions.

When people check in for the event, the manager, maitre d', usher, or organizer can check them in with an iPad or iPhone that provides real-time access to the networked database. This scenario plays to the strengths of both desktop and laptop computers, as well as to the strengths of mobile devices. To enter a reservation, a keyboard comes in handy, but to check in an attendee, the form factor of the iPad or iPhone works best. If the event is relatively crowded, networked iPhones or iPads can be used so that instead of people lining up at a check-in desk, the checkers-in can wander through the crowd and check in people wherever they are. Thus, the mobility of the iPhone or iPad comes into play; to check in someone, you just need to find the reservation and enter in the number of attendees. And if you need to verify attendance, the signature capture feature of FileMaker Go lets people sign in on an iOS device.

You see how to set up the FileMaker database in this part of the book. In Part II, you'll see how to build the software for iPhone and iPad using FileMaker Go. In Part III, you'll see how to build the software for iPhone and iPad using FileMaker Bento. And in Part IV, you'll see how to build the software for access over the web. This makes it possible to use any browser-equipped mobile device to update the FileMaker database.

Entering Events

Before processing reservations, people need to be able to enter events. In this example, the minimum information is provided:

- Event name
- Location
- Date
- Time

If you are building on this model, there are some other fields you might want to add, as follows:

- Event end time
- Total number of attendees allowed for the event

Note that this scenario envisions someone making a reservation through a phone call, email, or a walk-up transaction. You could expand this structure to enable people to make their own reservations on the web; if you do so, you'd probably want to provide additional information for them, such as a synopsis of the play, whether a walking tour includes hills or steps, and the like.

In addition to enabling people to create new events, the basic database enables them to delete them. If you flesh out this system, you would want to post a warning if events have attendees with reservations and someone tries to delete the events.

Entering Reservations

After events have been entered into the database, it should be simple to enter a reservation. Only two steps should be required, as follows:

1. Locate the event.
2. Enter the data for the reservation.

It can also be useful to capture the date and time of each entry as well as the name of the user who is entering the data. You can use FileMaker's auto-enter feature to have this data entered without the user having to do anything at all.

In this basic implementation, the running total of reservations for each event is displayed. If you enhance it, you can provide warnings when the event is near capacity.

Processing Reservations

Checking in attendees is simple, particularly with a mobile device. Even if a laptop is used at a formal check-in station, it's just a matter of locating the event and typing in the number of attendees. In fact, it's usually much simpler. When you're checking in attendees, it's usually for the current event, so you locate it at the beginning and then process all the attendees for that event. In fact, all you do is find the attendee and type in the number of people checking in.

Here's what each interface looks like. You'll find more details and how-to steps throughout the book.

Looking at Reservations on FileMaker Pro

Figure 1.1 shows FileMaker Pro running on Mac OS X (it appears almost identical on Windows).

The figures in this book that show screenshots are not to scale, so the relative sizes of the iPhone and iPad screens are not so noticeable as they are in real life. With Mac OS X, screen sizes and resolutions vary from computer to computer and monitor to monitor. The figures in the book have been resized for legibility and compatibility with the book's production process.

What you are looking at in Figure 1.1 is an example of a powerful and user-friendly interface of the type that has become so familiar, you might not notice many of the features. Here's a critical and descriptive look.

The window shows a record from a FileMaker database (there's more on what's underneath in Chapter 2). At the top of the window, a *toolbar* provides one-click access to buttons that perform functions such as finding and searching. At the left of the toolbar, controls let you step through records in the database.

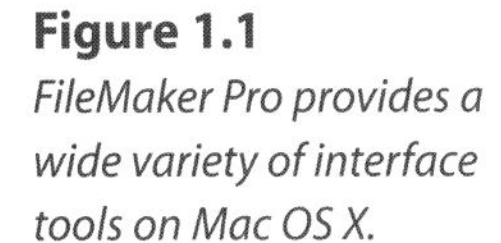

Figure 1.1
FileMaker Pro provides a wide variety of interface tools on Mac OS X.

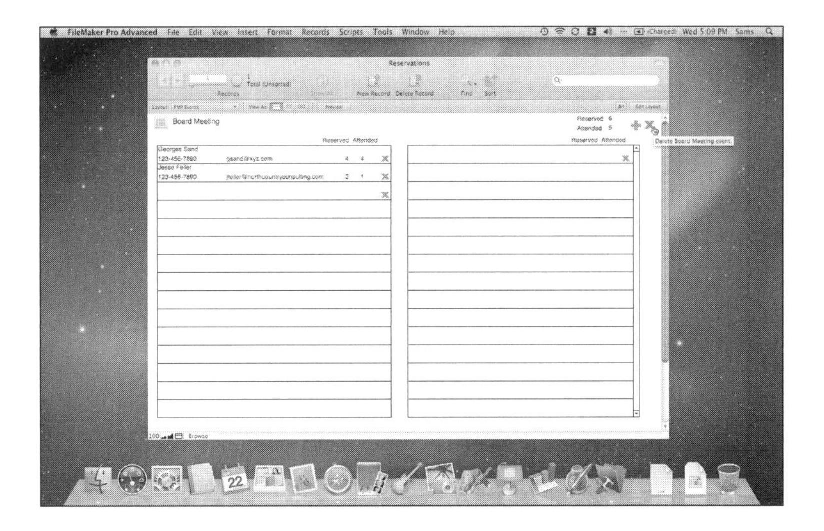

The body of the window below the toolbar is a *layout* that you construct with FileMaker Pro. You design the interface, adding buttons (such as the add and delete buttons—the + and X), text, and the contents of data fields.

When you start to edit (that is, when you click or tab into the first field), dotted outlines appear around all the editable fields, as you see in Figure 1.1. Some fields are not editable, such as the calculated fields for the number of reservations and the number of people who have showed up. These fields are updated automatically based on the data that is entered.

The main part of the window is taken up by a *portal*. This scrolling list shows all the attendees for this event: the *related records*. This FileMaker database has been set up to allow you to enter a new attendee just by filling in the blanks at the bottom of the list; note the dotted lines around blank fields in the new record at the bottom of the list. (FileMaker enables you to always have a blank record like this at the bottom of a portal or to require users to explicitly create it by clicking a button or running a script that you write.)

The interface includes *help tags* that provide assistance when you hover the pointer over an interface element. For example, the large X that lets you delete the event record provides a help tag that is calculated to include the name of the event.

What the user sees and refers to as *help tags* are often referred to as *tooltips* by the developer. They are the same thing.

The interface is a veritable treasure trove of functionality packaged up in FileMaker's powerful and concise interface. For example, at the bottom of the toolbar, you can choose to go to another layout, and you can choose to display the records as a list, as a form (shown in Figure 1.1), or in a spreadsheet-like table. You can switch into various modes, as follows:

- *Browse* mode lets you view and enter data.
- *Find* mode lets you search for data. You supply values for any of the fields on which you want to search, and FileMaker will do the rest.
- *Preview* mode lets you display the data as it will appear when it prints. Totals and subtotals can be calculated, and page numbers will be generated.

This database can be used either for a single user or with one of the FileMaker sharing options described in Chapter 2.

There's much more to FileMaker. You can find out more in the author's book, *FileMaker Pro in Depth* from Que Publishing.

Perhaps the most important things to look at in Figure 1.1 aren't in the FileMaker window itself. They are so integral to the way that you use a computer (be it Windows or Mac OS X) that you might not even notice them anymore. These critical interface elements include the following:

- The menu bar at the top of the screen (Mac) or the top of the window (Windows)
- The window's resize control and draggable title bar
- The dock at the bottom of the screen on Mac OS X and the buttons at the bottom of the Windows screen
- The help tag shown in Figure 1.1

Also, although only one window is shown in Figure 1.1, you can have many windows open.

When you go mobile with iPad or iPhone or even other devices from other manufacturers, you can say goodbye to menu bars at the top of windows or the screen, multiple and resizable windows, and system-wide controls at the bottom of the screen (this last point is true for iPad and iPhone but not necessarily for other devices)—and you certainly say goodbye to help tags. There is no such thing as hovering over an interface element on a touch screen; you tap it or you don't. Hovering doesn't exist.

As you see throughout this book, thinking about designing an interface for a mobile device often means thinking about accomplishing tasks in a different way than you would on a desktop-based computer.

WHAT'S THE OPPOSITE OF A MOBILE DEVICE?

A mobile device can be held in your hand and carried around. It has a touch screen, and it generally has communication features that typically include Wi-Fi and may include telephony.

What's the name for other devices? Some of them are laptop computers, others are desktop computers, and still others are tower computers that might live under your desk or table. Although you can carry around many of them, they are not portable in the way in which mobile devices are portable. And, as you will see, there are some very significant interface differences.

One of the most important differences is that with operating systems such as Windows and Mac OS X, there is a desktop image in the background of the screen. You can open windows in front of it and move documents and folders around on this virtual desktop. (This is the basic graphical user interface design.)

With mobile devices, the desktop disappears, and you work directly with one window at a time.

For that reason, in this book, the opposite of mobile device is referred to as a desktop-based computer. "Desktop" is used in the sense of the desktop that appears on the computer screen and not in the sense of the desk on top of which your computer might be placed.

Looking at Reservations on FileMaker Go

You can open the Reservations database shown in Figure 1.1 on iPhone or iPad using FileMaker Go. This section shows both devices. This is the same database that you saw previously on Mac OS X, but the layouts have been modified to accommodate the screen sizes of iPad and iPhone. Modifying layouts is a critical part of moving a database to those devices, and you'll see how to do it for FileMaker Go, Bento, and web browsers throughout this book.

FileMaker Go on iPad

Figure 1.2 shows Reservations running on iPad with FileMaker Go.

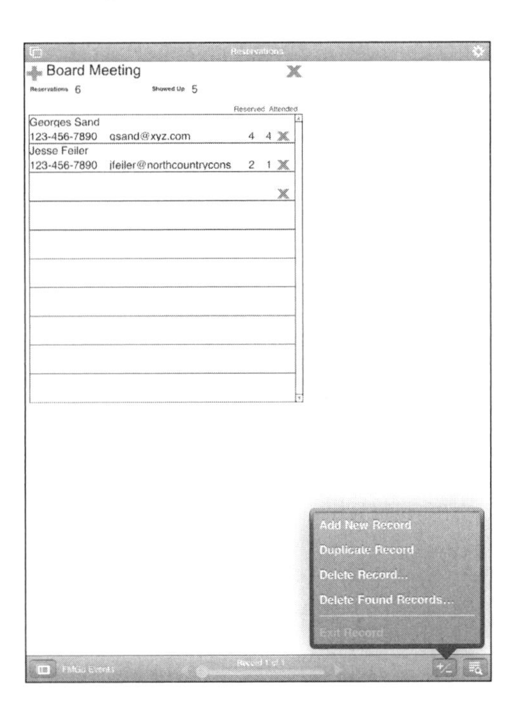

Figure 1.2
Open Reservations on iPad with FileMaker Go.

The status toolbar at the top of a FileMaker window on Mac OS X or Windows is replaced with a toolbar at the bottom of the screen, as you see in Figure 1.2. Buttons in that toolbar adhere to the human interface guidelines for iPad and iPhone; when you tap them, a *popover* with choices appears. A popover is similar to a dialog on a desktop computer; it presents information or choices for you to deal with now. However, unlike dialogs on desktop-based systems, there are no Cancel or OK buttons; to cancel a popover, tap the screen anywhere outside the popover. Tap a command inside the popover to execute it and close the popover. The operating system takes care of positioning these popovers, and there's always an arrow pointing to the button that you tapped to open it. (Popovers can appear in other places on the screen, as well as the bottom.)

One of the key features of desktop-based computers is not visible in Figure 1.1: the keyboard. When you tap a field in FileMaker Go that allows text input, the onscreen keyboard appears as you see in Figure 1.3. If you have defined the field as being of a specific type such as numeric, the numeric keyboard appears. In this case, the field that is being entered is a telephone number, and although the underlying data is numeric, phone numbers often are shown using the letter equivalents for the keys so it is defined as a text field.

➔ For information on using a wireless keyboard with iPad or iPhone, **see** Appendix A, "Using a Wireless Keyboard," **p. 359**.

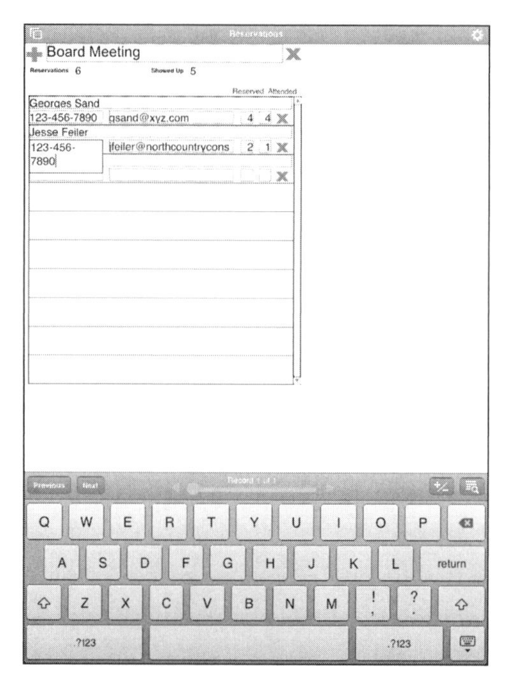

Figure 1.3
The onscreen keyboard opens automatically as needed.

FileMaker Go on iPhone

On iPhone, FileMaker Go appears very much the same. Compare Reservations on iPhone, as shown in Figure 1.4, with the iPad version shown in Figure 1.2. The popover is not visible in Figure 1.4, but only because it hasn't been opened with a tap.

When you open the popover on iPhone (see Figure 1.5), it is in approximately the same position. Because of the size of the screen, though, the effect is somewhat different.

The comparison between Figure 1.1 (Mac OS X) and Figures 1.2 and 1.3 (iPad) and Figures 1.4 and 1.5 (iPhone) demonstrates one of the key concerns involved in moving FileMaker databases to mobile devices. There are two columns of data in the portal on the Mac version, and there's only one column on iPhone and iPad. In addition, the running totals of reservations and attendees have moved from the right-hand side to the left-hand side on iPhone.

Modifying the layout to fit different screen types is not something that is done automatically for you by FileMaker Go. Rather, it is accomplished by the interface designer. The technique is to create multiple layouts for your data, with each one optimized for a specific type of screen. Users can use the control in the lower left of the iPhone or iPad screen to choose the layout they want to use. When you develop your basic FileMaker solution, you can automatically switch layouts depending on the device that someone is using.

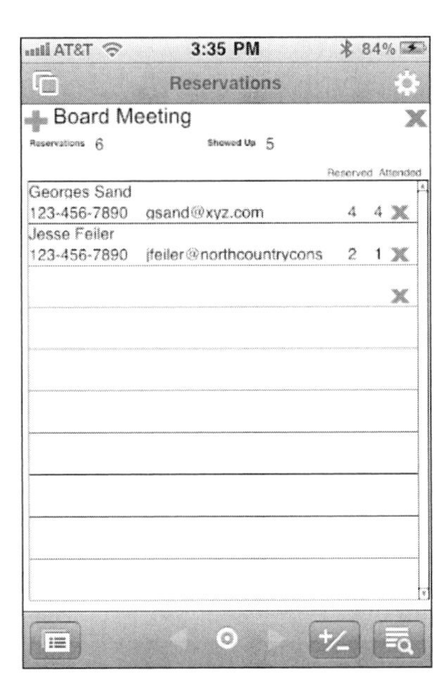

Figure 1.4
Use Reservations on iPhone with FileMaker Go.

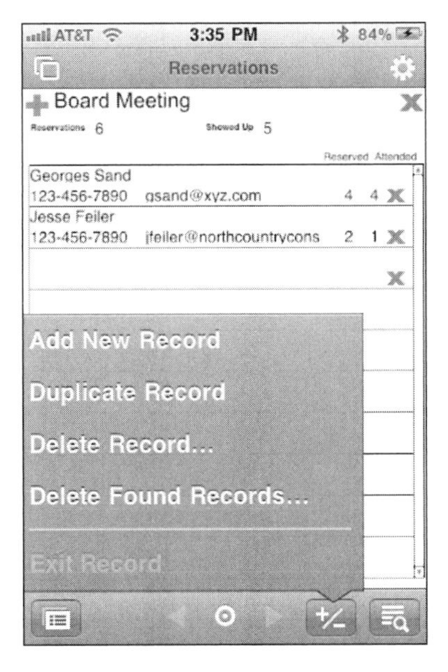

Figure 1.5
Choose additional commands on iPhone.

However, you can use FileMaker's autosizing feature to move and resize objects on the layout. Frequently you can use this to avoid having multiple layouts. For example, moving fields so that their size or position relative to an edge of their container (often the layout itself) remains constant can be done with autosizing. If you want to have a portal with two columns (as is the case on the Mac in Figure 1.1), that is a decision that you make with the layout. In practice, you may use both features: alternate layouts and autosizing for individual objects on the layout.

➔ You'll find more information about switching layouts and customizing them for mobile devices in Chapter 9, "Designing a FileMaker Go Solution," **p. 213**.

Looking at Reservations on Bento

FileMaker's personal database, Bento, provides a concise and powerful tool to use existing templates or create your own libraries for your own purposes. Bento databases can be shared and synchronized among five Macs on a local area network. Whereas FileMaker runs on both Mac and Windows computers, Bento is a Mac OS X-only product. However, just as FileMaker has FileMaker Gowith apps that run on iPhone and iPad, Bento has apps for iPhone and iPad.

You can build a Reservations library for Bento. Note the difference in terminology. FileMaker lets you build databases; each is stored in its own file, and each can contain a number of tables. With Bento, you have a single Bento database on your Mac, and you create *libraries* within it. Libraries in the Bento database are very much like tables in a FileMaker database.

→ Chapter 11, "Using Bento and Bento Libraries," **p. 251**, introduces you to Bento and shows you how to set up a Reservations library.

Bento on iPad

Given a Bento library that is structured much like the FileMaker Reservations database, you can use it with the Bento iPad app, as shown in Figure 1.6.

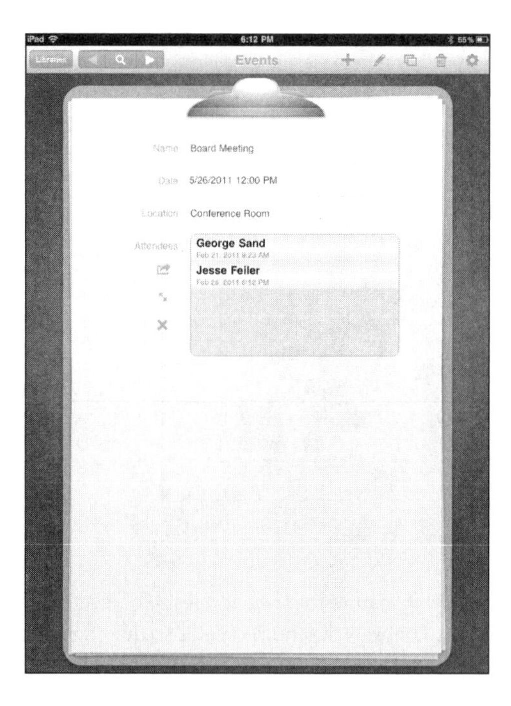

Figure 1.6
Use Reservations on iPad.

Bento on Mac OS X as well as on the iPhone and iPad apps provides a lot of the interface infrastructure that you can develop and customize on FileMaker. Notice that the + and X buttons to add and delete events are part of the Reservations database you build with FileMaker Pro. On Bento, the + at the top of the screen to add a new record and the trash can to delete the current record are part of the Bento app itself rather than your own library. From the user's point of view, it doesn't make much difference.

You can add related records as you see in Figure 1.7. Tap the Action icon (note the arrow from the popover to it) and you can either choose related records from another library or create new ones.

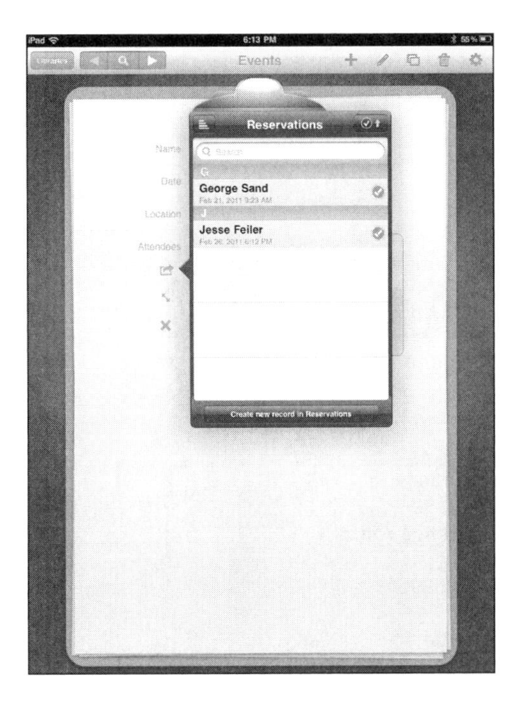

Figure 1.7
Bento formats related records automatically.

Bento on iPhone

On iPhone, Bento is able to provide a different interface. Whereas FileMaker and FileMaker Go rely on layouts that you create, with Bento a lot of that infrastructure is provided for you. One of the by-products of that infrastructure is that Bento for iPhone knows the logical structure of related records. This is how the Reservations Bento library functions on iPhone.

You choose the event you are interested in from an alphabetical list, as shown in Figure 1.8. On FileMaker Pro or FileMaker Go, you would implement this with a separate list layout.

When you tap an event, you see its details, as shown in Figure 1.9.

Figure 1.8
Select an event from the alphabetized list.

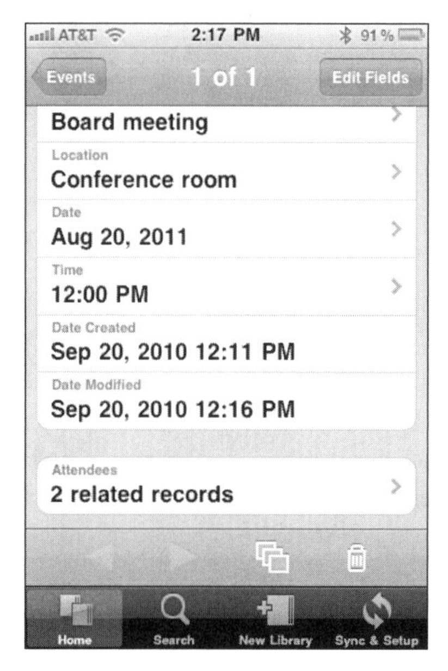

Figure 1.9
View an individual event.

Bento for iPhone indicates that there are two related records; tap the arrow to see them, as shown in Figure 1.10.

Tap one related record to see its data, as shown in Figure 1.11.

Figure 1.10
View the list of related records.

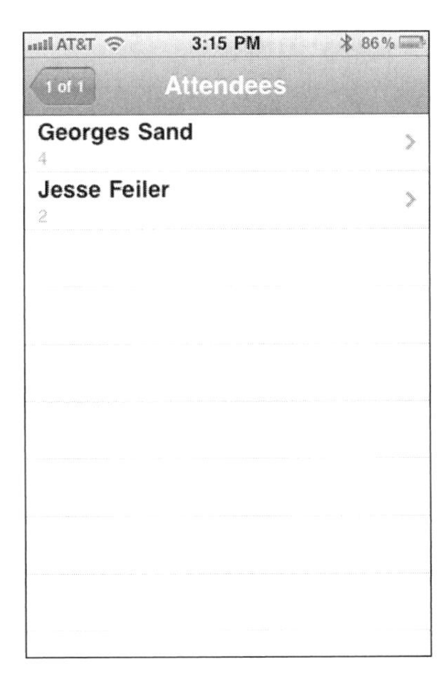

Figure 1.11
See a related record's data.

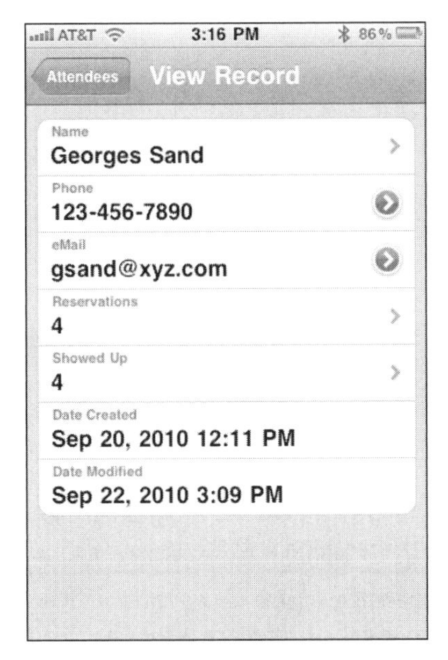

This standard iPhone interface is called a *navigation* interface. It consists of a navigation bar at the top of the screen that usually has a left-pointing arrow to take you back one level. For data in the main section of the screen, disclosure indicators at the right take you deeper into the data. You might have a button at the right side of the navigation bar such as the Edit Fields button shown in Figure 1.9.

Looking at Reservations on a Mobile Browser

Using the Bento and FileMaker Go apps on iPad and iPhone is a great way to access your data from a mobile device. However, there's yet another method. Because one of the characteristics of today's smartphones and mobile devices, such as iPad and iPhone, is that they incorporate browsers, FileMaker Server can serve your database up as a website. There's a library of PHP code that you can use to build your dynamic web pages so that they access a FileMaker database running on FileMaker Server.

There's even an interactive PHP Assistant available in FileMaker Pro Advanced that helps you build your website. You can choose from a variety of templates for sites that allow updating as well as just browsing. To no one's great surprise, one of those templates builds a site that is designed specifically for iPhone.

Figure 1.12 shows a default listing of records from the Reservations FileMaker database.

Figure 1.12
View the list of records on your iPhone.

As you can see in Figure 1.13, you can browse a single record's data. Remember that these web pages are built by the PHP assistant based on your database's fields. You have the option to choose which fields to include and exclude, and, of course, you can modify the resulting PHP files as you wish.

The interface to allow searching of the database is available to you and your iPhone users along with (if you want) the ability to update the database. The searching interface is shown in Figure 1.14.

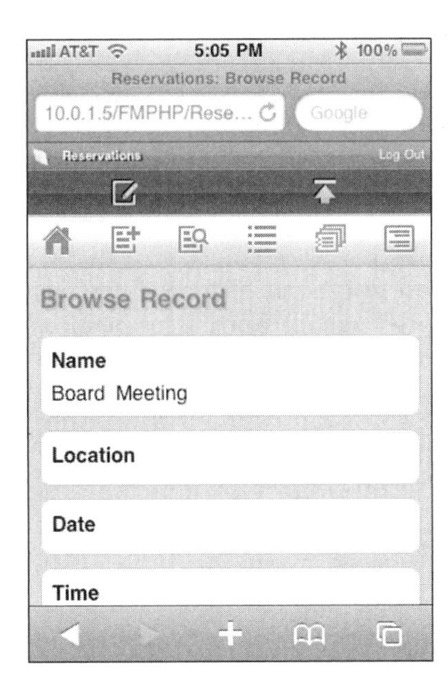

Figure 1.13
Browse a single record's data on iPhone.

Figure 1.14
Users can search your database from their iPhones.

Web publishing is nothing new for FileMaker. What is new now is that mobile devices support browsers, and because you can easily build a website designed for the size and shape of a mobile device's screen (although the iPhone template is the easiest to use), your web-enabled FileMaker database is automatically made mobile when accessed from a mobile browser.

➔ Find out more about web publishing for mobile devices with FileMaker in Part IV, "FileMaker Web Publishing: Instant Web Publishing (IWP) and Custom Web Publishing (CWP) ."

Further Steps

Here are some suggestions for exploring the concepts discussed in this chapter:

- As you use your iPad and iPhone, pay attention to how complex data is presented on them. Make a note of when you get lost and disoriented and when you suddenly notice that the data you want is right in front of you. What makes the difference?
- If you already have a FileMaker database or if you are thinking of creating one for mobile devices, think about what is necessary. There is never enough space on the computer screen for all the data, and that is doubly true for iPad and iPhone. What can you omit?
- One way of handling the limited screen space is to add tabs and other interface elements that can reveal data only when needed. Watch for these, and watch how they are implemented on mobile devices so that you can use those techniques yourself.

IN THIS CHAPTER

- Understanding database and FileMaker Concepts
- Using the Interface Tools
- Managing a Database with the Development Tools
- Structuring your Solutions and Sub-solutions

Introducing the FileMaker Architecture

If you already know FileMaker, this book can help you enable mobile access to existing or new databases from mobile devices. It also can help you learn enough about FileMaker in general and the specific issues involved with mobile devices even if you do not have a background in FileMaker. This chapter is key to both approaches, because even the most experienced FileMaker developer is likely to have to do a bit of rethinking (not much!) to get into the world of mobile FileMaker data.

In many cases, you will be modifying an existing database for use with FileMaker Go or FileMaker Web Publishing. That is why an overview such as this might be sufficient. You just need to be able to follow your way around existing code and databases. Even if you are starting from scratch, you can start from templates and starter solutions so that you are really not on your own and starting from a blank database.

A large part of FileMaker's success comes from the fact that it brings the power of relational databases to people who want to use them—people who might have knowledge of the data and the general subject area but who are not necessarily experts in data management. FileMaker provides an integrated environment in which you can manage a database and also provide a user interface for browsing, updating, and searching the data in that database. (In this regard, it differs from databases such as MySQL, which provides just the database storage and manipulation; if you want to add an interface to a database of that sort, you have to write it yourself in a programming language or you have to add on an interface product.)

To get started in the world of FileMaker, you need to learn or brush up on the vocabularies for talking about databases in general and FileMaker in particular.

Talking Databases

In today's world, databases are almost always relational databases. In the past, databases using different structures existed for a variety of purposes, but now the world is relational. Two sets of concepts are particularly important: the relational database vocabulary and SQL. They are described in this section.

These concepts are important for all relational databases, including FileMaker and Bento. As you will see, SQL functionality is provided in FileMaker and Bento through graphical tool. You do not write SQL code, but you should understand the concepts so that you can interact with people and databases from the SQL side of the relational database world.

WHAT IS SQL?

SQL was originally an acronym for Structured Query Language and was pronounced "sequel," but for various reasons—including the fact that SEQUEL turned out to be a registered trademark for Hawker Siddeley—SQL lost its acronym status and became simply three letters. For that reason, it is properly pronounced "S-Q-L," although many people still call it "sequel."

Using the Relational Database Vocabulary

The relational database vocabulary is very simple. A database is composed of *tables*. You can think of each table as being much like a spreadsheet: *Rows* contain data for a single observation, person, or event; the *columns* contain specific data, such as a name, the date of an event, and so forth.

Many people use an alternate terminology in which the rows of data are called *records* and the columns are called *fields*. And just to keep things interesting, some people use both sets of terminology.

Database software provides all the functionality needed to store, retrieve, delete, and update the data. Database software also takes care of optimizing the operations involved in manipulating the data. In some large systems, a single database might be spread over a number of separate computers, but in FileMaker, life is simpler—a database is usually contained in a single file. (There is more on that in the following section.)

Looking at SQL

Although FileMaker is a relational database, it does not use SQL syntax. However, it is important to know about SQL because you may be interacting with databases that do use it (FileMaker is great at using SQL databases as *external data sources*), and you also might have to talk to people who come from the world of strict SQL (as, indeed, you might yourself).

The heart of SQL is the syntax for a *query*. A query has the following basic form:

```
SELECT <column> FROM <table> WHERE <some condition applies>;
```

Thus, a SQL statement could be as follows:

```
SELECT name FROM employees WHERE department = "Sales";
```

SQL statements end with semicolons in many implementations, and capitalization does not matter. However, for ease of reading and by convention, the reserved words in SQL are often capitalized as they are in these examples.

Additional clauses can be added as in the following example:

```
SELECT name FROM employees WHERE department = "Sales" ORDER BY employment_date;
```

In addition, you can retrieve data from several tables at a time, as in the following query:

```
SELECT name, salary FROM employees, employee_terms WHERE department = "Sales"
AND name = employee_terms_name ORDER BY employment_date;
```

There is much more to SQL, but this gives you a flavor of the main features.

There is comparable syntax to insert data into tables, and there is a great deal of syntax enabling a database administrator to control access to the database.

Talking FileMaker

FileMaker represents two significant advances beyond a basic relational database that may implement SQL: interface tools that make it easier for users to work with their data, and GUI development tools that make it easier for you to structure databases.

It is important to note that these tools can be built only with FileMaker Pro or FileMaker Pro Advanced. After you have built these tools with FileMaker Pro, they can be made available to anyone who is using your database through FileMaker Server (including FileMaker Server Advanced), FileMaker web publishing, FileMaker Go, and runtime versions built with FileMaker Pro Advanced. A few of the features are not available on FileMaker Go and in runtime versions.

Each of these tools is discussed in greater detail later in this chapter.

Working with the Interface Tools

FileMaker not only provides the data manipulation code, but it also provides a set of tools so that you can develop an interface for users. With other relational databases, you often have to write your own code in a programming language or add a user interface package to the database.

There are three basic sets of interface tools, as follows:

- **Layout Mode.** Layout mode provides the tools you use to create and modify the layouts that the user sees.

- **Scripts.** FileMaker provides a scripting interface. It is based on its own commands, so you do not have to learn some new scripting language. The scripting language is driven by a GUI interface, so instead of typing script steps, you select them from a list and click a button (or double-click the script step) to move them into your script. Scripts are used both to support user interfaces and to implement functionality in the database side of things. Scripts can be attached to buttons (in FileMaker terms, almost any interface element can be turned into a button regardless of its appearance just by associating a script with it).
- **Triggers.** Triggers are used to link scripts to actions taken by users (or in some cases by scripts). The trigger consists of three parts. It specifies an event (such as a keystroke) that can happen to an object (such as a button or field in a layout, a layout itself, or a file). If and when that event happens to that object, a script is executed. In this way, you can create a trigger that causes a script to run when a layout is opened, when text is entered, or when any of a number of other conditions occurs.

You will find more details on these concepts later in this chapter, as well as throughout the book.

Working with the GUI Development Tools

FileMaker implements the functionality of SQL and the relational database model using a graphical user interface for you. Thus, instead of you having to write a SELECT statement in order to get data, you design the database and create relationships graphically so that the equivalent of the SELECT statements can be created automatically by FileMaker.

There are three areas where the GUI development tools assist you:

- **Tables and Fields.** This is the heart of the database: the tables and the fields (or columns) within them. Tables and fields are actually the heart of any relational database, and FileMaker handles their editing much as other databases do. Fields can store different types of data such as text, numbers, dates, and even binary data that could be an image or audio clip. In FileMaker, these are called *container* fields.
- **Relationships Graph.** This is one of the most important differences between FileMaker and SQL-based databases. As you saw in the syntax examples previously in this chapter, a SELECT statement can include a WHERE clause that joins two tables together (for example, WHERE employee_ID = person_ID, and in that case employee_ID is in the employee table and person_ID is in the address_book table). The WHERE clause need not be equality; it can ask for a less than or greater than relationship, as well as a number of others.

  In FileMaker, you use the *Relationships Graph* to draw the relationships, as you see later in this chapter. Thus, whereas the relationships are created in SQL as needed, in FileMaker, the relationships are created as part of the database design process instead of as part of the programming process. There are advantages to both methods, but in the case of FileMaker, the fact that the relationships are created with the database and not for each query means that the skills needed to create individual queries do not need to include the ability to think about relationships and their structures. It is just one more way in which FileMaker makes things easier for its database designers.
- **Calculations.** In addition to familiar field types such as text, numbers, and dates, FileMaker supports a *calculation* field type. This can be calculated from a combination of arithmetic and logical operators along with constants and the values from other fields in the same table or in a related table. The ability

to use related values in a calculation lets you create sophisticated calculations that do some of the work you might otherwise need to do with scripts. You'll find examples of calculations later in this chapter.

Talking Security

FileMaker uses a security system based on accounts and privilege sets. You typically create an account for each user of the database. Separately, you create privilege sets that determine what actions can be performed, such as which tables can be updated, whether data can be printed, and so forth. You normally have a few privilege sets and a number of users. For each user, you simply choose the privilege set that applies, and your security is in place.

Like the GUI development tools and the layout tools, security is created and edited only with FileMaker Pro (or FileMaker Pro Advanced). It applies to all the versions of FileMaker.

Security is set at the file level, not at the individual table level although you can specify settings down to individual fields, scripts, layouts, and other components of the file. User accounts and privilege sets apply to the file as a whole.

Working with the Interface Tools

There are three groups of interface tools, and they all interact with the underlying database. It is not uncommon to separate interface from data in developing software, whether it be application development, website design, or database solutions. Both the interface and the underlying data must come together in a complete and intuitively designed whole, but it is often the case that different people with different skills address each of those areas.

It is also common, particularly with FileMaker, that a single person does it all. If that is your case, you have the challenge of switching roles, but you also have the great advantage of being able to decide to make a change that affects both the interface and the database structure without the dubious pleasure of endless meetings.

This section provides an overview of the interface tools in FileMaker. It is relatively brief because the focus of this book is on the development and deployment of databases for mobile devices. Although FileMaker is a unique product, part of its attraction is that it combines intuitive functionality and standard interface elements so that once you learn the concepts, you can get to work.

It is also worth noting that most of the time, you are not starting from scratch in developing your FileMaker solutions. Many people start from the Starter Solutions that come with FileMaker Pro.

Using Layout Mode

Most of your users' interactions with your FileMaker database are through layouts. You can create your own, use layouts that are part of the Starter Solutions provided by FileMaker, or use the New Layout/Report assistant. Much of your work with layouts consists of making changes to existing layouts to customize them for your own use. When it comes to working with FileMaker databases on mobile devices, the most common task to do with layouts is to copy an existing layout and then modify it for the mobile device that you are targeting.

The database and its layouts are closely linked. For example, you can create a data entry field on a layout that displays data from a field in the database. The database field must be created first so that the layout field has something to display. However, many people find it easier to look first at the layout tools so that they can see how their database fields can be used. In practice, you will often jump back and forth between the two. In this chapter, you might want to jump back and forth between the database and GUI tool sections.

➔ This section shows you how to perform basic layout creation and editing. The specific tips for customizing layouts for mobile devices either with FileMaker Go or with FileMaker Instant Web Publishing are provided in Chapter 9, "Designing a FileMaker Go Solution," **p. 213**, and Chapter 15, "Deploying FileMaker/IWP with FileMaker Server Advanced," **p. 325**.

FileMaker Pro has four modes you can use, as follows:

- **Browse mode** is the default mode; it lets you enter, view, modify, and delete data subject to security concerns.
- **Find mode** lets you search for data.
- **Preview mode** presents data as it will appear when printed. Totals and subtotals are calculated. It is not supported on FileMaker Web Publishing, FileMaker Go, or FileMaker runtime solutions.
- **Layout mode** lets you create, modify, and delete layouts. It is not supported on FileMaker Web Publishing, FileMaker Go, or FileMaker runtime solutions.

  To start editing layouts, you need to switch to Layout mode. Enter Layout mode in any of these ways:

  - Choose View, Layout Mode from the menu bar.
  - Choose Layout from the pop-up menu at the bottom left of the FileMaker window frame. By default, it is set to Browse mode, so just click on Browse and choose Layout to switch modes.
  - In the Status toolbar at the top of the window, click Edit Layout at the right to enter Layout mode. When you are in Layout mode, the button changes to Exit Layout.
  - Use the keyboard shortcut ⌘+L (Mac) or Ctrl+L (Windows).

In Layout mode, the menu bar changes and the window, as shown in Figure 2.1, changes.

The window is customizable, so your version might look somewhat different. In particular, you can control what elements are shown from the View menu. For example, the View menu's Show submenu lets you see outlines for the data entry fields with the names of the database fields shown as you see in Figure 2.1. You can change that option to view sample text rather than the field names.

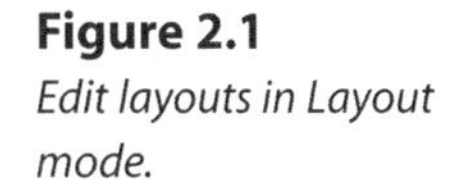

Figure 2.1
Edit layouts in Layout mode.

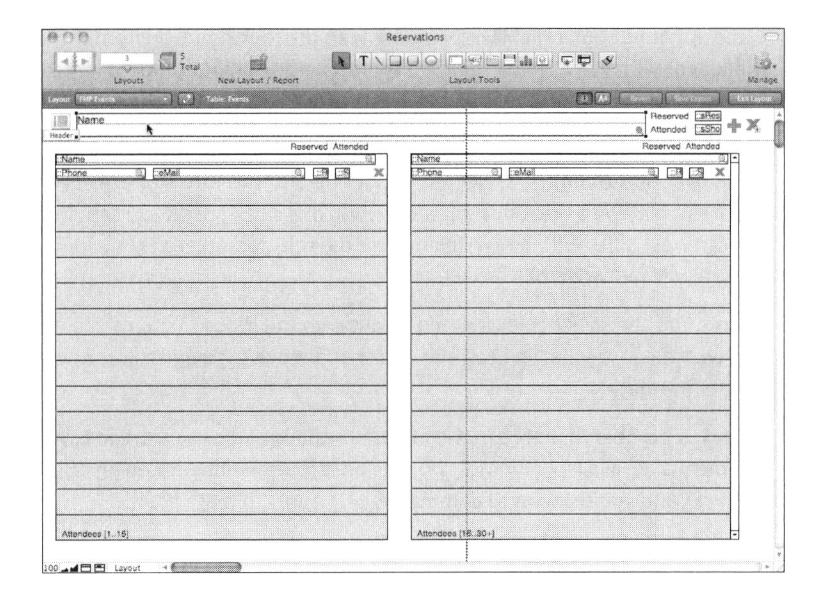

Note in particular that at the bottom of the Status toolbar are (or can be) two smaller horizontal toolbars, as follows:

- **Layout bar.** The Layout bar, shown at the bottom of the Status toolbar in Figure 2.1, lets you select a layout from the pop-up menu at the left. It is visible in all four modes, although it appears differently in each one.
- **Formatting bar.** The Formatting bar, which can be shown in Browse and Layout modes, provides quick access to formatting tools that let you select fonts, styles, and alignments. It is shown in Figure 2.2.

Figure 2.2
The Formatting bar might appear below the Status toolbar.

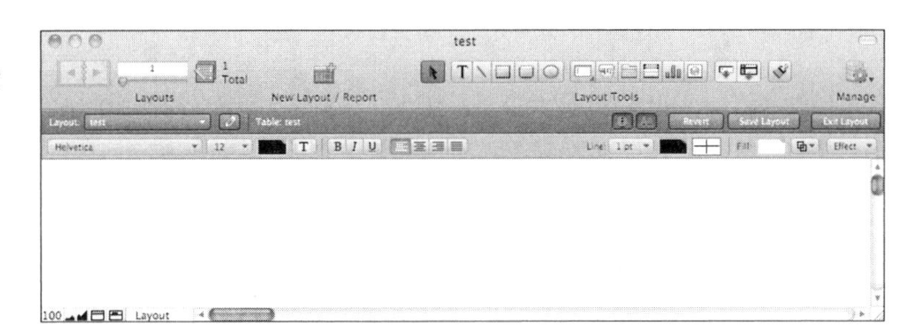

Introducing the Layout Mode Tools

In Layout mode, you have five sets of tools to work with. Each is described further later in this chapter:

- **Status toolbar.** The Status toolbar at the top of the FileMaker window changes to provide layout tools. Compare Figures 1.1 and 2.1 to see the difference between the Status toolbar in Browse mode (see Figure 1.1) and Layout mode (see Figure 2.1). Show and hide the Status toolbar with the oblong button in the upper right of a FileMaker window, with the fourth button from the left in the bottom frame of a window, or with View, Status Toolbar. (This applies to all four modes.)

- **Inspector.** The Inspector lets you view and set details about any selected element in the layout. Show or hide the Inspector with View, Inspector or ⌘-I (Mac) or Ctrl+I (Windows). The Inspector shows information about the currently selected layout object. It has three tabs (Appearance, Position, and Data) so that it can provide a great deal of information in a small space. You can open multiple Inspectors so that you can view different tabs in each one for the selected object. Open additional Inspector windows with View, New Inspector.
- **Mouse.** You can select objects in the layout with the mouse. When selected, you can view and set attributes with the Inspector; you can also resize and move the objects just as you would in any other graphical user interface.
- **Menus.** A number of menus in the menu bar change when you are in Layout mode. Shortcut and context menus are available for most objects on layouts. Right-click on an object (or, on a Mac, you can use ⌘+click), and you see a list of commands that apply to that object.
- **New Layout/Report assistant.** This assistant walks you through the process of creating a layout pretty much automatically. You can choose from various styles, and you can modify the resulting layout or report in Layout mode later on.

Using the Status Toolbar

The easiest way to find your way around the Status toolbar is to look at all the possible buttons that can be placed there. Do that by choosing View, Customize Status Toolbar, as shown in Figure 2.3.

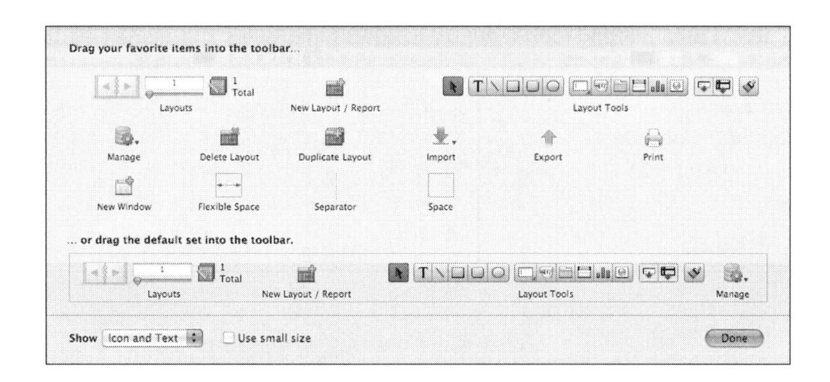

Figure 2.3
You can customize the Status toolbar.

At the left, standard controls let you go to the next or previous layout either with the forward or back buttons or with the slider; you can also type in a layout number above the slider. (This same interface lets you navigate through records in Browse mode, through search requests in Find mode, and through page in Preview mode.)

A single New Layout/Report button lets you create a new layout or report. (The distinction between the two is actually minor: Layouts are used in Browse and Find modes, and reports are used in Preview mode including printing. The elements and tools of each are basically the same.)

The layout tools are shown at the right of the top line of the Status bar (unless you have moved them) and at the top right of Figure 2.3, and they provide you with most of the tools you need for building layouts. There are five sections of the Layout tools. From left, they are as follows:

- **Arrow pointer.** At the left of the layout tools, the arrow pointer lets you select objects in the layout and move them around.
- **Draw graphic tools.** Next, five buttons let you choose what you will draw with the mouse. The T lets you click and type text into a text box that you can then move and resize. Next to it, you can draw a line, a rectangle, a rounded rectangle, or an oval. Holding down the shift key while you draw gives you a regular object (that is, a square instead of a rectangle, a horizontal or vertical line, or a circle rather than an oval).
- **Draw control tools.** These let you add objects to the layout. Just as with the drawing tools, you select one of these and then draw a rectangle where you want the inserted object to be placed. You can always move or resize the object (or even delete it).
    - **Field/Control.** This lets you draw a rectangle for a database field or control. After you have drawn it, you are prompted to choose the database field to be shown in the layout field. You can change this later on with the Inspector. Also, note that the same database field can appear several times in the same layout.
    - **Button.** You can draw a button to which you can attach a script or a standard action (such as going to the next or previous record). You attach the script with Format, Button Setup.
    - **Tab control.** You can create a tab control with the next tool. Tab controls are common methods of getting the most out of screen space because the same area can be used for several different sets of data depending on which tab is chosen.
    - **Portal.** A portal displays data from a related table and might allow editing of it. For example, in Chapter 1, "Making Data Mobile," you saw how an event record could contain fields with information about an event; in addition, a portal contains a scrolling list of attendee records. Each attendee record represents one attendee's data; those records are stored in a related table, but they are displayed in a single portal on the event record.
    - **Chart.** FileMaker now supports built-in charting. The next tool lets you draw a chart and then configure it as to chart type and the data that it displays.
    - **Web Viewer.** The final tool lets you draw a web viewer that can contain live data from the web. This data can be requested dynamically depending on the values of various database fields.
- **Insert tools.** The draw control tools work like the graphics drawing tools: click a tool and then draw the object where you want it. The two insert tools represent a different interface. You click one of these tools and then drag the tool to the approximate location you want on your layout. You can modify it later. The first tool lets you add a field, providing an alternate method to the field/control drawing tool. The second lets you add a layout part (see "Working with Layout Parts" later in this chapter).

- **Format painter.** Finally, the paint brush at the right lets you copy the formatting characteristics from one object to another. These characteristics do not include size, but they do include attributes such as text style and color. See "How to Standardize Layout Objects" later in this chapter.

Using the Inspector

One of the most important elements of modern interface design is the concept of inspectors; they were introduced to FileMaker in FileMaker Pro 11. Inspectors are small windows (sometimes called *windoids*). They typically appear when the application is active; when you switch to another application, its windows remain visible on the desktop, but its windoids and palettes are typically hidden until you return to the application.

Inspectors are dynamic: They present data for the currently selected object. In FileMaker, the Inspector has three tabs, as you can see in Figure 2.4.

Figure 2.4
The Inspector has three tabs.

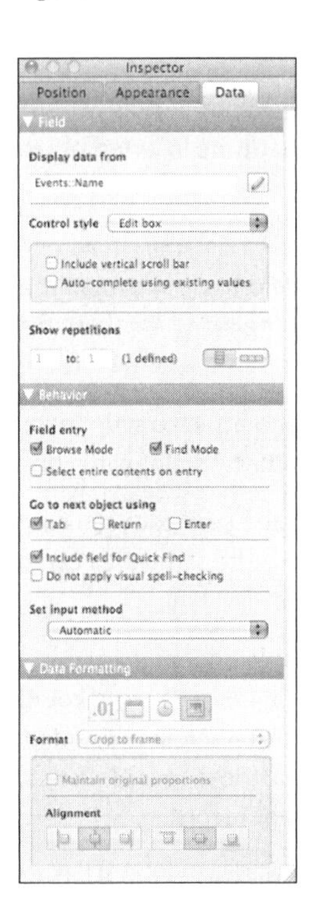

Instead of switching from tab to tab, you might want to open two or even three Inspectors at the same time and set them to different tabs. Then, as you switch from one object to another, you will be able to see all the information about the selected object in the tabs of the multiple inspectors. Open a single Inspector with View, Inspector (it alternates between showing and hiding the first Inspector opened); open multiple Inspectors with View, New Inspector.

The Data tab, shown in Figure 2.4, lets you configure an object with regard to its data. The most important setting is the first item in the first section (Field). Here is where you specify the database field that is shown in the layout field. Click the pencil button to bring up a dialog in which you can choose the field to use. (There is more on this dialog in the "Relationships Graph" section later in this chapter.)

You can choose which type of control style is used from among these built-in styles:

- Edit Box
- Drop-Down List
- Pop-Up Menu
- Checkbox Set
- Radio Button Set
- Drop-Down Calendar

Many people use an edit box and pair it with a text field so that any data can be entered. If you enter "next Saturday" in a text field, that is just fine; however, FileMaker cannot perform date validation on that data.

When you start thinking about mobile devices, the overuse of the edit box becomes increasingly problematic. When someone taps in an edit box on an iOS device, the on-screen keyboard appears, and they can tap out the data.

On a computer with a keyboard, it is often faster for an experienced typist to just type in the data; taking your hands off the keyboard to pick up the mouse so that you can click a radio button might slow you down.

On a mobile device, your hands are not on the keyboard (there is no keyboard), so being able to tap a radio button can actually save time and improve accuracy. On your mobile interfaces, look for opportunities to replace keyboard input with simple taps on interface elements, such as checkboxes and radio buttons.

Two other sets of options on the Data tab are particularly useful for mobile devices, as follows:

- Under Behavior, the option to include a field in Quick Find can be helpful. Quick Find enables users to do a search for data without specifying the field in which it is located. For many purposes, putting a name field into the Quick Find list might be sufficient for the bulk of user searches. Quick Find is fast, and it requires less input than a full-fledged find request. On mobile devices, the faster and easier you can make it for a user to accomplish a task, the better things are.
- The Data Formatting settings work for fields that are specified as being other than text fields (generally they are number fields). This can eliminate the need for a user to type in carefully formatted text. The formatting is handled automatically regardless of what the user has typed in. Again, minimizing input on a touch screen is a positive thing.

The Position tab of the Inspector is shown in Figure 2.5.

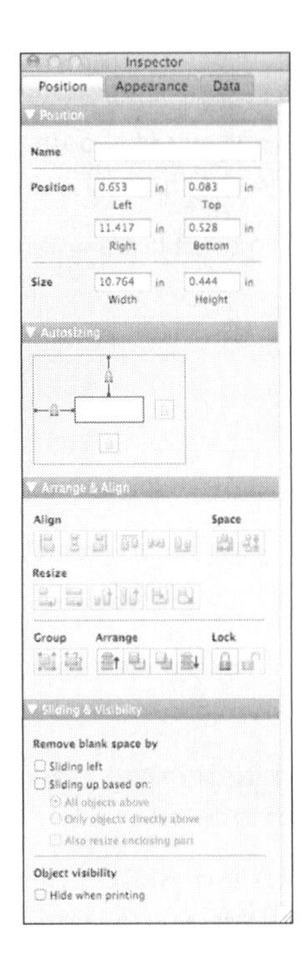

Figure 2.5

Use the Position tab to locate an object.

At the top, in the Position section, you can specify the position and size of an object with numbers. This is often more precise than working with the mouse. (See the "How to Standardize Layout Objects" section later in this chapter for more tips.)

At the very top of the Position tab, you can specify a name for the object. It does not hurt to name an object, and it can save you time later on. Many script steps require you to name objects, so if you have done so when you create them, you are ahead of the game. After a while, you will get a sense of when you need that name. But in the meantime, if you need to name an object while you are in the middle of writing a script, this is where you do it.

Autosizing lets you automatically position the object in relation to its container (usually the window in which it appears). The most common setting is to click two of the padlocks to control the autosizing. By default, most objects are positioned where you place them on the layout, even if the window changes size. However, if you have a button in the lower right of a window (such as a Submit button in a dialog), you typically want that button to be in the same position relative to the lower-right corner of the window when the window is resized.

In a related situation, if you have a large edit box in which a great deal of text can be entered, you might want to have that edit box expand or contract as the window is resized. Anchor it (with padlocks) to all four dimensions. Its upper-left corner will be in the same place no matter how the window is resized, and its lower-right corner will expand or contract with the window.

None of this matters on mobile devices: You cannot resize the window. However, the same logic that accommodates window resizing is necessary for handling the rotation of an iOS device from horizontal to vertical. In fact, setting the autosizing options might be even more essential on mobile devices than it is for desktop-oriented computers.

If you have more than one object selected on a layout, you can use the Arrange & Align tools to adjust them. It is common to use these settings to configure a set of related fields (possibly with their labels); when you are done, leave the objects selected and click the Group button so that you can move the objects as a single unit as you continue with your layout editing. (Note that you can ungroup objects with the second button in the Group section.)

Similarly, when you are happy with a single field, lock it so that it cannot accidentally be changed.

The Appearance tab, shown in Figure 2.6, lets you set attributes such as font, alignment, and spacing. Note that at the top in the Object section, you can set a tooltip for the object. If you click the pencil, you can edit this tooltip so that instead of text that you type in here, it can be the result of a calculation. Instead of telling someone they are about to delete a record, you can tell them that they are about to delete a specific record (this was shown in Figure 1.1).

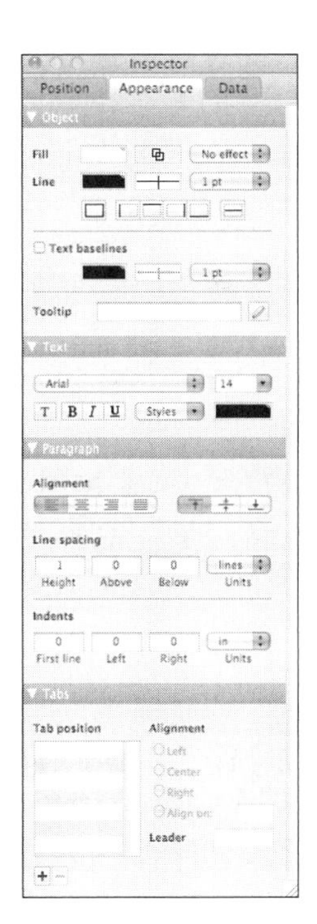

Figure 2.6

Set fonts, alignment, styles, and tooltips with the Appearance tab.

Tooltips are very useful in making your interface easier to use, but remember that they do not exist on mobile devices. The concept of hovering over an object applies only to mouse-based devices. On a touch screen, an object is tapped or it is not—there is no middle ground.

How to Standardize Layout Objects

Organizing a set of interface objects with consistent appearance can make the interface appear better and be easier to use. This is particularly important on mobile devices where the screens are smaller than on desktop-based systems. On iPhone and iPod touch screens, the issue is particularly relevant. Here is an easy way to standardize layout objects using the tools described previously in this section. If you are not familiar with the tools, try out the steps in this section. After you have gone through it once or twice, you will find that you can speed through the process in a few seconds and your layouts will look much better!

1. Create the interface data fields and their labels, if you are using labels. Without worrying about the exact placement, put the label fields next to their layout fields in two columns.
2. Select an interface field and set its attributes exactly as you want them using the Inspector.
3. In the Appearance tab, set the alignment to left-aligned. (See the note about numeric fields.)
4. Select all the other data fields and temporarily group them.
5. Select the fine-tuned field and click the Format Painter button at the right of the tools. That picks up the attributes.
6. Click any of the grouped fields. The attributes will be applied to all of them. Ungroup them.
7. Select the label for the field and set its attributes as you want them. They might be the same as for the data fields, but they might look better with a slightly smaller font size or with a gray instead of black font color. Experiment to see what you like. Set the alignment attribute to right-aligned.
8. Repeat the temporary grouping and use Format Painter to make all the label fields match.
9. Select the data field you have fine-tuned and lock it from the Position tab (or with Arrange, Lock).
10. Arrange the other data fields below the fine-tuned field (or above it). Do not worry about being precise.
11. Select all the data fields, including the fine-tuned one.
12. On the Position tab, select the left alignment for the selected fields or choose Arrange, Align, Left Edges. Note that alignment on the Position tab refers to the objects themselves. Alignment on the Appearance tab affects the alignment of text within a field, and you have already set that in Step 3.
13. On the Position tab, use the Distribute Vertically button in the Space section or choose Arrange, Distribute, Vertically.
14. Those fields are now set, so you can group and lock them.
15. Position the top label field where you want it next to the top data field. You might want to use the Align command in the Arrange menu or the align buttons in the Position tab to help you. Lock this label field.
16. Position the bottom label field where you want it next to the bottom date field. Select it and the locked label field at the top of the column. Align the right edges of the fields (because the top label field is locked, you can move only the bottom one). Lock that field.
17. Select all the label fields, including the locked top and bottom ones. Align their right edges and distribute them vertically from the Position tab, as you did in steps 12 and 13.
18. Group and then lock everything.

Note that numeric fields often look best with their data right-aligned in a column. For data entry, many people prefer left-alignment. Also, for iPhone screens, you might want to experiment with small labels in a gray font color placed above the data entry fields.

Mouse

When selected, most objects have handles in each corner so that you can reshape the object. Dragging from any part of an object except a handle moves it around on the layout. This is standard graphical user interface practice. You can select multiple objects by holding down the Shift key just as you normally do on Mac or Windows. You can also draw a box around several objects to select them all.

FileMaker now relies heavily on shortcut and contextual menus; select an object and click the right mouse button or ⌘+click (Ctrl+click for Windows) on a one-button mouse to bring up the menu.

Menus

The FileMaker menu structure varies depending on your mode. The menus and the right and left of the menu bar are constant, but in Layout mode, the intermediate menus change. The View, Insert, and Format menus have commands relevant to Layout mode in them. In addition, an Arrange menu to let you manage layout objects appears. The most commonly used menu commands are also available in the Status toolbar and in the Inspector.

Layouts Menu

In addition to the redefined menus in Layout mode, you have a Layouts menu. The Layouts menu lets you manage your layouts. You also can indicate what part of the Relationships Graph a layout displays. Because of this close interaction, the Layouts menu is described at the end of this chapter following the description of the database design.

Scripts

Scripts in FileMaker use a graphical user interface for their creation. The first item in the Script menu is Manage Scripts, and it opens the interface shown in this section. Below that, scripts are shown in a list; they can be executed from the Scripts menu. Often, scripts are attached to buttons (and remember that in FileMaker, almost any interface object can be a button by attaching a script to it), or they are triggered as described in the next section. Scripts can also be run using script triggers as well as the scheduling mechanism in FileMaker Server.

This section is designed to give you a brief overview of scripting. It is covered in more depth in the author's *FileMaker Pro in Depth* from Que Publishing, as well as in other books and in FileMaker's own documentation.

Scripting has evolved with FileMaker over the years. New script steps have been added, and the pop-up menu at the lower left of the scripting window now lets you see which script steps work in which environments. Perhaps the most important changes are the addition of variables, parameters, and results. These three changes let you use FileMaker scripting as a traditional and powerful programming tool.

Using Script Variables

You can now set a script variable using the `Set Variable` script step. Variables are preceded by a $ and can be accessed throughout a script. If you use $$, variables remain available until the user quits out of FileMaker Pro, and they can be used in a variety of scripts. This concept is called *scoping*: the scope in which a variable is defined and can be used. With FileMaker, you do not have the scoping issues that apply in a more complex programming environment. With FileMaker, a variable is named and able to be used either within a single script or within any script during your FileMaker session. There is nothing in between.

Using Script Parameters

When you specify a script in a trigger or with a button, you can pass parameters into it. Typically these are values such as a record number, a field value, and an option to control the script's execution in some way. You can then retrieve the parameter in the script and process it. This replaces some old coding processes in which one script copied a value and another script pasted it, assuming that nothing had changed. Now you can explicitly pass the value.

Using Script Results

The Exit Script script step now allows you to return a script value. You can then pass it along to another script as a parameter.

The simplest way to get an overview of scripting (or to get a refresher if you already have used it) is to jump right in and write a script. If you have not used FileMaker Pro for a while, you might notice some differences in the interface, but you will soon feel at home.

How to Use a Script to Automatically Switch to a Mobile Layout

Anything that you can cause to happen automatically makes it easier for your users and reduces the possibility of error. One common script can be one that is triggered when a file is opened so that the user is sent to a known condition: perhaps a standard layout and a given record that automatically shows all the records in the file. It really does not matter too much what that known condition is: The user will expect to be there when the file is opened. It can be very distracting to wind up on a different layout each time and with a different record selected. This problem can be exacerbated in an environment where several people are using the same computer. It is sort of the FileMaker equivalent of "who didn't refill the paper tray on the photo copier?"

On mobile devices, you may want a FileMaker database to open not to a standard layout but to whatever was last used (that is a principle of Apple's Human Interface Guidelines for these devices). The reason for this distinction has to do with the different types of user experiences on desktop and mobile devices. In some ways, starting or resuming an app on a mobile device is more akin to switching from one app's window to another app's window, and that's why preserving the last selections in the interface (*state* to be precise) might be the best choice on mobile devices.

Nevertheless, you might want to automatically switch to a specific layout when you open a FileMaker database. That standard layout may be a layout specifically designed for a mobile device if that is what you are running on. This section shows you how to do that.

1. Enter the script editor by choosing Scripts, Manage Scripts or (starting in FileMaker Pro 11) File, Manage, Scripts. This opens the dialog shown in Figure 2.7. You can search scripts to find the one you want—the list can quickly become lengthy. To help you organize your scripts, you can use the New button in the lower left to create folders; you can then drag scripts into and out of them, as well as open or close each folder with the disclosure triangle. Finally, note that you can control the order of scripts in the Script menu and, by using the checkbox next to each script, you can specify if each script is shown in the menu.

Figure 2.7
Review your scripts.

2. Create a new script with the New button in the lower left or open an existing one by double-clicking it. The script will open as shown in Figure 2.8.

3. Add the script steps from the list at the left by double-clicking a script step you want to use. Alternatively, highlight it and click Move at the bottom left of the main part of the window. Some script steps require additional data and others provide you with a pair of script steps. Start by moving `If` from the control steps at the top of the left side of the window. It will give you an `If` and an `End If` line.

4. The `If` has square brackets, which means there is more data to be provided by you. Highlight the `If` statement, and click Specify in the lower right of the window. (The Specify button appears only if you highlight a script step that requires more data.) The Specify Calculation dialog opens, as shown in Figure 2.9.

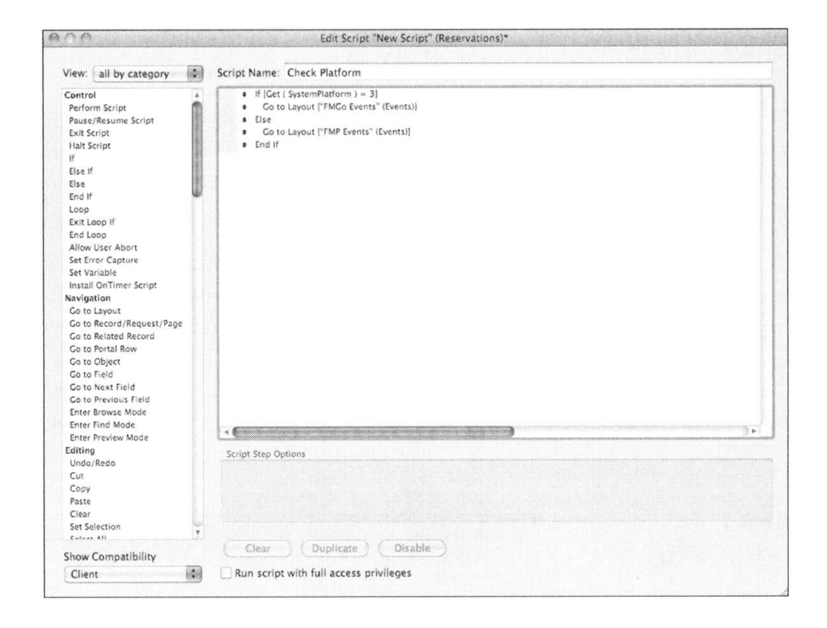

Figure 2.8
Start to edit a script.

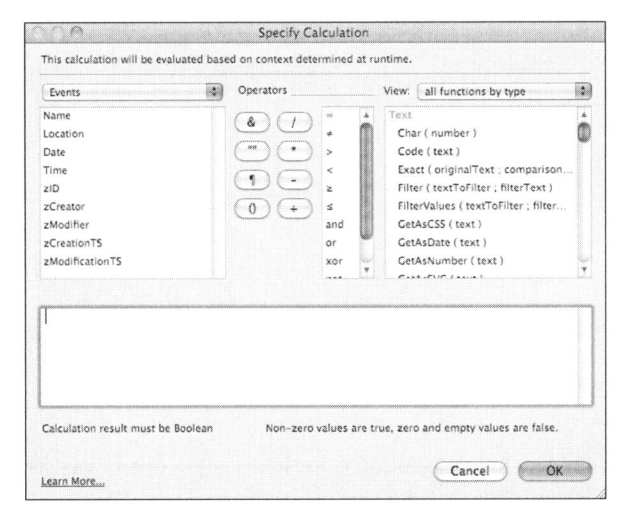

Figure 2.9
Start to specify the calculation for the If statement.

5. This dialog works much the same way as the script dialog does. You can double-click a field from the list at the top left to place it into the calculation. You can also double-click a function from the list at the upper right to place the function into the calculation. You can type intermediate characters. For now, choose the `Get (SystemPlatform)` function and double-click it so that it appears in the body of the calculation, as you see in Figure 2.10.

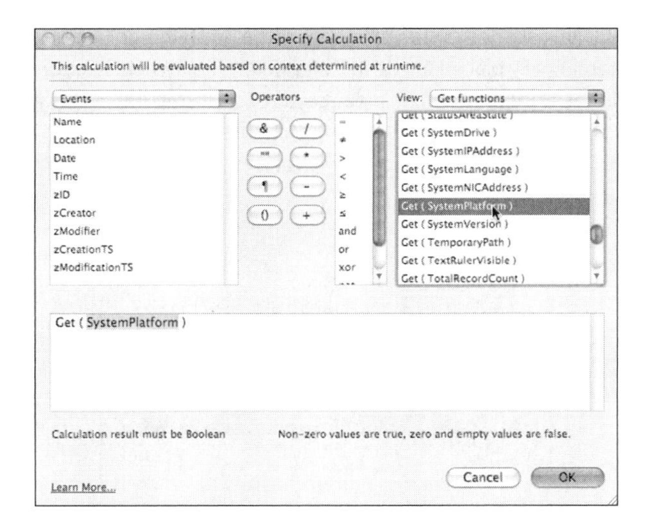

Figure 2.10
Test the system platform in the If statement.

6. You need to provide the actual test. If the system platform value is 3, you are on a mobile device. You can improve the readability of your calculations by adding a comment preceded by two slashes at the end of the line, as you see in Figure 2.11.

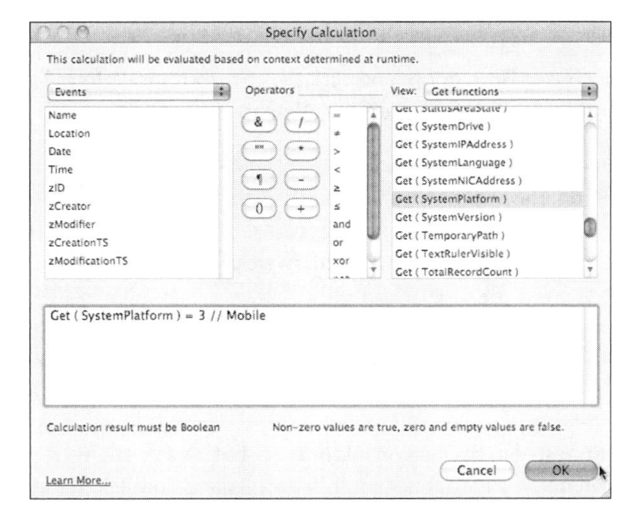

Figure 2.11
Use comments in your calculations.

You can make this test more complex if you want. Many databases that can run on mobile devices in FileMaker Go may have multiple layouts for the mobile device just as desktop-oriented solutions may have multiple layouts. You might want to test to see if you are on a mobile device as described in this step, but you might want to add a test to check to see if the current layout is a mobile layout. This ensures that you have a mobile layout on the mobile device which might be the last layout in use for this database on that device. To perform this test, you would add `Get ( LayoutName )` to test for specific layouts. Adding the word mobile to your mobile layout names makes this strategy even simpler to implement.

Here is the code snippet that tests for the presence of the word mobile in the current layout name:

```
PatternCount ( Get ( LayoutName ) ; "mobile" ) > 0
```

7. Add the `Go to Layout` script step to your script. If the `If` statement is selected, the `Go to Layout` script step is automatically placed inside it. If it appears somewhere else, drag it with the two-headed arrow until it is in the right place. The `Go to Layout` script step has a Specify pop-up so you can choose the layout. You can calculate the layout or simply choose one, as you see in Figure 2.12. Click OK to close the layout selector dialog and then close the script and agree to save it in the dialog that will appear.

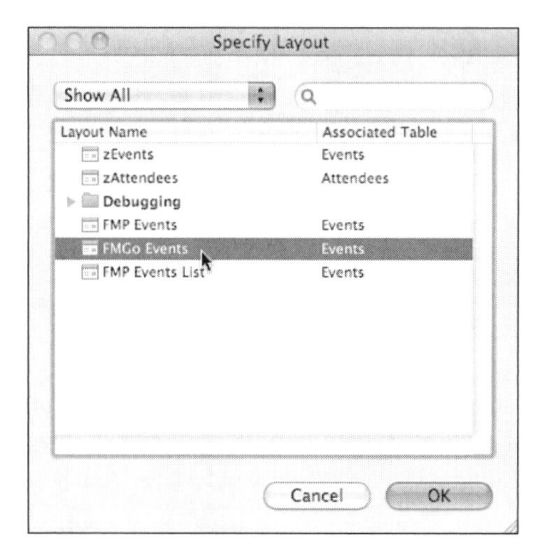

Figure 2.12
Select the layout to go to.

8. In the File menu, choose File Options and click the checkbox for Perform Script. Select the script you have just written, and it will be executed each time the file is opened.

This is almost all you need to know about making your FileMaker solutions behave appropriately for whatever device they are running on. The only other point to consider is that a mobile device might or might not be what you want to test for. Particularly when it comes to layouts, you might want to use a different function—`Get (ApplicationVersion)`—for the test. ApplicationVersion returns strings such as

```
Go x.x.x
```

and

```
Go_iPad x.x.x
```

You are normally not interested in the version information, but it's sometimes very important to know if you are on an iPad or on an iPhone or iPod touch (both return Go and only iPad returns Go_iPad). There is more on this in Part II, "FileMaker Go."

Working with the GUI Database Development Tools

FileMaker implements the functionality of SQL and the relational database model using a graphical user interface for you. Thus, instead of you having to write a SELECT statement in order to get data, you design the database and create relationships graphically so that the equivalent of the SELECT statements can be created automatically by FileMaker.

In the world of databases, the structure of a database is generally referred to as a *schema*. The processes described in this section fall into the topic of editing your database schema.

Just as the layout tools are available only in FileMaker Pro and FileMaker Pro Advanced, so, too, the GUI database development tools are only available in those products. You access these tools using File, Manage, Database. They are available in all four modes. When you are in Layout mode, a Manage button might appear at the right of the Status toolbar; it contains the Manage submenu, including the Database command. In this section, it is assumed that you are working in the Manage Database window.

Tables and Fields

The Manage Database window, shown in Figure 2.13, has three tabs at the top to let you manage tables, fields, and relationships.

The figures in this section show FileMaker Pro Advanced. There are some small differences between it and FileMaker Pro, including the options to copy and paste elements. For serious FileMaker database development, it is usually worth using FileMaker Pro Advanced. Check out filemaker.com to see special offers on the product. There often is a discount on FileMaker Pro Advanced if it is purchased with several licenses for FileMaker Pro.

Figure 2.13
Use Manage Database.

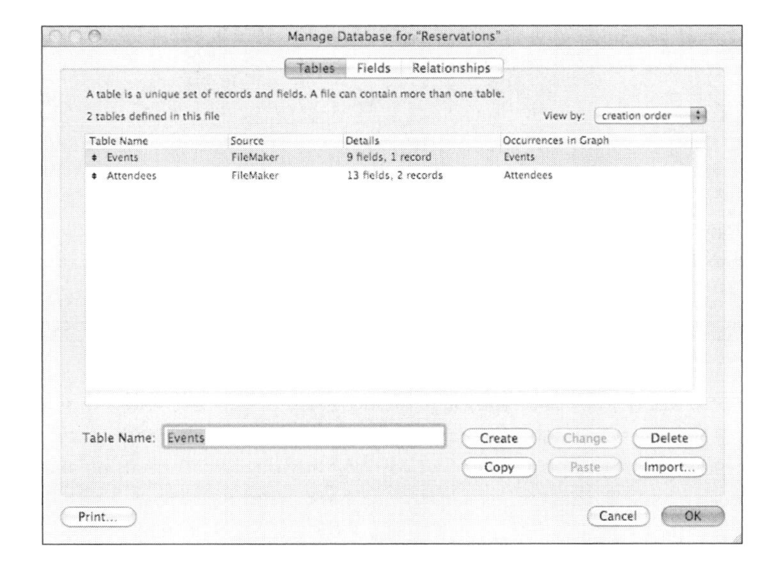

Managing Tables

The tables section of the Manage Database window is basic and simple. Simply identify the table you want to create and click Create. Table names should be clear and relatively short. Users never see them, but you will need to quickly be able to identify the tables. The pop-up menu at the upper right of the window lets you specify the order in which the tables are shown in the Manage Database window—it has no effect on the database itself.

The number of fields and records in each table is shown in this listing. The number of records can be very useful as you are debugging and developing a database.

Managing Fields

Figure 2.14 shows the Fields tab. Use this tab to modify, add, and delete fields.

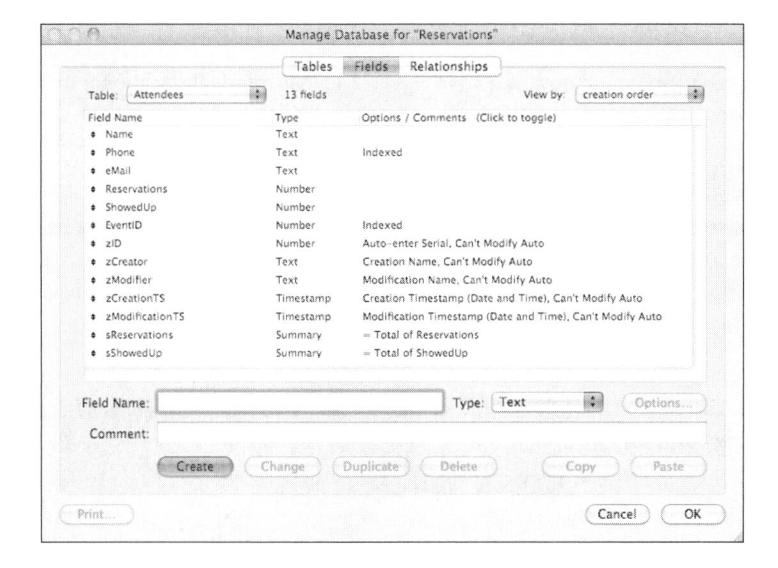

Figure 2.14
Modify, delete, and add fields.

The field names are in some ways more important than the table names. As you will see in the next section, the actual table names are not always shown. Field names should be clear and descriptive without being too lengthy. Special characters are not allowed, but underscores are. This lets you construct a field name such as private_phone_number. Alternatively, you can use what is called *camel case* to construct a field name such as privatePhoneNumber. Whatever you do, make it clear.

FileMaker provides built-in field types. If you make everything a text field, things will probably work without problems—FileMaker automatically converts data from one type to another as it needs it. However, making the field type as specific as possible can let you take advantage of formatting and validation tools that are available only for specific data types, such as numbers, dates, and times.

The FileMaker data types are as follows:

- Text
- Number
- Date
- Time
- Timestamp

- Container
- Calculation
- Summary

Most of these are common field types, but four of them deserve special attention, as follows:

- **Timestamp.** A timestamp is a composite data type that contains both a date and a time. You can choose to display it in various ways; for most purposes, using timestamps rather than dates or times will serve your purpose. Because the timestamp data type is relatively new in FileMaker, you will often find databases that do not use it; instead of a single timestamp field, separate date and time fields may be used.
- **Container.** A container field contains digital data (much like a BLOB field in other systems). That can be audio, video, images, or anything else—you can even put a spreadsheet into a container. FileMaker containers also have the ability to store references to files. To the end user, it frequently appears that a FileMaker field contains an image or other digital data, but you can organize it so that it contains only a reference to a file that contains the data.
- **Calculation.** Calculation fields are specified using the same Specify Calculation dialog you saw previously in Figure 2.11. They can include fields from the same or related tables, constants, and the results of functions.
- **Summary.** A summary field can contain a total, count, average, minimum, maximum, standard deviation, or fraction of the total for the values in the table. Summary fields are used in conjunction with layout parts (described later in this chapter). If you place a summary field in the subtotal for a layout part, it provides the relevant summary. If you copy that field and place it in the grand total part of a layout, it displays the relevant summary for all records.

After you have created a field, the Options button at the lower right of the fields tab of the Manage Database window lets you set additional options. For example, you can create a timestamp field and set an option so that it automatically is updated with the timestamp for the record's creation or last update. You can also declare a number field and set an option so that it is automatically filled with a unique serial number. Anything that you can make automatic is a benefit to you and others working on the database.

➔ There is more on the Options button in the "Using Auto-Enter Options" section in Chapter 5, "Preparing FileMaker for Mobile Use."

Many people preface field names with special characters so that when the field list is alphabetized (using the pop-up in the upper right of the window), certain fields are shown together. One of the most common practices is to prefix internal fields with z or zz. (You can see this in Figure 2.14.)

Most fields have separate values for each record of the table. However, you can specify global storage for any field. This means that it will have the same value for each record. There are two principle reasons for doing this:

- If you need to set a field using a dialog from a script step, you can set a global field. Such a field might be called `gTemp1`, `gTemp2`, and so on (g is a common prefix for global fields).
- If you need to set a value and refer to it in scripts or calculations, a global can work for you. Note that FileMaker now supports script variables, so in many cases, globals can be removed from old code and replaced by script variables.

Relationships Graph

The Relationships tab lets you get to the heart of your relational database. It provides a visual representation of your database tables. As you create a table, you see it added to the Relationships Graph, as shown in Figure 2.15.

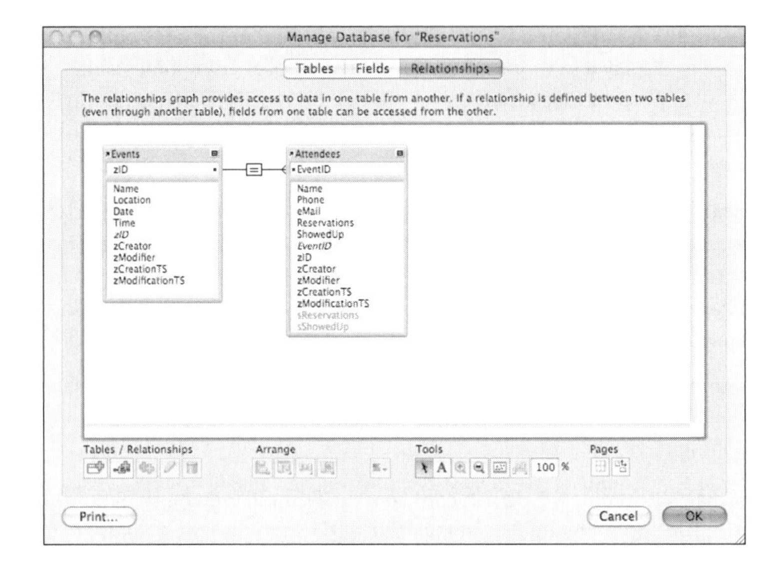

Figure 2.15
Tables are automatically added to the Relationships Graph.

You see the fields in the table by default. The small widget in the upper right of the tables lets you cycle through the full fields view to a summary view and to a one-line view with only the table name. At the upper left, the curved arrow lets you identify the underlying table. Double-click on the table name, and you can change the name of the table in the Relationships Graph. This is why the table name that you set in the Tables tab might never be seen. What you usually see is the name of the table in the Relationships Graph.

To create a relationship, draw a line from a field in one table to a field in another, as demonstrated in Figure 2.15. Note that one end of the line has multiple lines (like a crow's foot); this indicates that it is a one-to-many relationship. In this case, FileMaker has sensed that one event can have multiple attendees.

In the middle of the relationship line, a small box described the relationship. It starts as equality, but you can change it. Double-click that box to open the window shown in Figure 2.16.

You can change the fields that are related, change the type of relationship, or even add additional criteria for the relationship.

Figure 2.16

Edit a relationship.

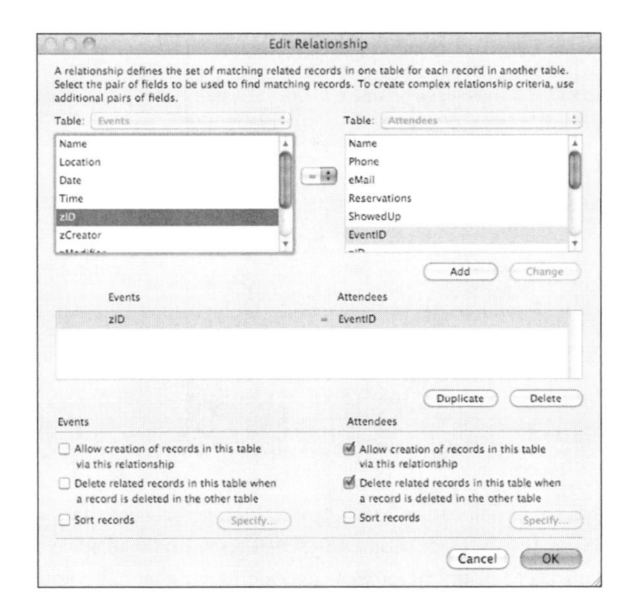

Putting Together Layouts, Relationships, and Files

A given table can appear in a number of different relationships. Just click the new table button in the lower left of the Relationships tab, select the base table, and give it a new name. (These are called *table occurrences or aliases* because they are not the actual tables themselves.) You can construct relationships among table aliases just as you do with tables themselves.

FileMaker provides you with a number of ways to focus on parts of your database, and the Relationships Graph is one of them. After your tables are defined (in the Tables tab), you can create aliases for some of them and draw the relationship that you want to use in one context or another. As you will see, when you create a layout, you specify the table it uses. However, when you are using a script step such as Go To Related Record (GTRR), you go to a related record in a table alias, and therefore, in the underlying table. The relationships defined for the table alias apply in this case. In that way, you can isolate various sets of relationships that make sense among themselves and then take advantage of them when you need them.

Another way in which FileMaker helps you focus on parts of your database is by the use of layouts. So far, layouts have been discussed in terms of the user interface, but a layout has another purpose: it identifies a set of fields that are displayed in the layout, and it describes the layout's context—its location in the Relationships Graph.

You add fields to a layout using the Insert menu or the buttons on the Status toolbar. When you do so, you are given the dialog shown in Figure 2.17 to select the field you want to use from the database.

Note the pop-up menu at the top of this dialog. It lets you move to a related table in the Relationships Graph. (If you have more than one occurrence of a table in the Relationships Graph, this is where the naming of those occurrences becomes important so you can tell them apart.) In this way, you construct your user interface for the layout.

Figure 2.17

Select a database field for a layout field.

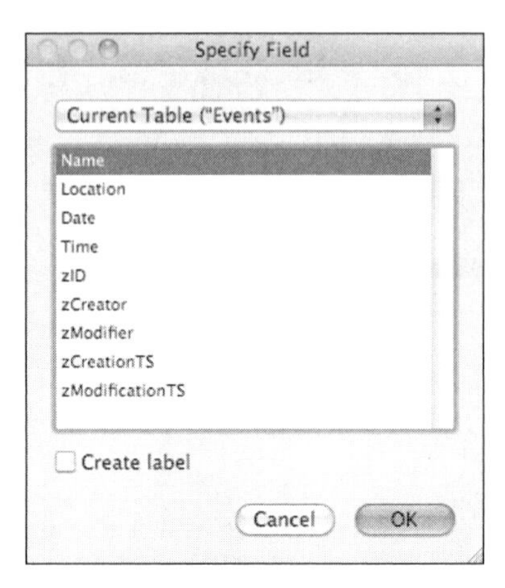

You are also constructing a set of fields from a variety of table occurrences that are bundled together in that layout. As you will see in Chapter 16, "Deploying FileMaker/CWP with FileMaker Server," the PHP Assistant uses layouts in this way. As you walk through the process of building your website, you specify a layout, and all of those fields are provided in the web solution.

In fact, it is not unusual to construct a layout and add fields to it solely for the purpose of creating a layout to be used by the PHP Assistant in building the website. (You can get rid of the layout after the website is built, but it takes up little space; it is usually a good idea to keep it around in case you need to modify the website, which is almost always the case.)

Creating a Layout

The simplest way to create a layout is to use the New Layout/Report assistant. There are three types of layouts you might want to build, as follows:

- **Interface layouts.** This is the most common and intuitive use of a layout: Build it for users to use in working with their data.
- **Web layouts.** Build a layout to use in the PHP Assistant. You do not care what it looks like; as described in the previous section, all you want is a layout that serves as a collection of fields.
- **Debugging layouts.** It can be useful to have a layout for each table that you can use in debugging. Some people do this automatically, and they label these layouts with the name of the table preceded by z, so that they are zNames, zInvoices, and so forth. FileMaker now lets you group layouts together, so you can create a Debugging folder. Use File, Manage, Layouts to create folders and rearrange layouts just as you do for scripts.

How to Create a Layout with the Layout/Report Assistant

The Layout/Report Assistant is the simplest way to create new layouts and reports. Here are the steps you need to follow.

Before starting to create a layout, make certain that you have created the tables that will be used in it. Also, ensure that the Relationships Graph is updated properly. You might need to add table occurrences or relationships. You will have a single base table occurrence, and you must be able to get to all the related tables for the layout by following lines in the Relationships Graph. If you want to create a separate set of table occurrences to support the new layout, this will work well (some people prefer to work this way, whereas others use existing table occurrences).

1. Go into Layout mode.
2. Choose New Layout/Report from the Layouts menu or click the New Layout/Report button on the Status toolbar. This opens the window shown in Figure 2.18.

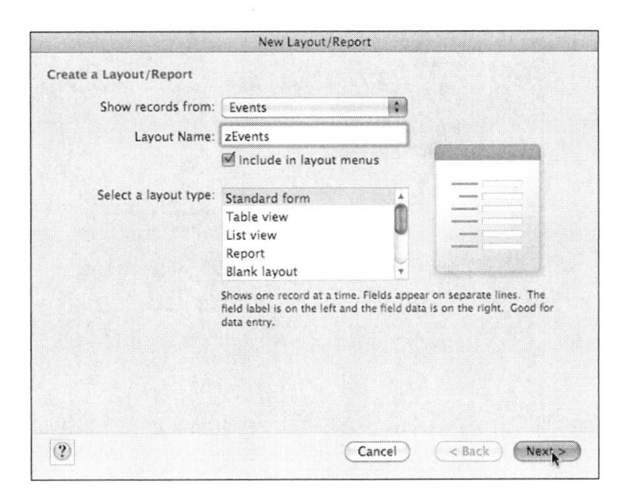

Figure 2.18
Start to create a new layout or report.

3. Select the table occurrence and name the table. For a debugging table, you might start the name with a z. Select the layout type. For a debugging table, the standard form is a good choice. Do not worry: Until the end of the process, you can always come back and change your mind. Click Next.
4. Specify the fields as shown in Figure 2.19, and click Next.
5. Select a theme as shown in Figure 2.20. You might want to use a distinctive theme (perhaps one you do not like) for debugging layouts and layouts designed just for the PHP Site Assistant so that you do not confuse them with true interface layouts.

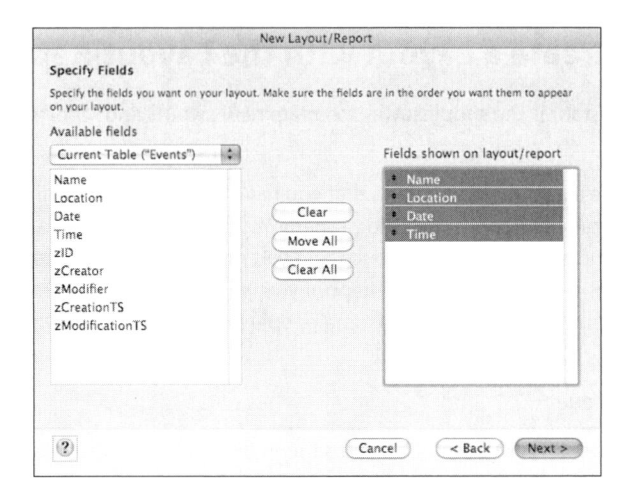

Figure 2.19
Specify the layout fields.

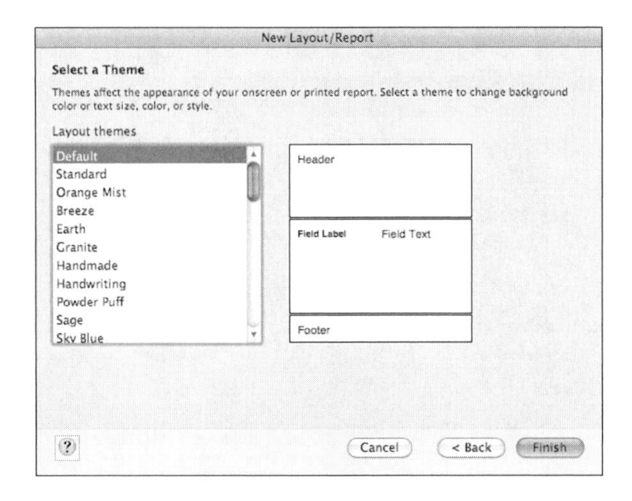

Figure 2.20
Select a theme.

6. You can choose to sort the records and to have a script built automatically for you. If you have chosen a report layout, you can specify totals and subtotals using summary fields, but for basic layouts, you are through.

When you are in Layout mode, the Layouts menu lets you change layout settings, including the name of the current layout and the base table occurrence, by using Layouts, Layout Setup. You can also manage parts by using Layouts, Part Setup. That opens the window shown in Figure 2.21.

If you choose to create a new part, you can provide its details, as shown in Figure 2.22.

Sub-summaries are very powerful, but they rely on the table being sorted in a specific way. As a result, it is almost always essential to invoke the layout from a script that performs the sort. If you are not familiar with parts and subsummaries, use the New Layout/Report assistant to develop a report and choose the option at the end to create a script. Then you can examine the layout and the script to see how they work together.

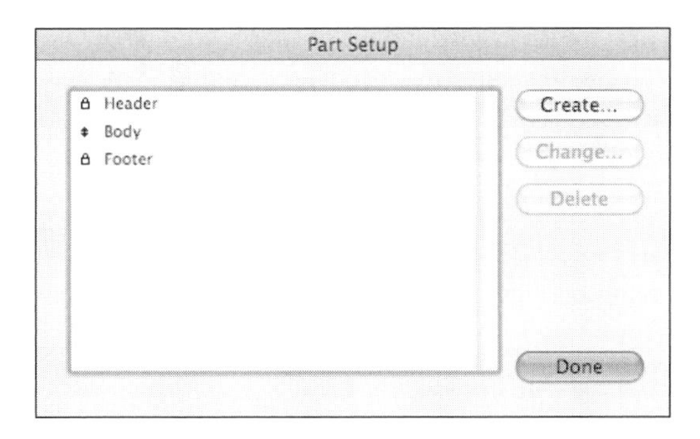

Figure 2.21
Manage layout parts.

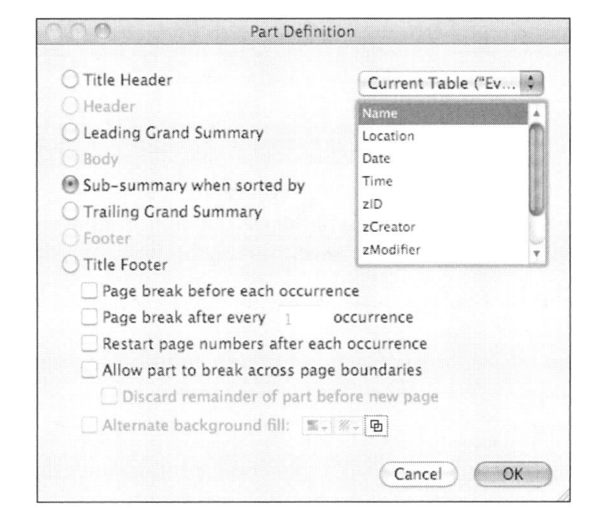

Figure 2.22
Specify details for your parts.

Structuring Solutions and Sub-Solutions

This chapter has provided a quick and high-level review of many of the developer features in FileMaker. If you have used FileMaker in the past, some of them might be new to you, because the last several versions of FileMaker have added significant new features. They can come together to make it easier for you to create mobile database solutions whether you are using FileMaker Web Publishing or FileMaker Go.

Figure 2.23 provides a schematic view of a FileMaker file as described in this chapter so far. It contains several tables that can be related to one another, as well as a number of scripts and layouts. The scripts and layouts can access any of the tables in the file through their occurrences in the Relationships Graph.

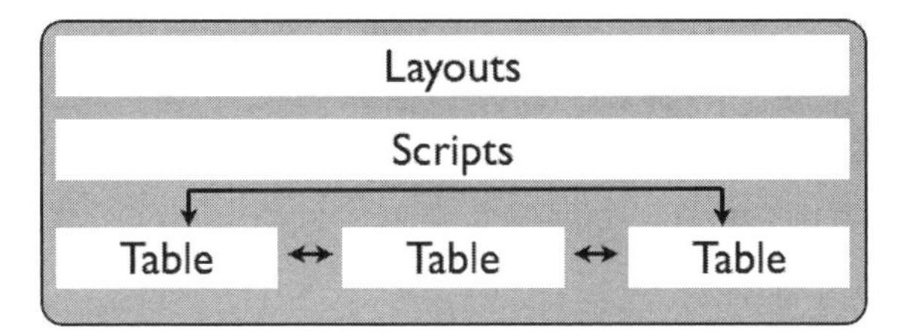

Figure 2.23
Scripts and layouts in a file can access any table occurrences in that file's Relationships Graph.

When you create a new table occurrence in a Relationships Graph, you specify the table on which it is based. You can also use that window to create a new *external data source* that, in many cases, is simply another FileMaker file (it can also be an external SQL data source). This new table occurrence is used in your Relationships Graph just as any other table occurrence is used. It does not matter that the underlying table is in another file.

Many FileMaker solutions have multiple files, each of which has its own tables, scripts, and layouts; they can be linked together with table occurrences. You do not need all three types of components. In fact, many solutions consist of two files, as follows:

- **Data file.** This file contains the actual tables.
- **Interface file.** A second file contains layouts and table occurrences based on the first file's tables.

Experienced developers in any modern language will recognize this as a classic separation of data from interface, a known way of simplifying application development and maintenance.

But where do the scripts go? The answer to that becomes simple when you consider that one file is primarily the interface and the other is primarily the data. Scripts that interact with the user go in the interface file. Scripts that are only used internally go in the data file. Because the interface file tends to be much smaller than the tables file, you can often slip the interface file in and out of the finished solution, even emailing it to users whose data files would never fit through an email gateway.

When you get around to thinking about putting FileMaker on mobile devices, a new architecture can emerge, as shown in Figure 2.24.

Figure 2.24
Use three levels of files for complex solutions.

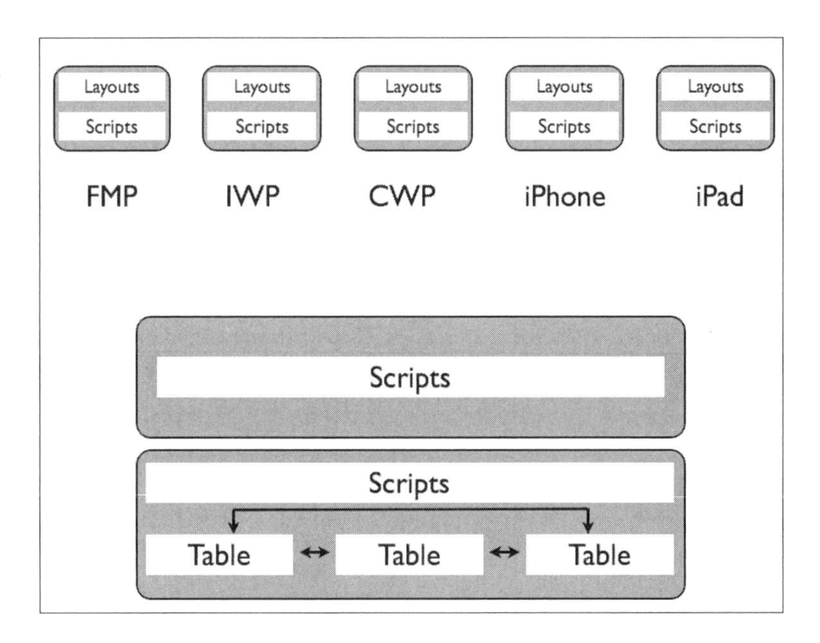

The data file still contains the tables and all the data. It might contain scripts that are used for data modification—no user interaction is used here.

The next file contains utility scripts that might be used by interfaces in general. You would normally think of putting them in an interface file, but this architecture envisions a variety of interface files for each type of platform. Some of them will need utility scripts, so you can put them in the intermediate file along with the table occurrences necessary to support them.

If you start thinking about an architecture such as this as you begin building FileMaker databases, it will be easier to add new platforms to your solution.

Further Steps

Here are some suggestions for exploring the concepts discussed in this chapter:

- If you have a FileMaker database, make a copy of it an experiment with restructuring it along the lines of Figure 2.24. You might want to start with two layers (data and interface) and then refine it.
- Whether you are creating a new database or using an old one, start going through the fields and ask yourself if they are true data fields. Particularly with old databases, you might find globals that are no longer necessary (you can substitute script variables).
- Because you can have separate interface files, start thinking about what functionality you want on mobile devices. Throughout this book, you will be prompted to ask and answer that question.

3

IN THIS CHAPTER

- Explore the Different Kinds of Database Sharing
- Share a Database with Others
- Access a Shared Database from Someone Else

Managing Data on the Move

Being able to access your data on the move can be an incredibly empowering experience. By the same token, when both you and your data are moving around, you can open a wide array of issues that can quickly get the better of you and more than offset the increases in productivity you had anticipated.

This chapter explores the concepts and terminology involved in managing data on the move. It also shows you specific steps and procedures you need to know to work with FileMaker Pro, FileMaker Server, FileMaker Go, and Bento. The basics are simple.

Synchronizing, Copying, and Sharing Data

When people start to think about accessing databases from mobile devices, they often envision a world in which their device can access their data from wherever they happen to be. Just grab your iPad or iPhone, turn it on or wake it up, and presto! Your database is in your hand.

That can certainly happen. You can create a new Bento library on your iPad or iPhone using the Bento for iPad or Bento for iPhone app, and you can do just about everything with it that you could do with Bento for Mac OS X. Your Bento library is right there in your hand.

Remember that Bento has a single database, and you create libraries within it. When people say "database" informally, they often mean something with the characteristics of a Bento library. In fact, most of the time when people say "database," they mean nothing more than something that stores some data.

In this simple scenario, you have a computer (iPhone, iPad, or Mac OS X computer) and your database is stored on that computer.

As soon as you start thinking about moving around, this simple model usually needs some revision. Many people use several computers. They might have an iPhone, an iPad, an office desktop computer running Mac OS X, and a home laptop running Windows. It is not unreasonable for them to think that they can access their database from any of those devices.

But in the simple scenario, the database is stored on the computer on which you created it. How do you access it from the variety of computers and mobile devices that you use in the course of a day? There are three ways of accessing your data from a variety of devices—three different architectures for managing your data. You will see in this chapter how to implement these three architectures with FileMaker Pro, FileMaker Server, FileMaker Go, and Bento, but the basic architectures have nothing to do with these products. The three architectures for sharing data predate all of these products, and they remain unchanged since the first days of databases back in the 1960s. The three architectures are synchronization, copying, and sharing.

Synchronization

With synchronization, you have the database on each of the devices that you use to access it. The database is on your iPhone, another copy of it is on your iPad, a third copy is on your laptop, and so forth. Each of those databases can be accessed on its own device; you can make changes on any of the devices.

Periodically, the databases are *synchronized*. In this process, a pair of databases are compared. When discrepancies are found between them, the later data is selected, and it is copied to the other device where it overrides the older data. Sometimes during synchronization, it becomes obvious that data has been changed on two devices, and it is not possible to definitively and automatically determine what the best data is. In these cases, the user is normally asked to intervene.

Synchronization works best when it is done automatically and relatively frequently. If you have a month's worth of updates on two separate copies of the database, chances are that you will have more conflicts and confusion than if synchronization is performed daily (or even hourly).

Furthermore, the longer the interval between synchronization, the greater the possibility that you have misleading data. Consider, for example, a calendar database. If you schedule a meeting for 3:00 p.m. on your iPad version, it will not show up on your desktop version until after synchronization. This means that you can schedule another meeting for 3:00 p.m. on your desktop calendar. When synchronization does take place, the two appointments are noted, and you have to untangle them.

Synchronization is implemented on personal digital assistants (PDAs) and smartphones precisely for this purpose. By now, the implementation of a synchronization process is not terribly complex because developers have been dealing with it for a long time.

In the case of more than two copies of a database, synchronization is often performed on pairs of the databases, with an extra pass through the databases scheduled automatically so that data is propagated throughout the set of databases. At the end of a synchronization process, the data is normally identical on all the database versions, and then the process of updating and synchronizing can proceed again.

One of the disadvantages of synchronization is that for it to take place, both databases must be available so that the synchronization software can look at both of them at the same time. For larger numbers of databases, each pair must be available at a certain time; the next pair can be synchronized later when those two databases are available, but that allows a period of time when you are working with unsynchronized and possibly inconsistent databases.

Sometimes, one database is selected as the central database, and all synchronization happens between that database and any one other database in the group.

Copying

Instead of using synchronization in which changes to each of a pair of databases are compared, you can simply copy databases data back and forth. There are two ways in which this can be implemented.

Copying Entire Databases

You can designate one database as the current database and put it on the device you are using. Then, you can copy it to another device and use it there. This strategy works very well for a personal database that you keep with you. You can keep it on your computer at work, copy it to a disc, and then put it on your home computer. Many people are used to packing up a briefcase, backpack, or purse when they travel between home and work, so this architecture fits comfortably into that behavior.

Copying Updates

Instead of copying an entire database, individual changes to a database can be copied back and forth. Instead of waiting for a synchronization to pick up all changes, they are processed one by one. This keeps two databases more closely synchronized.

This method works particularly well with Internet connections, including the connection that may go over a mobile phone network to an iPhone or iPad. It is used to synchronize MobileMe calendars and contacts if you choose to do so. That is how you can enter a new appointment or change a phone number on your iPhone and see the revised data a minute or two later on your Mac or PC.

Sharing

Both synchronizing and copying are strategies for keeping two copies of the same database more or less consistent. You can eliminate the need for consistency checking by using a simple strategy: Share the database. In this architecture, there is only one copy of the database, so it cannot be out of sync. Whatever is in that one copy of the database is the data.

Managing FileMaker Pro

If you are using a FileMaker database on a single computer and not sharing it, you do not have to worry about synchronizing, copying, or sharing. However, you can use FileMaker Pro to access databases that are shared by other people, as well as to share your own databases with other people. Here is how you do both of those tasks.

Accessing Databases

You have two choices when you are trying to access databases: You can open local databases or shared databases.

Opening Local Databases

When you open a local database, you use File, Open to select the file. You can navigate to any FileMaker file to which you have access on your own computer or on any computer to which your computer is connected. The basic File, Open command uses the Windows or Mac OS X file system.

You get FileMaker databases onto your computer and its networked disks in two ways:

- You can copy them from a CD, DVD, portable ("thumb") drive, or website.
- You can create them with FileMaker Pro or FileMaker Pro Advanced.

Opening Shared Databases

When you want to access a shared database, you use the command just below File, Open. It is File, Open Remote, as shown in Figure 3.1.

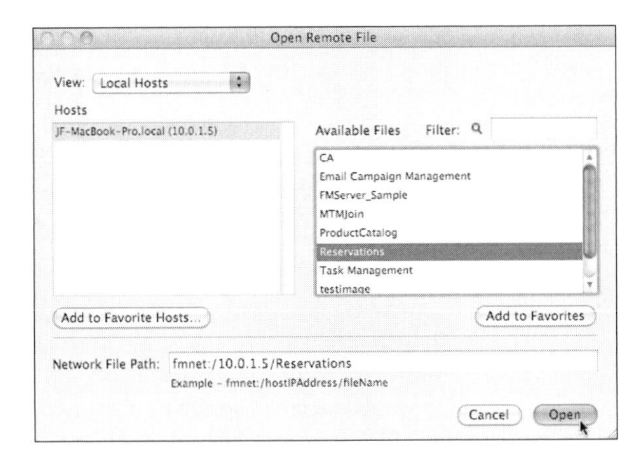

Figure 3.1
Use Open Remote to open a shared database.

The pop-up menu in the upper left of the dialog lets you select the type of computer you are looking for, as follows:

- **Local Hosts** automatically finds servers on your local network. These servers have names (provided by their owners) and network IP addresses. Select the server you want to use; in the right-hand pane, you see a list of the available databases on that server. Note that "server" can mean FileMaker Server, as described in the following section, or it can mean FileMaker Pro, as described later in this section.
- **Favorite Hosts** is a list that you can create with the names and addresses of your favorite servers. Notice in Figure 3.1 that at the bottom of the window is a URL that will take you to a specific server (it is filled in as you select a server and a database). You can copy that URL and then create a favorite so that you can easily go back to it.
- **Hosts Listed by LDAP** lets you connect to FileMaker databases through an LDAP server. If this is part of your configuration, your support and help desk will show you how it works. It is not specifically a FileMaker technology.

If you are connected to a shared database fairly frequently, it makes sense to store its address in the Favorites list. If you use a database fairly frequently and do not use a wide variety of other databases, you can bypass the Open Remote dialog and choose File, Open Recent which simply reopens a recent database whether it was on a server or your own computer.

When you are connected to a database that is not on your own computer, the name of the computer that is hosting it appears next to the database name at the top of the window, as shown in Figure 3.2.

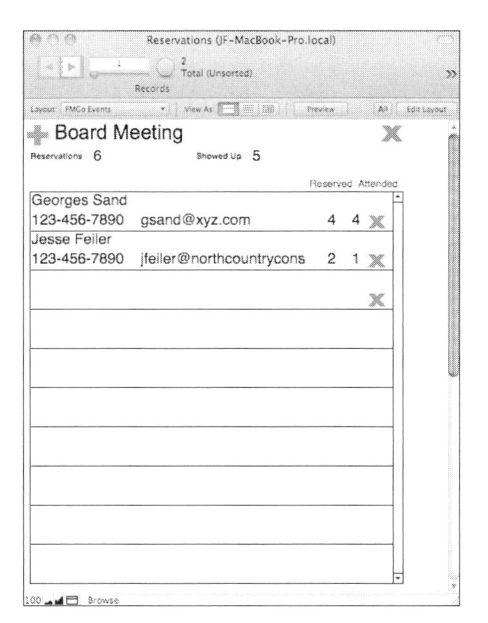

Figure 3.2
Remote databases show their host computer name next to the title.

Sharing Your Databases

With FileMaker Pro, you can not only open databases that other users have shared, but you can also share your own databases with up to nine people. In order to share a database, open it as you normally do (by double-clicking it or by using File, Open).

Make certain that the privilege sets that will be used by people who will access the database over your network allow them to do so. Choose File, Manage, Security; click the Edit Privileges tab, and double-click the privilege set to open the window shown in Figure 3.3. Check that Access via FileMaker Network is turned on in the list of Extended Privileges in the lower left.

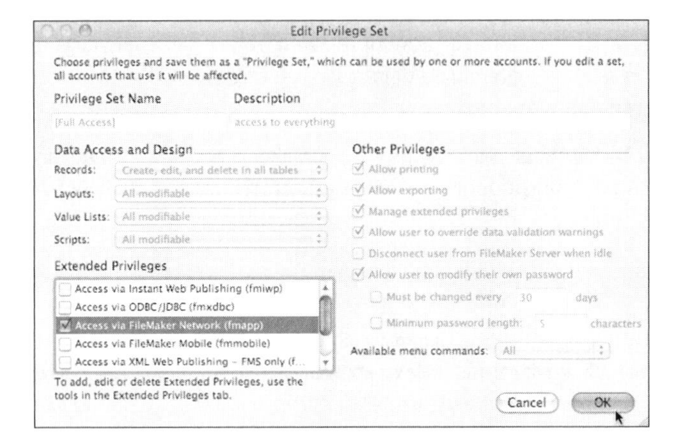

Figure 3.3

Allow network access to the database.

Remember that you only have to set privileges for privilege sets and then allow specific users to use the privilege set. In that way, you do not have to redo the settings for each individual user.

Choose File, Sharing, FileMaker Network to open the dialog shown in Figure 3.4.

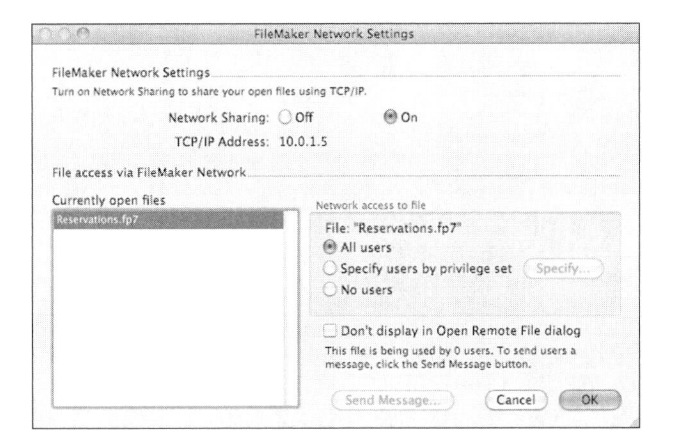

Figure 3.4

Share your own databases.

If Network Sharing is off, turn it on at the top of the window. If you cannot turn it on, look at the next line to see if you have a TCP/IP address. If you do not, your computer is not yet set up for sharing over a network. If you cannot access a website or your email, that is the most likely case. If you can get to websites and your email, you do have a network connection, and you should be setting a TCP/IP address. If you do not, check the FileMaker Knowledge Base at www.filemaker.com/kb/?nav=support-kb.

Select the database from the list at the lower left; then choose the settings at the right. You might want all users to have access if you are sharing a database over a local area network. Alternatively, you might want to use privilege sets so that some databases are visible only to specific groups (such as managers using a management privilege set).

The checkbox in the lower right lets you control whether or not the database is shown in the Open Remote dialog shown previously in Figure 3.1. If all of your databases are not available to everyone, it makes sense to not put them in the list most of the time.

If there are people using the shared database at this time, the lower right is where you can broadcast a message to all of them. Many people forget about this feature and where it is because most of the time, this dialog is used just to set up sharing and not to communicate with the users.

After you have set the extended privilege for the database and turned on sharing with File, Sharing, FileMaker Network, up to nine users can connect, as described in the previous section.

Remember that everyone is connecting to this one database, so there is no question of synchronization or copying from one computer to another. However, because everyone is connecting to that one computer, it must be powered on, and the database must have been opened in FileMaker before people can connect to it with Open Remote.

This means that this type of sharing works well in a controlled environment where there is someone around to do a little troubleshooting if someone cannot connect to the database. An office is a great example because there is usually someone there. People can connect to the shared database even if they are outside the office (even halfway around the world) as long as it is running and as long as they know the IP address of the computer on which it is running.

If people will be connecting from the outside, you might have to take additional steps to allow them access to your network, particularly if it uses IP addresses that change from time to time. This is common in networks served by a cable or DSL modem. The address of the modem can change from time to time, and the addresses on the local area network might also change. You can assign static IP addresses to the computers on your network so that you know what they are, but in most cases, the only way to know the address of the modem is to arrange with your Internet provider for a static IP address.

Managing FileMaker Server

FileMaker Pro enables you to share your database with up to nine people, but it is not designed for heavy-duty work. That is provided by FileMaker Server, a separate product.

FileMaker Server has only the server functionality. You cannot use it to browse FileMaker databases, and you cannot use it to create or modify them. To actually use the database in the ways that most people do, you need a copy of FileMaker Pro.

There are two versions of FileMaker Server. In the basic version, access for up to 250 users is provided. In addition to access through FileMaker Pro, you can turn on PHP web publishing so that PHP code in web pages can get to the database and retrieve data or update it.

FileMaker Server Advanced adds support for Instant Web Publishing for up to 100 users (with FileMaker Pro, you can use IWP for up to five users; it is not available on FileMaker Server). Furthermore, with FileMaker Server Advanced, the only limit to the number of direct users (FileMaker Pro) is the hardware configuration. (Compare these limits to sharing supported in FileMaker Pro for up to nine users.)

All database servers do the same basic job and use the same basic structure, although the terminology varies. This section shows you the basics of sharing a database with FileMaker Server. It is the FileMaker Server version of the previous section, "Sharing Your Databases."

The first step is still the same: Make certain that the extended privilege for FileMaker Network is set for the database file. (If it is set for sharing on FileMaker Pro, it is also set for sharing on FileMaker Server.)

Now you have to open the database in FileMaker Server and leave it open. FileMaker Server is most often deployed on a computer that is on all the time. Another common configuration is a computer that runs during the workday. When it is turned on, you can configure FileMaker Server to start up automatically and open its databases for you. Ideally, this computer runs nothing other than FileMaker Server so as to get the best possible performance. (For small installations, this is not a great concern, but if you are thinking about scores or even hundreds of users, you do have to worry about performance.)

The installation process for FileMaker Server is highly automated and easy to follow. During that process you will provide an administrator name and password so that you can log into FileMaker Server Admin in the future. There is more information on the disc or disk image that contains FileMaker Server software. Also, the FileMaker website has further information.

The deployment process starts after you have installed FileMaker Server on a computer to which you have access. It can be on your own computer, another computer on your local area network, or a computer anywhere else on the Internet that you can access. As part of that process, you need to open some ports for use by FileMaker Server. Here are the port numbers so you can get started quickly:

- **Open firewall ports:** 5003, 16000, and 16001
- **Open internal ports (not necessary on the firewall):** 16004, 16006, 16008, 16010, 16012, 16014, 16016, 16018, 50003, and 50006
- **For web publishing:** 80

How to Manage Databases on FileMaker Server

The standard installation leaves FileMaker Server running and set so that it starts up each time you reboot the computer. This enables you to start to deploy databases right away:

1. Connect to FileMaker Server from a browser. Use the IP address of the computer where FileMaker Server is running. If you are working on the same computer on which FileMaker Server is running, you can use `localhost` for the IP address. FileMaker Server is listening on port 16000, so the URL will be something such as 10.0.1.5:16000, localhost:16000, or myserver.com:16000. If FileMaker Server is running, you see the screen shown in Figure 3.5 in your browser. As long as you have an Internet connection, FileMaker Server checks to see that you have the latest software, as shown in Figure 3.5 at the bottom.

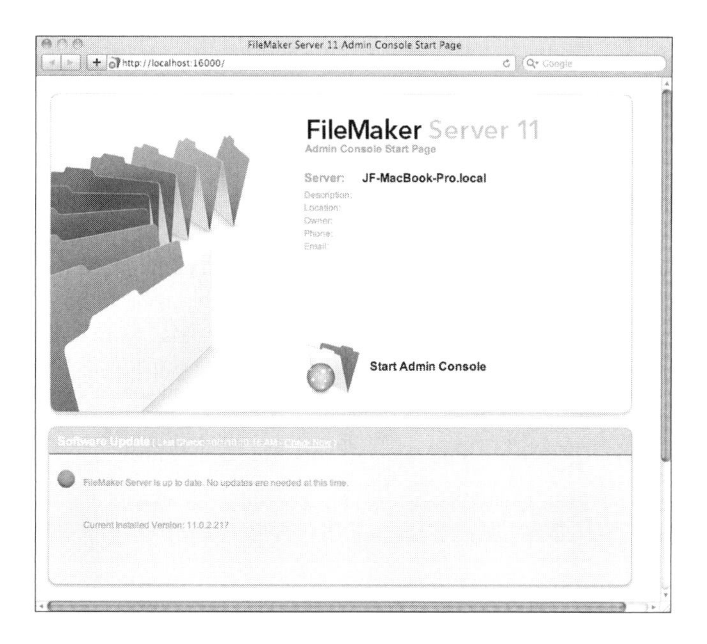

Figure 3.5
Connect to FileMaker Server on port 16000.

2. Scroll down to see links for documentation and testing, as shown in Figure 3.6.

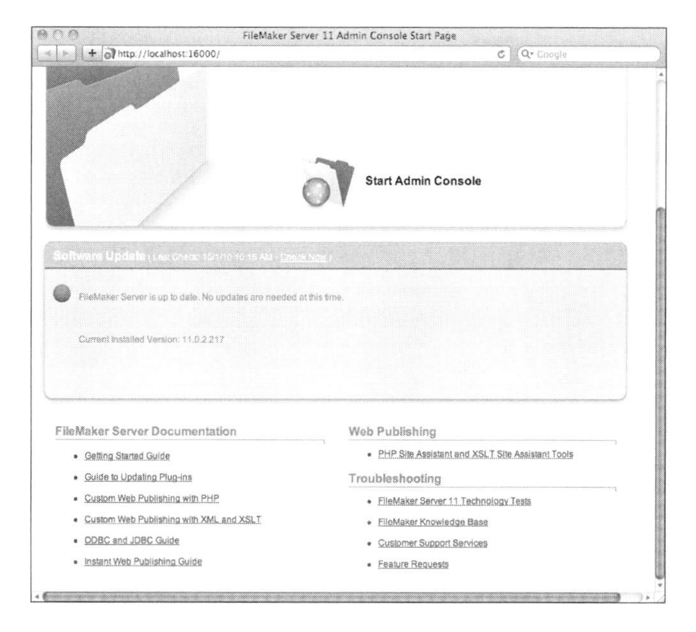

Figure 3.6
There are links to additional resources at the bottom of the page.

3. Click Start Admin Console to begin to administer FileMaker Server. Admin Console is written in Java, so you might be asked if you will allow it to have access to your disk, as shown in Figure 3.7. You might also be asked if you want to put an icon on your desktop (or anywhere else) so that you can easily launch Admin Console in the future, but steps 1 and 2 will work even if you do not have a desktop icon.

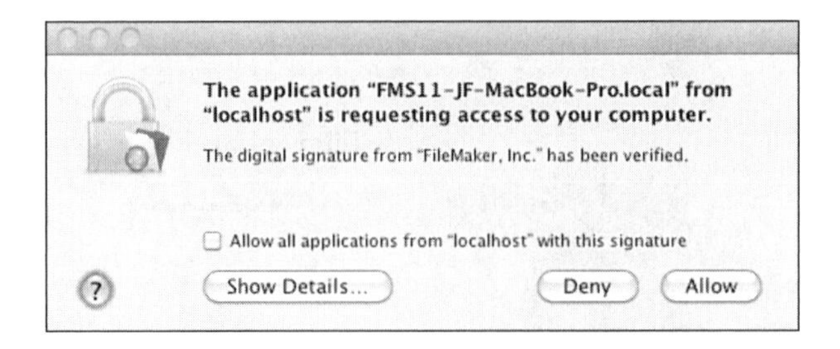

Figure 3.7
Grant access to Admin Console.

4. When asked, provide your FileMaker Server login name and password.

5. The Admin Console window opens, as shown in Figure 3.8. You can select any of the items in the list at the left to see details. You start with the FileMaker Server Overview, as shown in Figure 3.8. Just check that everything is OK (you should see green dots and no red ones).

   You can go to this overview whenever you have doubts about FileMaker Server so that you can check the status of all the components.

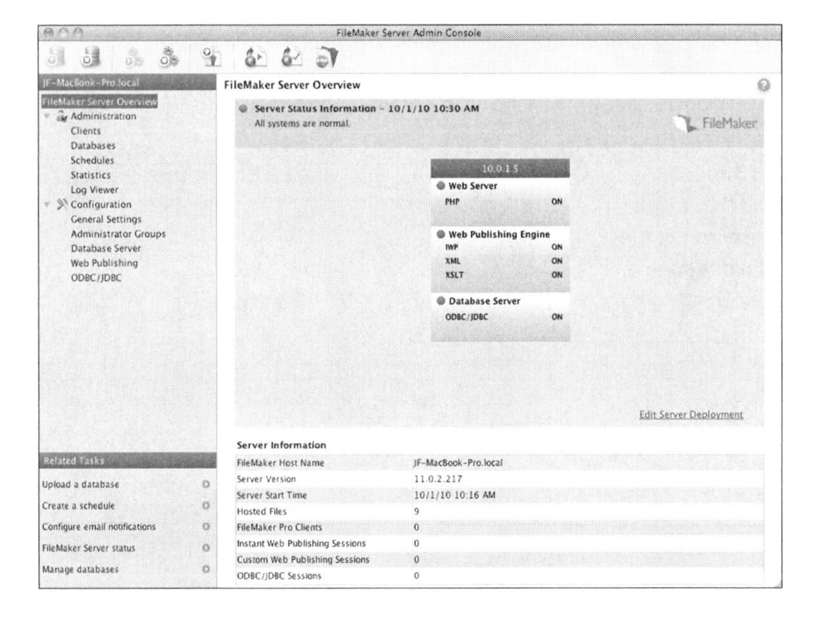

Figure 3.8
Check the server status.

6. Go to the Databases view, as shown in Figure 3.9. This is where you close and open databases (or, in traditional database-speak, stop and start them). Select one or more databases, and then choose an action from the pop-up menu, as shown in Figure 3.9; then click Perform Action. The window shows the status of each of the databases.

7. To add a database, use the tasks in the lower left of the window. Click Upload a Database to select a file and upload it. Do not simply drag the file into the list of FileMaker's databases.

8. Finally, click General Settings at the left and confirm that FileMaker Server is set to start automatically, as shown in Figure 3.10. After you have done this, you do not have to worry about it again.

Figure 3.9

Select actions for the databases.

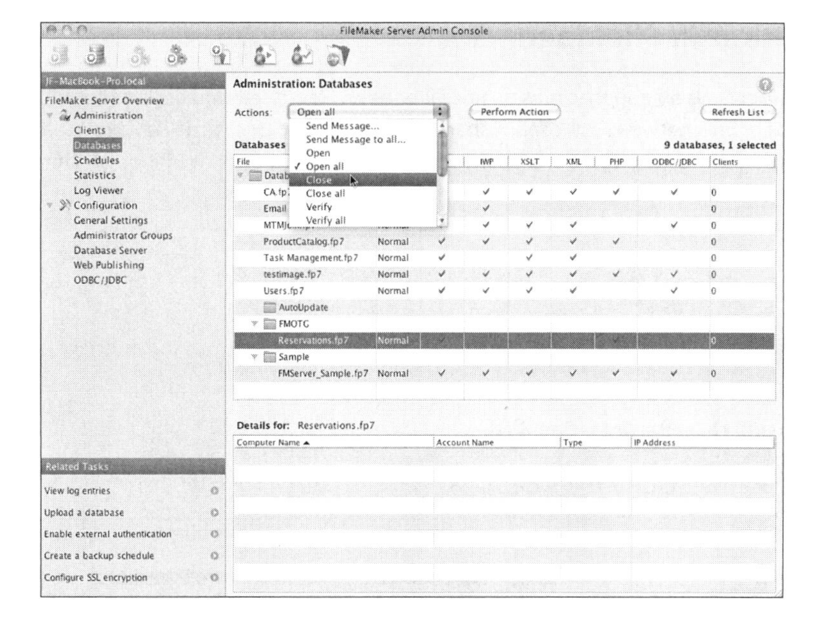

Figure 3.10

Configure FileMaker Server to start automatically.

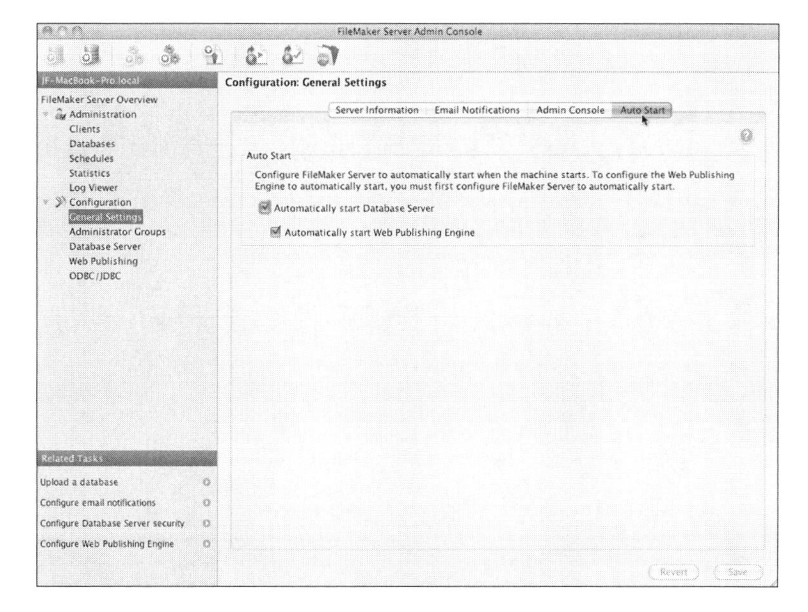

4 1 1

There are hosting services that provide FileMaker Server either on dedicated or shared computers. Search www.filemaker.com for hosting to find links to various vendors. Prices generally depend on the number of files you need to have hosted. Using a hosting service can make deploying a shared database with FileMaker Server much easier than doing it yourself.

Managing FileMaker Go

FileMaker Go is an app that runs on your iPhone or iPad. It lets you access FileMaker databases on the mobile device or on a network. You cannot share your own databases with FileMaker Go, so this section is about accessing databases. As is the case with FileMaker Pro, there are two kinds of databases you can access: local databases and shared databases.

Accessing Local Databases

The functionality of FileMaker Go is similar on iPad and iPhone, but the interface is different.

Accessing Local Databases on iPad

When you launch FileMaker Go on iPad, you might see the screen shown in Figure 3.11. On the left are the databases on your iPad. Tap the one you want to open, and you are ready to go.

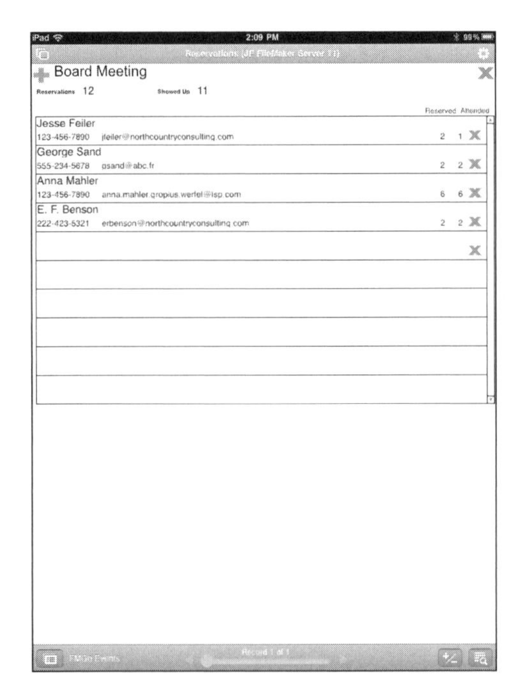

Figure 3.11
Open a local database on iPad.

The database opens, as shown in Figure 3.12.

→ There is more on how to use the FileMaker Go interface in Chapter 7, "Introducing FileMaker Server," **p. 151**.

Figure 3.12
Use the iPad database.

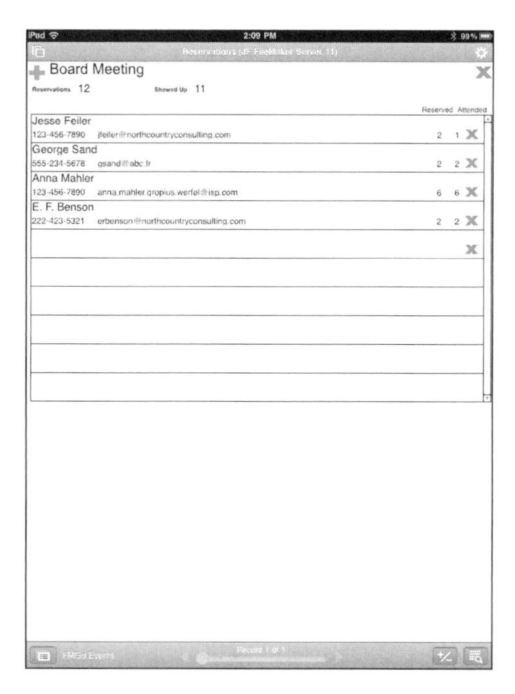

Putting Local Databases onto Your iPad or iPhone with iTunes

It is easy enough to put a local database onto your Mac or Windows computer by using a CD, DVD, or copying it from a website (or from an email message). But the process is different for iPhone and iPad.

You use iTunes to install software and files on your iOS devices. Using a cable or dock, connect your iPad or iPhone and open iTunes (it might open automatically if you have that preference set). Select your device from the pane at the left of the window, as shown in Figure 3.13. Then, select the Apps tab at the top of the window as you see in the figure.

Scroll to the bottom of the window, as shown in Figure 3.14. As you can see, each of your installed apps can have its own files. (Not all apps will have files.) Use the buttons in the bottom right of the window to add new files to your device or to save files from the device to another location. If the file does not transfer immediately to your device, click Sync in the lower right of the window.

If you want to remove a file from the device, the standard delete gesture does the trick. In the list of local databases shown previously in Figure 3.11, just swipe across a database and then tap the Delete button that appears to the right of the file name.

Figure 3.13
Go to your device's Apps tab in iTunes.

Figure 3.14
Manage files from iTunes.

Putting Local Databases onto Your iPad or iPhone with Email or the Web

You (or a friend) can email a FileMaker database to you, and you can open that email message on your iPhone or iPad. iOS recognizes FileMaker files (with the .fp7 extension) as belonging to the FileMaker Go app. When a message arrives with a file type that iOS recognizes, it displays an icon for the file and provides a button that lets you open it with the appropriate app. For FileMaker databases, the button is labeled Open in FileMaker Go.

If you access a FileMaker database on a website, the same process occurs. iOS recognizes the file type and provides you with a button to open it in FileMaker Go. You can navigate to the file on a website by using a URL such as http://mywebsite.com/mydatabase.fp7 or by clicking a link that contains that code.

At that point, it will be added to your list of recent databases, and you can reopen it even after the email message has been deleted. You can remove the database from your mobile device by swiping across its name in the local databases list. You can also use iTunes as described in the previous section to copy that database onto your desktop computers disk.

When using email to transfer a FileMaker database, remember to check the maximum file size that your email software allows you to send or receive.

Accessing Local Databases on iPhone

If you are using an iPhone, the single screen shown in Figure 3.11 is replaced by three screens, as shown in Figure 3.15.

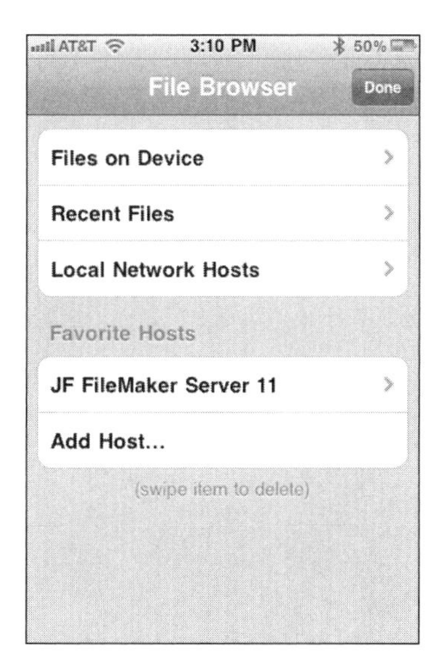

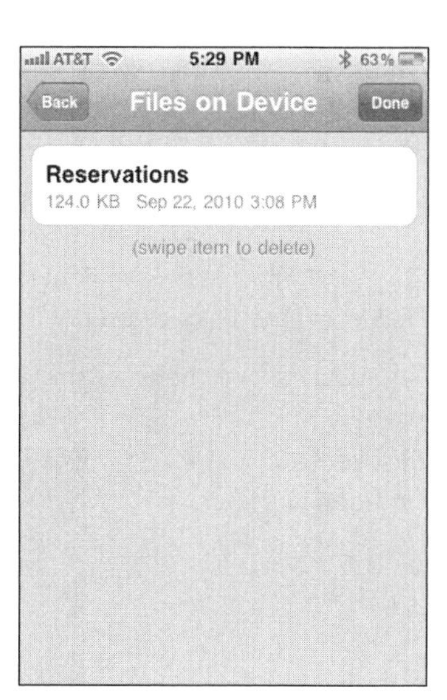

Figure 3.15
Open a database on iPhone.

From the main screen, tap Open File Browser. On the File Browser screen, tap Files on Device, and then tap the database you want to open.

Accessing Shared Databases

Both iPhone and iPad let you access a shared database over a network, which might be the data side of the phone network, or it can be a Wi-Fi network that you have joined. In either case, you can get to a FileMaker database as long as it has been shared and you have the right security access.

→ To review how to share a database from FileMaker Pro or FileMaker Server, **see** the "Sharing Your Database" sections previously in this chapter.

Accessing a Shared Database from iPad

You previously looked at the main FileMaker Go screen to see how it lists local databases at the left. Figure 3.16 focuses on the right-hand side, where you can see recent hosts and favorites.

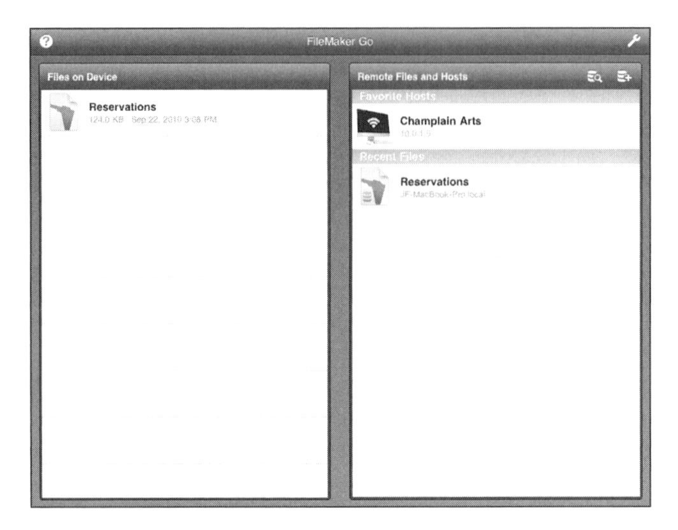

Figure 3.16
Connect to shared hosts.

The two buttons at the top right let you search for a FileMaker host on your local network and add a FileMaker host to your Favorites list.

Figure 3.17 shows the list of any FileMaker hosts on your local network.

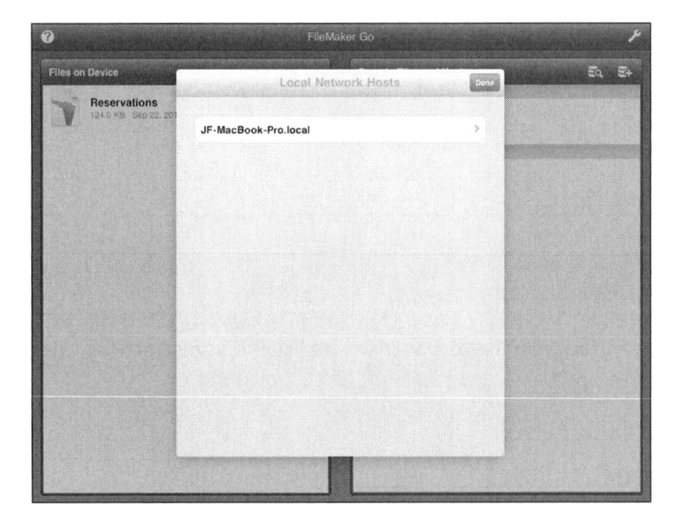

Figure 3.17
Find local FileMaker hosts.

If you want to add a host, you need to provide its address and, optionally, a name for it, as you see in Figure 3.18.

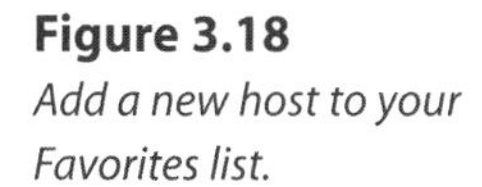

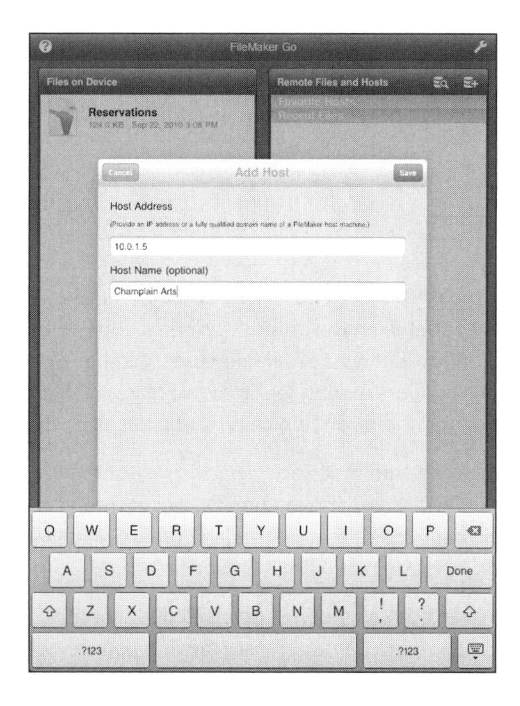

Figure 3.18
Add a new host to your Favorites list.

Connect to a host by tapping it in the Recents or Favorites list. You see a list of its databases, as shown in Figure 3.19.

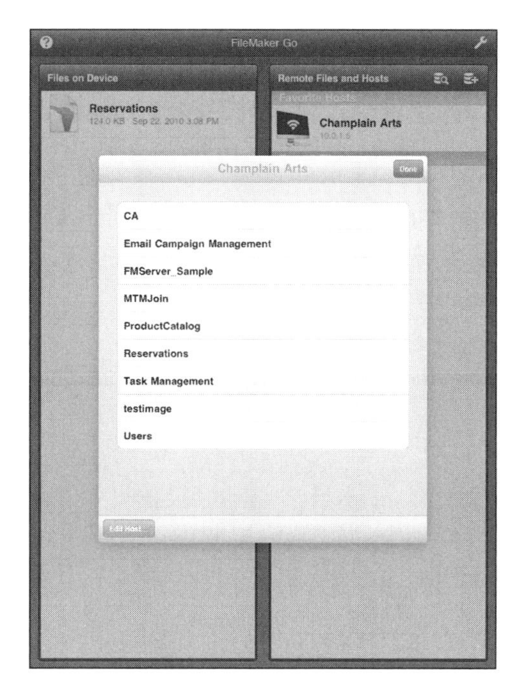

Figure 3.19
Select a database from a host.

Accessing a Shared Database from iPhone

On iPhone, the process is very similar. The only difference is that you step through several screens to do it, just as you did to open a local file (refer to Figure 3.15).

Managing Bento

Bento is a powerful and focused database. It lets you do many of the things that you can do with FileMaker Pro, but there are very clear differences between the products. When it comes to sharing your libraries (remember that Bento uses "library" in much the way FileMaker uses "database"), Bento does not allow real-time sharing with iOS devices. It does support sharing for up to five Macs on your local area network, so it works very well for a small office or a home network, which is what it has been designed to do.

Although Bento does not support sharing with mobile devices, if does something that FileMaker cannot do: It supports synchronization with a mobile device. As noted previously, synchronization can quickly get complicated if it needs to manage multiple devices. For that reason (and others), Bento supports synchronization between one iPad and one Mac and between one iPhone and one Mac.

This can be the same Mac. You can have both an iPhone and an iPad synchronized with Bento on a Mac, but you cannot have two iPhones synchronized to Bento on that Mac, and you cannot have an iPad or iPhone synchronized to Bento on two Macs.

Because there is no real-time sharing, you do not worry about connecting to shared libraries or sharing your own libraries directly with other users. However, you do accomplish many of the same goals by setting up synchronization.

How to Set Up Synchronization Between a Mac and an iPad or iPhone

The process is the same for both devices, but the screens are slightly different because the iPhone is smaller:

1. Start by making certain that Bento is running on both your iPhone/iPad and your Mac and that both are connected to the same wireless network.
2. On your mobile device, open Sync & Setup from the Actions button, as shown in Figure 3.20. Tap Sync with a New Computer. You are warned if you have already paired the device with another computer.
3. You are given a four-digit passcode, as you see in Figure 3.21.

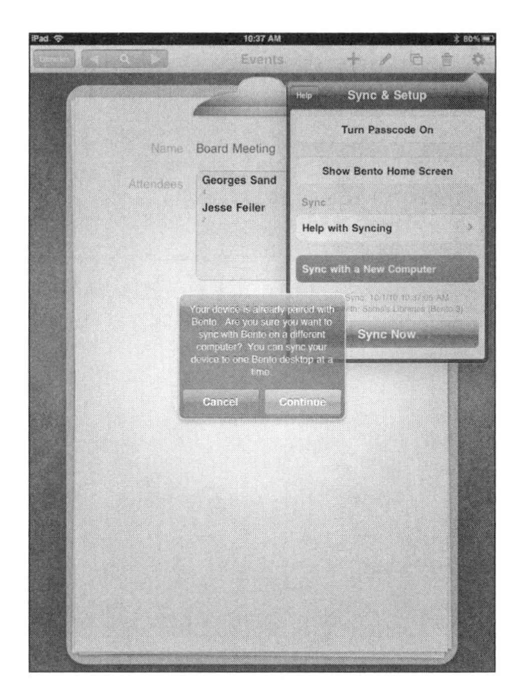

Figure 3.20
Set up syncing with a Mac.

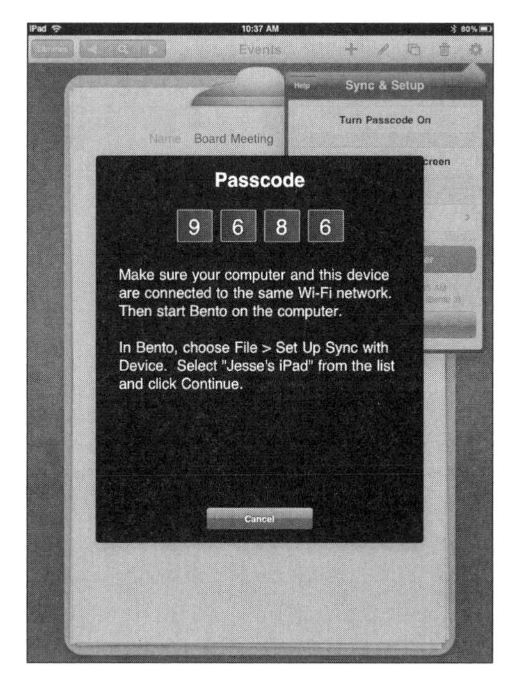

Figure 3.21
Get a passcode.

4. As soon as Bento on your Mac notices a computer that is trying to pair with it, you are asked to choose which device you want to pair with, as shown in Figure 3.22. (This matters because there might be several mobile devices on the wireless network.)

Figure 3.22
Choose the device to pair with.

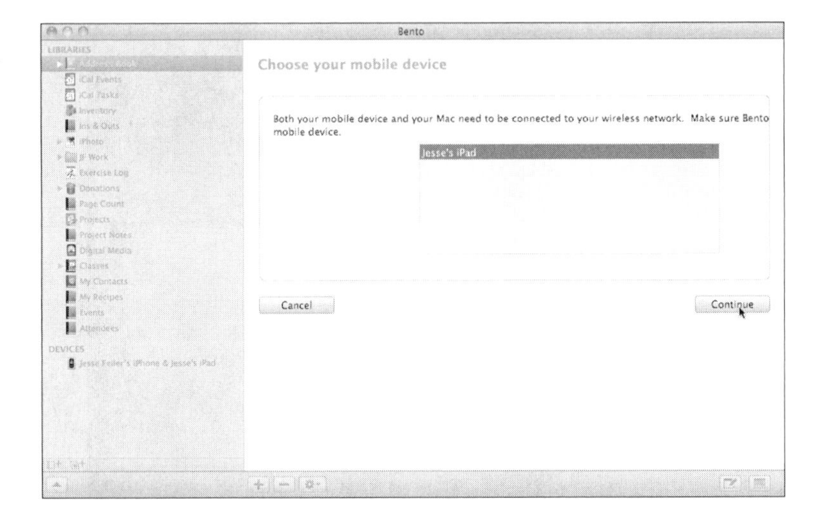

5. You get a confirmation that the pairing is complete
6. Continue on to select the libraries you want to sync (or all of them), as shown in Figure 3.23.

Figure 3.23
Choose the libraries to sync.

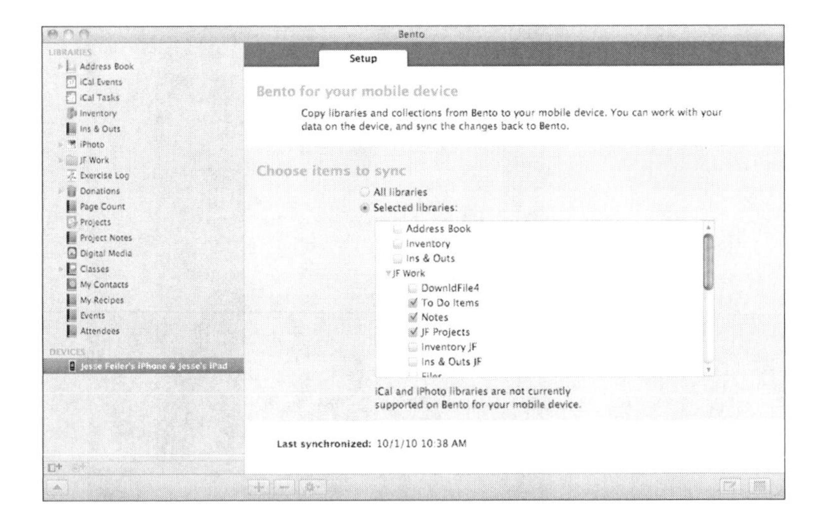

After you have set up synchronization, all you have to do is to remember to do it periodically. You initiate synchronization from your mobile device at a time when both are connected to the same wireless network and both are running Bento. Use Sync & Setup, as shown previously in Figure 3.19; this time, tap Sync Now.

The only other thing you might want to do is to modify the libraries that are synchronized. You can do that from Bento on your Mac. Choose Bento, Preferences and use the Sharing tab, as shown in Figure 3.24.

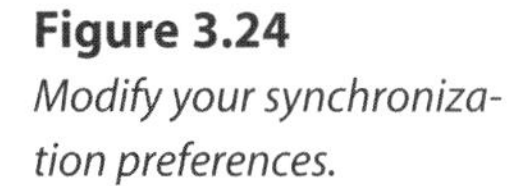

Figure 3.24
Modify your synchronization preferences.

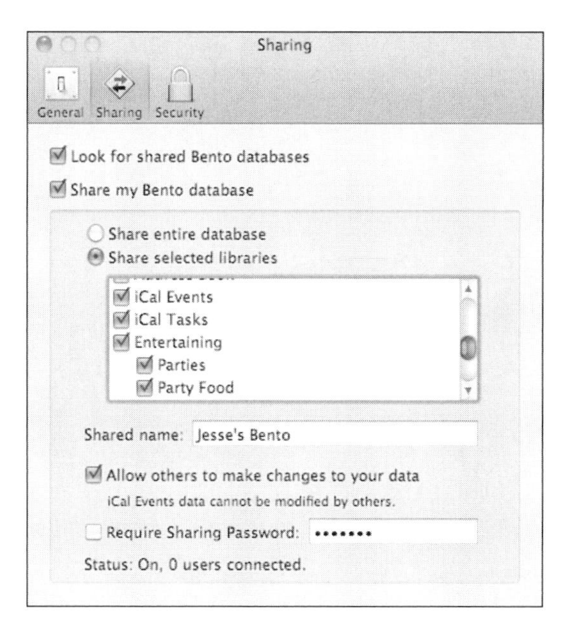

You can save a little bit of synchronization time if you do not sync all of your libraries, but the process is normally so fast that you might want to just sync all of them.

One of the main reasons that you might want to limit your synchronization is if you want to make certain that some data does not move to your mobile device or to your Mac. If you are the sole user of the mobile device and the desktop Mac this is not an issue. But if you are sharing either one with other people, you might want to limit the data that is passed along.

Further Steps

Here are some suggestions for exploring the concepts discussed in this chapter:

- As you start to enable remote wireless access to FileMaker Pro databases, review your security settings. Particularly look at automatic login set in FileMaker Pro with the File, File Options dialog. By default, you automatically log into the Admin account with all privileges. Chances are, you want to log into another account with only some privileges if you are exposing your database on a wireless network.
- Think about how you will handle data sharing. Remember that iPads and iPhones are mobile devices, so you cannot assume that they are always able to connect to a shared database. One strategy for dealing with this type of situation is to have updates stored separately from the main database and then merged into it periodically.
- Even if you are not planning to share a new FileMaker database with mobile devices, plan how you would do it. Chances are that one of these days, you will need that option.

4

IN THIS CHAPTER

- Using Your Fingers
- Working with Text
- Managing Mobile Files
- Printing with a Mobile Device

Working with Mobile Devices

Although the world of mobile devices has been expanding over the last few years, a pattern is taking shape. That pattern is centered around four types of devices:

- Laptops of various screen sizes
- Netbooks, which generally have smaller screen sizes than laptops
- Tablet computers, such as iPad
- Smartphones, such as iPhone

Laptops and netbooks generally have keyboards built into them; tablet computers and smartphones might have small keyboards, onscreen keyboards, or the ability to connect wireless keyboards to them. Many (including iPad and iPhone) have several of these options.

The combination of the relatively smaller screen size than a desktop-based computer and the use of a smaller keyboard for input is at the heart of what you have to think about when you are developing for a mobile device. Of course, at the same time, you have to consider that these mobile devices and their owners can go many places where traditional computers cannot go. The marketplace seems to be deciding that whatever drawbacks a smaller screen size might have (and, in the case of iPad, many people would argue that the screen is just fine as is) and that the smaller keyboard might pose, the greater mobility is far worth it.

The issues of screen size and keyboard, as well as what you have to think about when you are working in a world of touch control, recur throughout this book. Anyone who develops for mobile devices has to be aware of them because it is a new way of thinking about interfaces.

This chapter deals with issues common to most mobile devices. In the next chapter, you see some of the issues that are common to both mobile devices and the FileMaker database tools.

Working with Your Fingers

Pointing with your finger and pointing with a computer are two very different experiences; neither is intrinsically better than the other. It is your job as a designer of a database solution to handle both appropriately. There are two primary issues to consider:

- A computer mouse with its pointer on the screen can point much more precisely than a fingertip can. This means that clickable items on a user interface can be much smaller than tappable items on a touch interface: Fingertips are enormous compared to the tip of a pointer on a computer screen.
- For most people, a fingertip can move farther and faster than a computer mouse can. Even a wireless mouse needs to be picked up and moved away from the edge of a desk when you need to move the pointer on the screen a bit further.

There is another point to consider when comparing your fingertip to a computer mouse: Although you can watch the mouse pointer move along the screen as you drag the mouse on the desk, there is no comparable behavior with a touch interface. A computer mouse can participate in mouse-up tracking (that is, the movement of the mouse and its pointer on the screen without the button being down), but if your finger is not touching the screen, its movement cannot be tracked.

This means that every aspect of an interface that relies on mouse-up tracking just does not work on a touchscreen—there are no tooltips or help tags to guide people along. (Apple uses the latter term; the tooltip term and functionality were introduced in Microsoft Word 95.)

TOUCHING AND TAPPING A MOBILE SCREEN

Mobile devices such as iPad and iPhone use *touchscreens* (one word). The technology is referred to as *touchscreen technology* and there are many variations on that phrase (people sometimes talk about a touch environment). Touch is the basic technology.

The actions that the user performs are called *gestures*. The two most common gestures are taps and touches. There is a distinction (and it is a distinction that you will find in a standard dictionary, although the nuances of the distinction matter more in the world of touchscreens).

A *tap* is a down-and-up action. For programmers, in many cases it is the up part of the action that they care about. The *touchup* event is what triggers an action similar to a mouse click in many applications.

A *touch* is the down part of a tap. It comes into play in cases where the finger remains in contact with the touchscreen. You frequently find instructions such as, "Touch and hold the image and then drag it to your document." The hold part of that command can be a very short period of time, but the touch-and-hold action is just enough to let the device understand that something else such as the drag is going to happen.

You can see how touch-and-hold works on iPhone or iPad when you touch and hold your finger over some text, as you see in Figure 4.1.

Figure 4.1
Touch and hold over some text to bring up the magnifying glass.

The text is enlarged so that you can see where the insertion point is. As you drag your finger along the screen, the *magnifying glass* moves so that you can see exactly where the insertion point is.

When you lift your finger up, the dragging of the magnifying glass ends, and the *selection buttons* appear, as shown in Figure 4.2.

Figure 4.2
Lift your finger to show selection buttons.

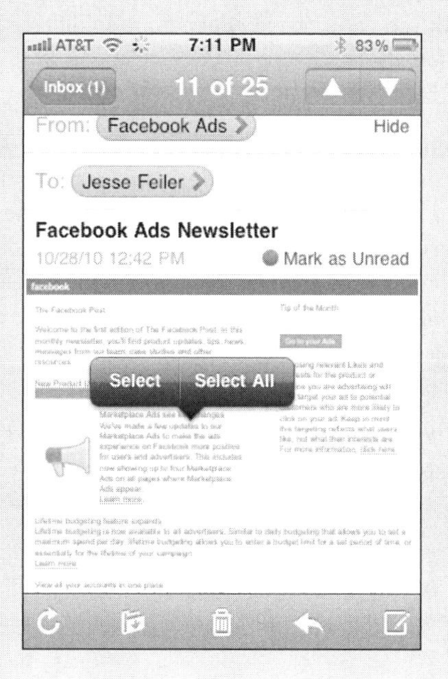

What to Do Without a Keyboard

Although iPad and iPhone have onscreen keyboards and the ability to use a wireless keyboard, as the designer of a mobile database solution, you should recognize that particularly for speedy touch-typists, the onscreen keyboard slows them down. For that reason (as well as the fact that not everyone carries a wireless keyboard with them), you can improve your mobile database solutions by constantly looking for what you can do without a keyboard.

When you don't have a keyboard, you can present tappable objects, such as radio buttons and checkboxes, wherever possible. These work very well in a touch environment, and, in fact, most people can handle them as quickly or even faster than a skilled touch-typist can enter data with a keyboard.

But there is one point to remember: In a desktop environment where a typist has a keyboard available as well as a mouse, the physical transition from hands on the keyboard to a hand on the mouse can slow people down. What this means to you is that you might have to think about having two interfaces for heavy data-entry portions of a database. You use one version for a desktop environment with a keyboard and mouse, and you use the other for a touchscreen environment.

Remember that this really becomes an issue only for intense data entry and for speedy typists. But even if it is an issue, FileMaker helps you out a good deal: Because the interface (*layouts* in FileMaker terms) is relatively independent of the data that it handles, you can switch layouts quickly and easily (and even automatically).

So the point remains for a mobile database designer: Look for every opportunity to convert data entry into touch.

And while you are looking for new opportunities to include a touch interface, there are two other points to consider. The first is worth mentioning, but it rarely causes issues for users. When you add touch to your interface for a database that is also shown in FileMaker Pro, you wind up with two ways of accomplishing the same task: a mouse-driven way and a touch-driven way. The experiences are so different (and people are now so used to them) that this just is not an issue.

What can be an issue is that because people are usually holding a mobile device in their hand, they can easily tap the screen inadvertently. Just as a stray mouse click on a desktop-based interface usually does not pose a problem, a stray tap also does not hurt in most cases. On the desktop, a stray mouse click that just happens to click on a Submit or OK button can do damage. On a mobile device, you close the mobile equivalents or dialogs or alerts by tapping somewhere outside the alert; there is normally no Cancel button. People soon learn this, but you might incorporate it into your design by remembering that dismissed dialogs might have been inadvertently dismissed, so you might want to re-confirm any action that is dramatic.

One terrific way to speed keyboard input is to use the auto-complete option. As a user begins to type, those characters are used to suggest a word or phrase that can be completed automatically. Unfortunately, this feature is not supported on FileMaker Go.

What to Do About Text

When you are designing a FileMaker solution for mobile devices, you move into a world that you cannot control as tightly as you can with desktop-based systems. If you use web publishing, you are at the mercy of the specific browser and operating system installation when it comes to available fonts. For iPhone and iPad, you are still at the mercy of the devices, but you have a bit more control in that you know the available fonts.

On the web, designers have gotten used to using fonts that they know will be available in all environments. This provides predictable stability and, in some cases, a bit of boredom when it comes to typography design. Although the collections of fonts on iPhone and iPad are likely to change over time, the current rosters of fonts on those devices are included later in this section. You can safely use those in your FileMaker layouts for FileMaker Go; layouts presented on those devices through Safari and Bento often work properly with those fonts. So, go ahead and design the interface you want with a bit more visual excitement than Helvetica.

These are the fonts available for iPhone and iPod touch:

- Arial
- Arial Rounded MT Bold
- Courier
- Courier New
- Georgia
- Helvetica
- Helvetica Neue
- Times New Roman
- Trebuchet MS
- Verdana
- Cochin

These are the fonts for iPad:

- Arial
- Arial Rounded MT Bold
- Courier
- Courier New
- Georgia
- Helvetica
- Helvetica Neue
- Times New Roman
- Trebuchet MS
- Verdana
- Academy Engraved LET
- Baskerville
- Chalkduster
- Optima
- Palatino
- Gill Sans
- Futura
- Cochin
- Snell RoundHand
- Didot

What to Do About Graphics

iPad can present a surprising challenge to you when you move a FileMaker database solution to it. This applies to any of the tools you might use: FileMaker web publishing, Bento, and FileMaker Go. The challenge stems directly from one of the great achievements of iPad: its remarkably clear screen. (That screen also accounts for the crispness of the installed fonts.)

The issue is inherent in iPad and iPhone 4 and later. Apple promotes its Retina Display (first launched on iPhone 4) as using pixels so small that they cannot be seen by the eye: Only the overall effect is visible. The flip side of that statement is that images stored in databases also are shown in amazing detail. Just as with the advent of high-definition TV, flaws and imperfections are noticeable. (More than one TV station had to rebuild or at least seriously touch up its news and other local programming sets as part of the transition to high-definition.)

If you have an existing database that contains graphics, it is time to start thinking about what you are going to do about the quality of those graphics. The first step is probably to recognize the issue and to move to higher-quality images for all future data entry into the database. Depending on your database and your users, you might want to launch an ongoing project to upgrade the existing images. In some cases, that is a matter of converting images, but in many more cases, it is a matter of reshooting photographic images or otherwise starting from scratch. Because this can be a long-term process, and because it is inherent in the use of high-resolution devices such as iPad and iPhone (in other words, it is not just a FileMaker issue), it makes sense to formulate a strategy and get started on it as soon as possible.

If you store images in your database, consider adding an image quality field with a companion value list (displayed in radio buttons or a pop-up menu). The values for the value list can be terms such as OK, Reshoot, Retouch, Hi-Def, or Lo-Res depending on your circumstances. It is often not enough to just look at the resolution of the embedded images: You need subjective human judgment about whether or not a speck of dust in the background of the image is now visible enough to ruin the image or to require manipulation with an image-editing tool. Furthermore, you may need to formulate standards and procedures for retouching images if you do not have them already so that it is clear what types of adjustments require new artwork and what types can be retouched.

Integrating Without a Visible File System

People have gotten used to organizing their files on their desktop-based computers. Although some files are hidden, and some files are protected from modification, the vast majority of files and folders on a desktop-based computer are your data, and you can use them as you see fit. You can rename and reorganize them until the structure makes sense to you. The standard File Open and File Save dialogs let you locate the files wherever you want them on your computer, as well as on any connected servers. If you want to put your accounting files in a folder called Photos, no file-organizing police will stop you.

Designers of operating systems have made this possible, and some of them have had second thoughts. There is something attractive to OS designers in having separate areas for the files of each application. Most operating systems implement some concept of a file's owner (the application that will be used by default to open it), but even in those cases, users can often change the file's owner so that, if you want, you can open your vacation photos with AccountEdge or Quicken (or at least try to).

If you are deploying a FileMaker database for use on iPad or iPhone, you have to think about two types of files:

- The database files need to be available to users.
- Graphics and other files used by the database need to be available.

Your FileMaker database files can be served up from FileMaker Server; you can access them using a local Wi-Fi connection or by connecting to them over the Internet. You use FileMaker Pro and the File, Open Remote command to connect to these databases; on mobile devices, you can use FileMaker Go.

→ There is more on FileMaker Go in Part II, "FileMaker Go," **p. 151**.

You can also publish databases on the web using FileMaker Server; in that case, people connect to your database using a browser. The browser option is available for mobile devices such as iPhone and iPad.

→ There is more on FileMaker Server in Part IV, "FileMaker Web Publishing: Instant Web Publishing (IWP) and Custom Web Publishing (CWP)," **p. 325**.

As for the files that your database needs to access, these can be inserted into the database directly; you also can insert them by reference. Those techniques are also described later in this section.

Moving FileMaker Databases to Your Mobile Device

Each application on your iPhone or iPad has its own area for files. You place files in that area either by moving them to your device or by creating them with an app. As you see in this section, you move them back and forth between your computer and your mobile device with iTunes or email.

One consequence of this is that if you remove an app from your device, all the files you have stored in that app's area are also removed; you should back them up to another device before removing the app.

Using iTunes

iTunes is the basic tool you use for moving files to and from your mobile device. The device normally has to be connected to your computer with a cable in order to move files back and forth. (Apps such as Contacts, Calendar, and Mail move data over wireless connections, but that data consists of relatively small items, such as individual phone numbers, appointments, and the like.)

To move files from your computer to your mobile device, connect the device. iTunes might open automatically; if it does not, launch it yourself. In the left-hand side of the window is a section for devices; your iPad or iPhone should appear there after a few seconds. If you do not see your device, check the cable connection and verify that it is turned on. Figure 4.3 shows an iPad connected through iTunes. Tabs at the top of the window let you see various parts of your device's storage: Click Apps to manage storage.

Scroll down to the bottom of the Apps tab, as shown in Figure 4.4.

Here is where you can see the file storage system on your iOS device. Each app has its own set of files. Click on an app, and you see the files (and, as noted, remove the app, and its files are removed, too).

Developers see more of the file structure, but basically this is it. People can discuss and argue about whether it is better or worse than the totally user-controlled file structure on personal computers in general, but this is what is provided on iOS.

If you want to add a file to your device so that an app such as FileMaker Go can access it, you connect your device, launch iTunes (if it does not launch automatically), and scroll down to the app in question.

Figure 4.3
Connect a mobile device using iTunes.

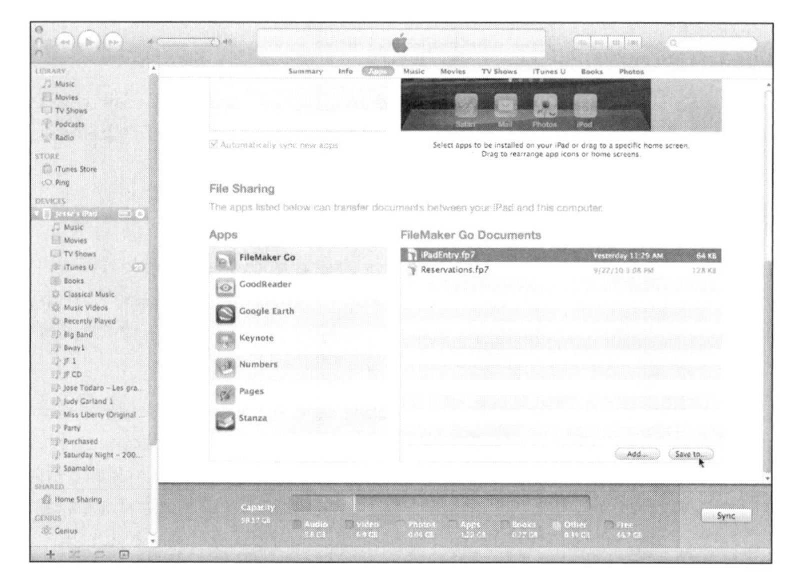

Figure 4.4
Scroll down to see file storage.

As Figure 4.4 shows, there are two buttons in the bottom right of the window that let you move files in and out. Click Add, and a standard File Open dialog opens that lets you select a file from your computer or available network locations and add it to your device. The file is immediately sent down to your device. You can also drag the file from your disk into this window; just make certain that you have selected the proper app.

With any of the files on your device selected, the Send button in the bottom right of the window opens a File Save dialog. This lets you choose a location and name for the file on your device. The file is saved on your disk (or network locations) with the name that you provide. It is not removed from your device. If you want to do that, you do it on the mobile device as you see later in this chapter (see Figure 4.9).

Note that this is not anything special about FileMaker Go. Files for the iWork apps (Pages, Numbers, and Keynote) work exactly the same way as do files for any other app that supports its own files.

The FileMaker Go files on your iPad are shown at the left of the screen you see when you launch the app, as shown in Figure 4.5.

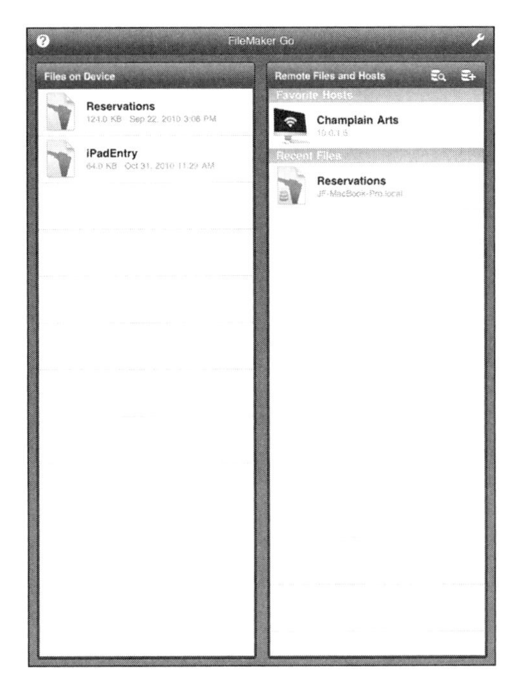

Figure 4.5
FileMaker Go files are shown when you launch FileMaker Go.

Using a Non-Paired Copy of iTunes

An iTunes library is paired with a specific iPad or iPhone (it can be paired with one of each). That enables you to sync your apps, music, and videos between your computer and your mobile devices. If you change computers or mobile devices, you can move your apps, music, and video to the new device or the new computer, but you only can have at most one iPhone and one iPad paired with your computer. (This is part of *digital rights management*—DRM. It protects copyrighted material from unauthorized distribution and duplication.)

Because most of the synchronization between your mobile device and your computer has to do with copyrighted material, the security mechanisms are activated when you connect the device to your computer.

However, the files that belong to you can be moved to and from the mobile device without running afoul of intellectual property laws. (That is, of course, unless you are trying to move a database that belongs to someone else to your mobile device or from it to your computer. That is against the law, but iTunes does not prevent you from doing so. The owner of the database or other copyrighted file must have it out with you.)

Because it is perfectly legal to move your own files between your iPad or iPhone and your own computer, iTunes lets you do so. However, it does remind you about the copyrighted material that is involved and asks you if you want to synchronize that material. Remember that most people are used to just plugging in their iPad, iPhone, or iPod and having their music sync automatically.

Here are the steps to take if you want to move your own files between an iPhone or iPad and your own computer. When you connect your mobile device, iTunes might launch automatically. Depending on your settings, it might start to synchronize your music, movies, apps, and photos automatically.

If you connect your mobile device to a computer that is not paired with it, iTunes might still launch automatically. If it does not, launch it yourself. You see a warning like the one in Figure 4.6.

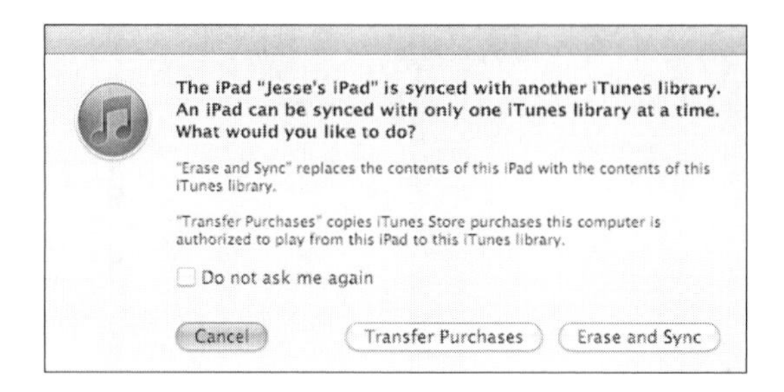

Figure 4.6
Cancel iTunes library synchronization.

The text of the dialog box presents you with two choices for your iTunes library (that is, your music, movies, and apps—copyrighted material that you have downloaded from the iTunes store whether or not it is a free app or a movie you have paid for). You can choose to erase the iTunes library from the device you have just plugged in and replace it with the contents of the iTunes library on the computer to which it is now connected. On the other hand, you might want to transfer the iTunes library content from the mobile device to the computer. Either way, at the end of the operation, the iTunes library on the mobile device and the computer will be synchronized.

If you click Cancel, neither of those things happens. If you are interested only in moving your own files between your mobile device and your computer, Cancel is your option.

You move on to your photos, as shown in Figure 4.7.

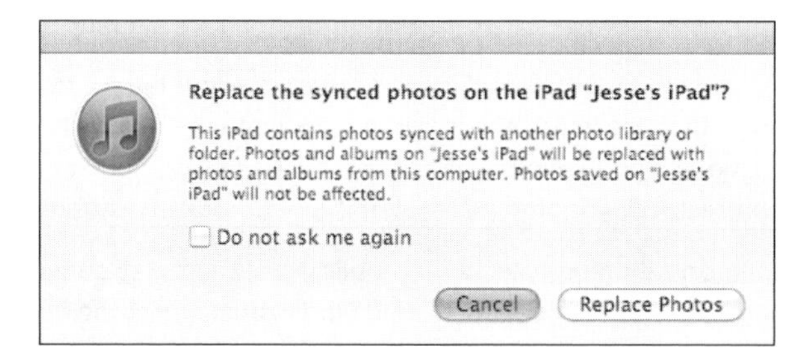

Figure 4.7
Cancel photo synchronization.

As you can see from the text in the dialog in Figure 4.7, you have an option to replace synced photos on your mobile device with those from the computer you are using (you do not have the reverse option as you did with the iTunes library).

If you are interested just in transferring your files, click Cancel again here.

In this way, all that happens during the connection is that you synchronize files and, perhaps, change some mobile device settings and recharge it as necessary.

If you refer to Figure 4.3, you can see that iTunes does not recognize any apps from the attached mobile device because it is not being synced for apps. At the right of Figures 4.3 and 4.4, you can see a dimmed version of the device's home screen (you can normally rearrange the app icons here). That image is dimmed and not modifiable unless you have decided to pair the device with the computer.

Using eMail

You can also move files to your iPad or iPhone using email. In many ways, this is simpler because you do not have to connect your mobile device directly to your computer.

In order to move a file to an iPad via email, just send it as an attachment to an email message to the iPad's owner. It arrives in the same way that any email message with an attached file arrives. Figure 4.8 shows such an incoming message.

Figure 4.8
Send a database in an email message.

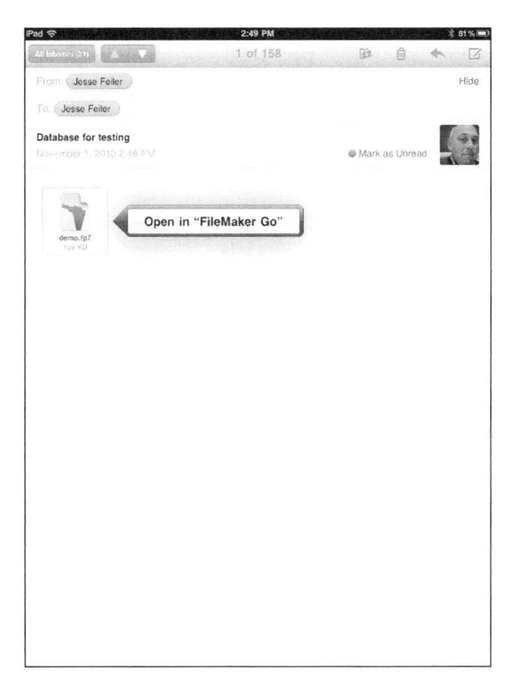

When you tap an enclosed file, you are given the opportunity to open it in the application associated with it, as you see in Figure 4.8. (If no such application can be found, the attachment icon simply is displayed in the message.)

After you have tapped a FileMaker database from an email message, it is opened in FileMaker Go and it is moved to the FileMaker Go file storage area. From now on, it shows up in the Files on Device (left-hand) column of the FileMaker Go home screen, as you see in Figure 4.9.

Figure 4.9
Review the FileMaker files on your iPad.

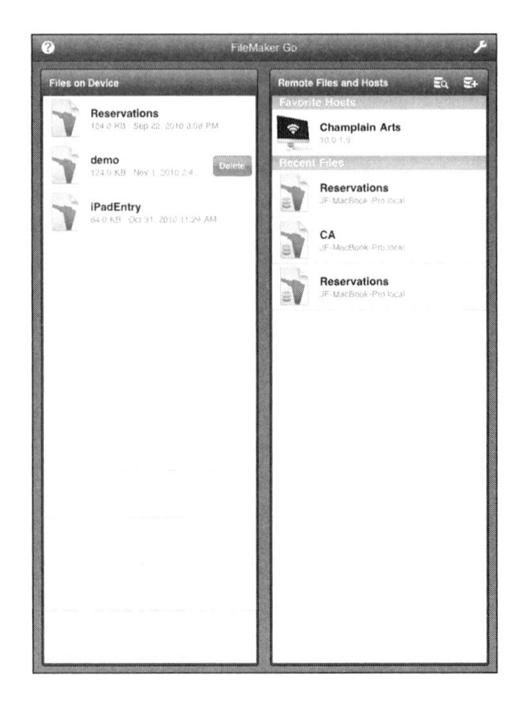

Figure 4.9 also shows how you can remove a file from your device: the standard left-to-right swipe brings up the Delete button, as you see in Figure 4.9.

On iPhone, the screen shown in Figure 4.9 is replaced by a file browser. By default, you see recent files on iPhone, as shown in Figure 4.10.

Figure 4.10
Start from recent files on your iPhone.

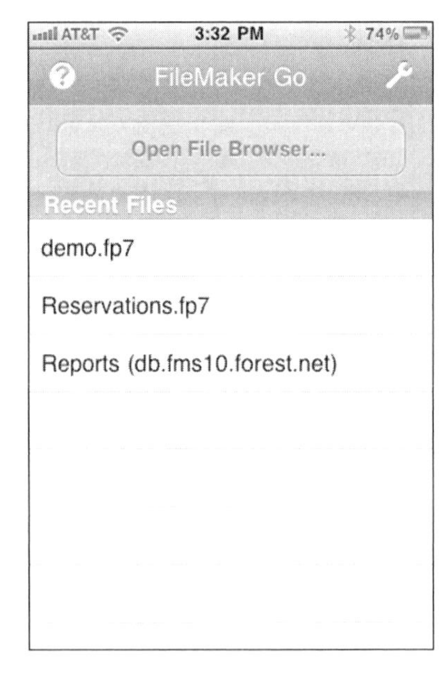

If you tap the File Browser button, you can look at the different types of files (Recent Files, Files on Device), and from the lists in those categories, you can open the files directly. Those screens are shown in Figure 4.11.

Figure 4.11
Browse and open the files you want.

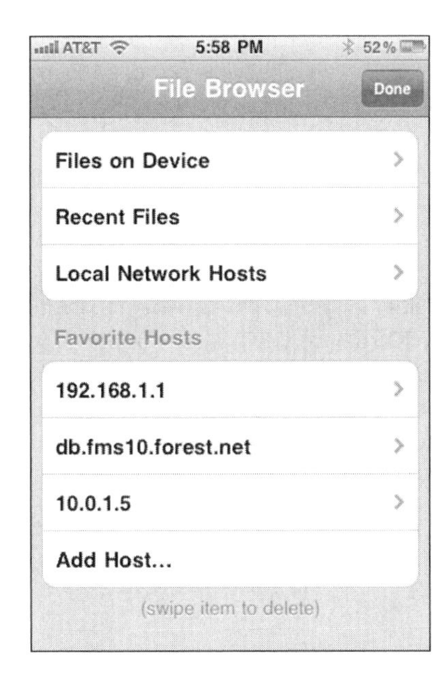

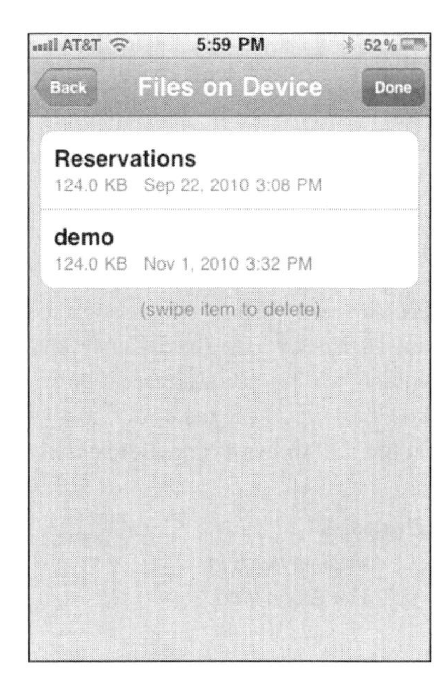

It is important to note the iTunes display of files for apps on your iOS device (shown previously in Figure 4.5). With iTunes, you can add files to your device or save them from your device to a location on your computer's disk or network. Remove files from the list on your device or select them on the Apps tab in iTunes (see Figure 4.4) and use the Delete command. This is standard practice for most apps.

Inserting Files and File References into FileMaker Databases

FileMaker databases can contain *container* fields. These are fields into which you can place references to files or even the contents of files. (In many ways, they are similar to *blob* fields in databases such as MySQL.)

When you place binary data into a container field (that is, data that could be the content of a file), its structure is preserved. FileMaker cannot display it or manage it in most cases, but it does store the data and can present it as needed. Another application can then process the data.

The most common use for container fields is to store images, movies, spreadsheets, PDFs, and even other FileMaker databases. You can do this in either of two ways:

- **Inserting files.** You can select a container field and then choose a command from the Insert menu to insert a Picture, QuickTime, sound, or file. In cases where you specify the type of content, FileMaker might be able to display it in the field. By inserting the file, you make your FileMaker database bigger (because it now contains movies, images, and the like).
- **Inserting file references.** A checkbox on the dialog that opens from the Insert menu lets you choose to insert a file reference rather than the file itself. This keeps your database's size manageable; the file

and all of its data remains on your hard disk or network. However, if you move your FileMaker database to another location, you must move the referenced files in such a way that FileMaker can find it.

Thus, you have a clear tradeoff: more portability of your FileMaker database versus larger database size. This matters when you are publishing a database on the web because you need to move the database and its referenced files together, and it also matters if you are moving a database to a mobile device (for the same reason).

Using Signature Capture

FileMaker Go introduces a new use for container fields: You can use them to capture signatures from an iOS device. When you tap a container field on an iOS device, you have the option to insert a photo from your library or to capture a signature directly into the container field (after all, these are *touch*screens). In addition, on iPhone you have the option to take a photo and automatically have it inserted into the container field. Figure 4.12 shows a container field in action on iPad.

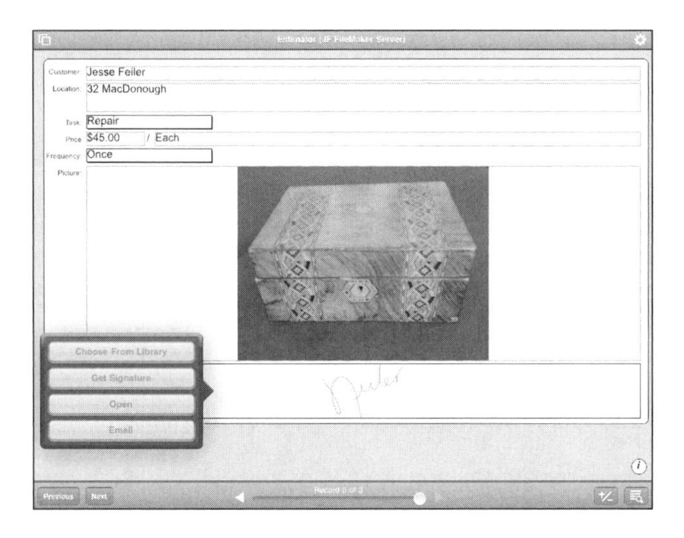

Figure 4.12
Use container fields in FileMaker Go on iPad.

→ You see how to build the Estimator example shown in Figure 4.12 in Chapter 9, "Designing a FileMaker Go Solution," **p. 213**.

Using Multiple Files in FileMaker Solutions

In Chapter 2, "Introducing the FileMaker Architecture," the section "Structuring Solutions and Sub-Solutions" described how you can use multiple files to construct your solution so that files focusing on interface elements are separate from those supporting the data itself. When it comes time to think about deploying your solutions and sub-solutions, you might have to address the issues relating to those files.

The issues center on the fact that you do not have the kind of access to the files and file system on the mobile device that you have on a desktop-based computer. All of your FileMaker files will be together in the FileMaker Go section of the device. Therefore, if you have constructed a complicated file structure (perhaps by connecting files in several folders or even several disks), this structure might not work easily on a mobile device. (In fact, this is very much the issue that you can encounter when deploying files for FileMaker Server. In that case, files are placed in a Databases folder for FileMaker Server, and so complex file/folder relationships are very much as simplified as they are for FileMaker Go.)

The simplest way to organize your files in a multi-file FileMaker solution is to place all the files in a single folder; you can then move that folder around (to the FileMaker Server deployment or to the FileMaker Go deployment) and the files remain together. This means that the links among the files are always within that one folder, with each related file being no more than one step away from any other file.

Before you can use an external file in a multi-file solution, you need to define it to FileMaker. To do so, you use the File, Manage, External Data Sources command, as shown in Figure 4.13.

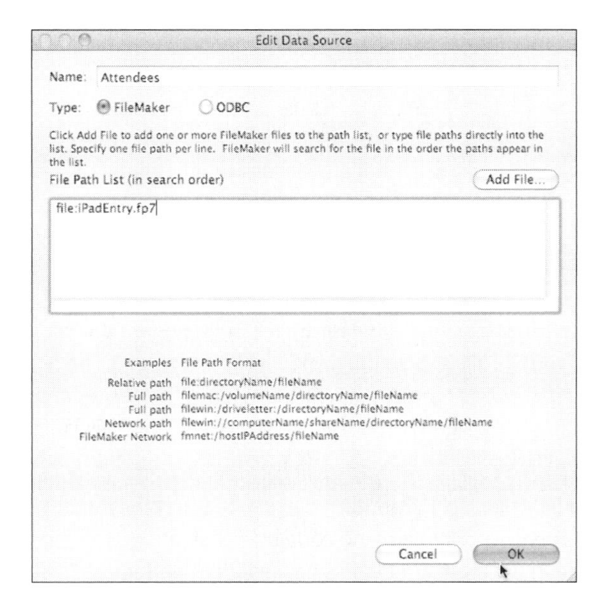

Figure 4.13
Specify external data sources.

➔ Figures 4.13 and 4.14 show parts of the Reservations example which you will see how to build in Chapter 7, "Using FileMaker Go," **p. 151**.

In this dialog, you can name an external data source and specify its location. As the dialog shows in the examples at the bottom, you can use files in other directories and on other servers as external data sources. You can also use non-FileMaker files if they conform to ODBC protocols.

If you are placing a file on a mobile device with FileMaker Go, it is easiest to use the single folder structure described in this section. If you are accessing a FileMaker Server database, you can deploy it as you normally would and simply connect to it from the mobile device.

After you have specified an external data source, you can use it in your relationships graph. Simply click the add table button at the bottom left as shown in Figure 4.14, and you can choose an external data source. (If it is not already set up, you can add it from the dialog, as shown in the figure.) After you have selected the data source, its tables are displayed, and you can select the one you want to use. From then on, everything is the same as if you were using tables within your main database file.

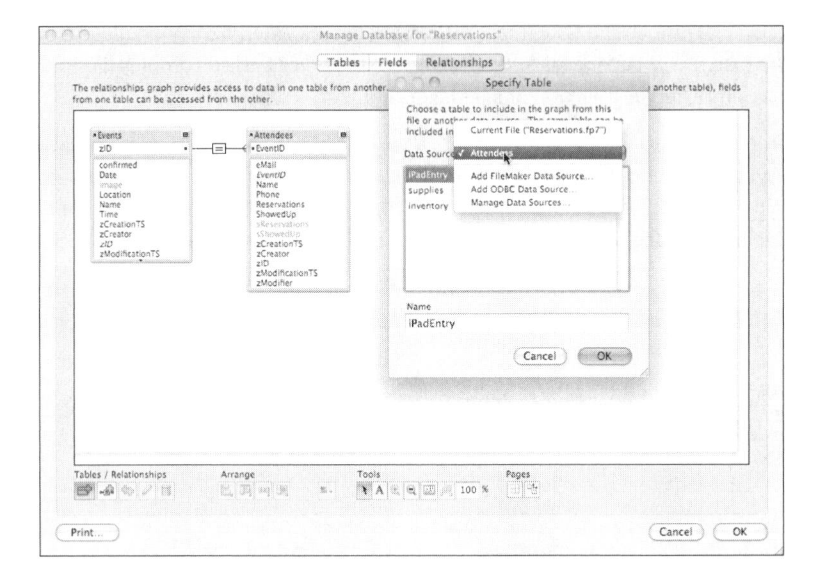

Figure 4.14
Select the table from an external data source.

Printing

Some special-purpose mobile devices incorporate printers. Examples are devices that print receipts, such as for car rental returns. Most mobile devices do not print, and for many people and many applications, this has posed a problem: People often need records of their transactions. The desire for printed output from mobile devices has spawned a variety of solutions right from the start. Initially, the iOS devices supported printing with third-party apps; however, within a relatively brief period of time, printing became an integral part of iOS.

➔ With the release of FileMaker Go 1.2, printing is expanded even beyond the basics described here. FileMaker has very extensive print handling code that enables you to dynamically reduce the size of some fields and even move them around on the layout depending on the data that they contain. This is so important that it receives its own chapter: Chapter 10, "Using Printing and Charting with FileMaker Go," **p. 237**.

Actually, printing in iOS was there from the start (indeed, from before the start), although not all of the pieces of the implementation were complete. Printing is usually a part of an operating system; you can notice this by observing that the print dialog for each of your applications on a desktop-based computer typically uses the same dialog. Often, an individual application adds certain special features to the common dialog, and printers, too, add certain options. However, the dialog basically remains the same. In the NeXTSTEP operating system (which became OPENSTEP, then Rhapsody, and then Mac OS X), built-in printing supported printing to a printer as well as faxing through a connected fax modem. This heritage of being able to print through a fax modem—that is, being able to print to a printer that is not physically connected directly to the computer—has been important in the development of iOS printing.

There are three ways of printing with FileMaker Go (and many other apps). The first way is to generate a PDF file that can be sent via email, and the second is to use the printing features that have been built into iOS since version 4.2. (In fact, deep down beneath the user interface, the two methods are actually very much the same.) The third way is the FileMaker-specific method described in Chapter 10.

Just as is the case with desktop-based computers, printing is supported and implemented in the operating system, but it is up to individual applications to enable printing of their documents and data. With FileMaker, printing is one of the security settings that you can control with FileMaker Pro using the Manage, Security dialogs.

How to Print with a PDF File

FileMaker Go provides an example of printing through a PDF file:

1. You start by navigating to the data that you want to print.
2. Tap the wrench in the upper right of the screen to open the menu of commands, as shown in Figure 4.15.

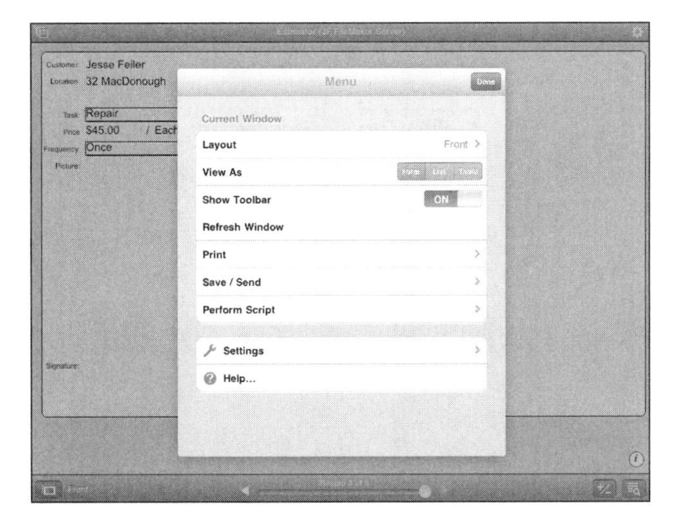

Figure 4.15
Tap to show the menu of commands.

3. **Tap Print to show the options you see in Figure 4.16.** This enables you to choose the page range, orientation, and so forth as shown in Figure 4.16.

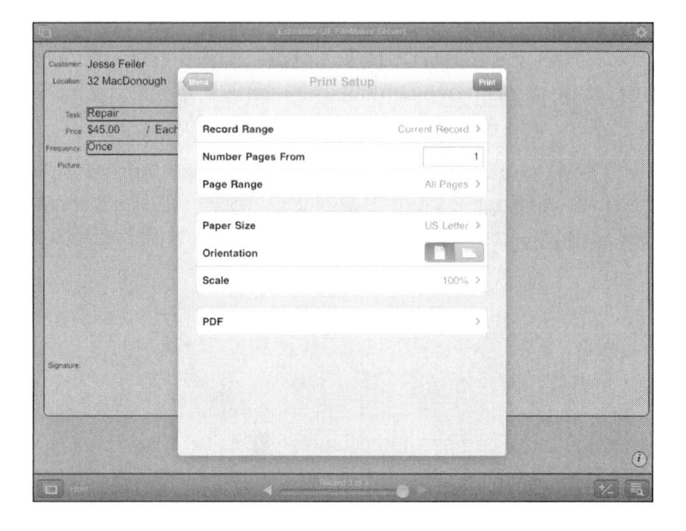

Figure 4.16
Select your printing options.

4. **Tap PDF to select the final options as shown in Figure 4.17.** As you can see, there are three basic choices for saving and sending; not all of them are available in all cases. Your choices might be limited by security (not all users can print, for example), or they might be limited by the current selection (you can only save or send field contents if you have selected a field...and if your security level allows you to do so) .

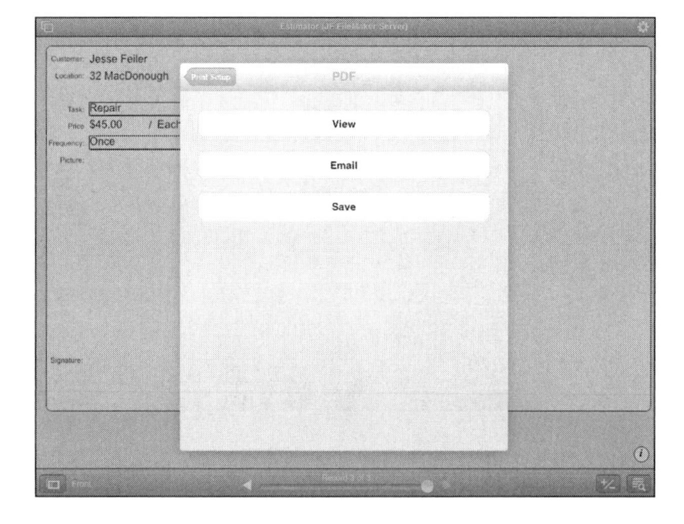

Figure 4.17
Select your PDF options.

As you see in Figure 4.17, you have three choices:

- You can view it on your device. If you choose this option, you can then choose further actions, as shown in Figure 4.16, so that you can open the file in iBooks or another app that handles PDF files (the choices depend on the apps you have installed), you can select from your apps, or you can print it as described in the following section (see Figure 4.17).

- You can email the PDF. If you have a printer that supports web printing, you can email the document to your printer. Web printing is supported on various printers that are connected directly to the Internet and that, therefore, can have their own email address. (The HP ePrint feature is a web printing feature.) You can use the email option shown in Figure 4.16 to send the PDF directly to a printer if it is configured properly and supports web printing.
- You can save it. If you choose to save the PDF, it is saved to your device, and you are able to view it or download it in the FileMaker Go section of iTunes, as described previously in this chapter in the "Using iTunes" section.

How to Print to a Printer for FileMaker Web Publishing

Beginning with iOS 4.2, Apple introduced its AirPrint feature that builds on the web printing and PDF emailing features described in the previous section. There is a subtle difference between AirPrint and the web printing technology described in the previous section. Web printing (such as HP's ePrint) sends print jobs to addressable printers over the Internet using email. AirPrint locates printers on the local area network where your computer or mobile device is located, and it sends the print job to the local printer. To print to a printer that is not on your local area network, use web printing and send the print job (a PDF file in most cases) via email.

This technique works with any app that supports printing. If you are using FileMaker's web publishing tools as described in Part IV, this is how you (or your users) can print out the results using Safari on iPad.

Here is how to print using AirPrint:

1. Go to the data you want to print. Various apps on iPhone and iPad support AirPrint (Safari is one of them).
2. Printing is provided in the Actions menu (the box with the arrow coming out of the top). It is located to the left of the address field in the toolbar on iPad; on iPhone, it is located at the center of the bottom toolbar. Figure 4.18 shows both interfaces.

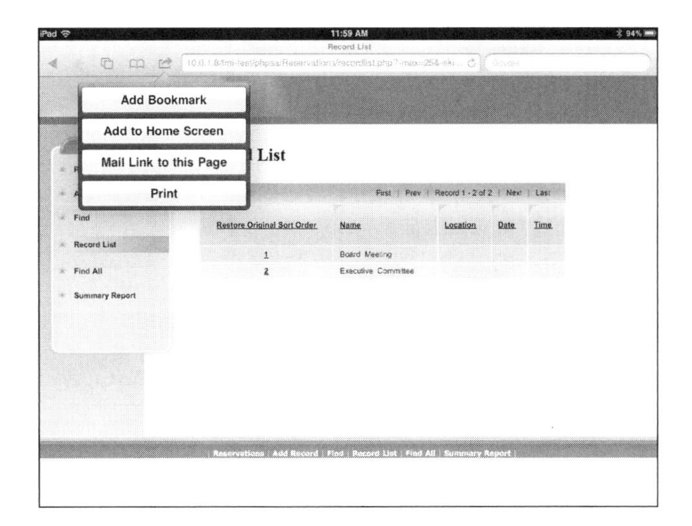

Figure 4.18
Print from Safari on iPad.

3. **Tap Print.**
4. **Choose the printer and the number of copies that you want, as shown in Figure 4.19.** That is all you need to do.

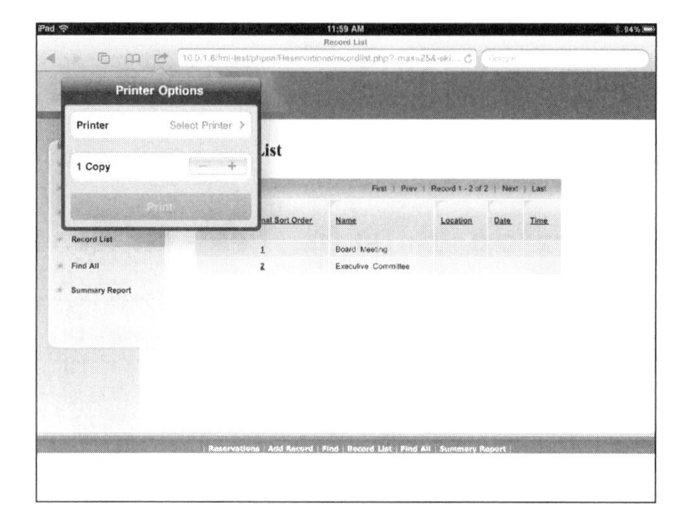

Figure 4.19
Set up the print options.

4. If you want to manage the print queue (for example, cancel a job or change the order in which jobs print), you can use Print Center. Double-tap the Home button to open the multitasking control at the bottom of the screen, as shown in Figure 4.20. Find Print Center at the left. (It only appears when a job is printing.) A badge on the Print Center shows you the number of jobs in the queue.

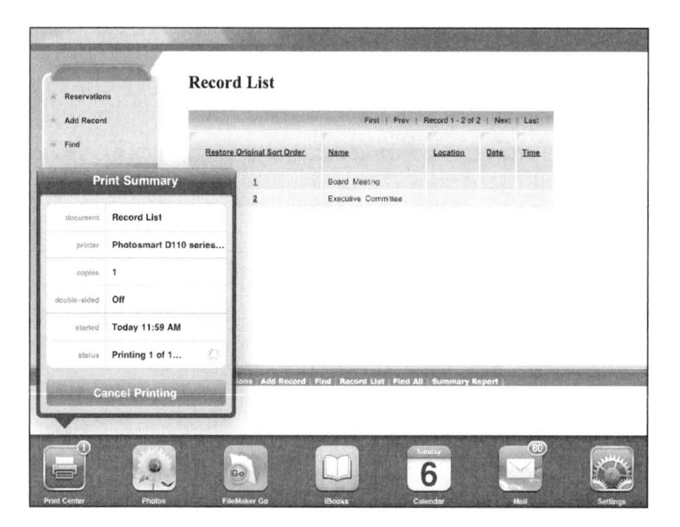

Figure 4.20
Use the Print Center.

5. If there is more than one print job in the queue, tap Print Center to see a list of the current jobs, as shown in Figure 4.21. You can reorder the jobs in this list just by touching and holding a job and moving it up or down.

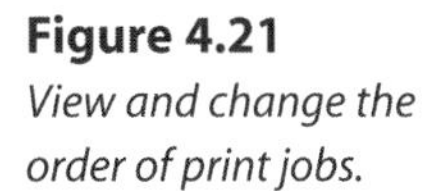

Figure 4.21
View and change the order of print jobs.

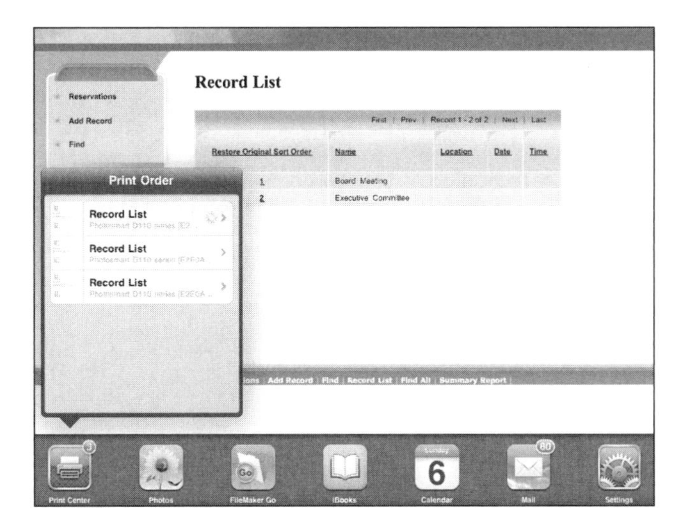

6. Tap a job to get its details, as shown previously in Figure 4.20. As you can see, this is where you can remove a job before it has printed (this is useful if you know that you have made a mistake and you do not want to waste paper and ink).

Further Steps

Perhaps more than any other topic, printing from a mobile device gets to the heart of how mobile devices differ from desktop-based computers. If you use a desktop-based computer, look around at the cables; they are everywhere. Most long-time computer users have a carton of old cables stashed away somewhere. Cables are not cheap, and users often keep old cables around so that they have them on hand in case they need another one. When the need arises, you search out the carton, and upend its contents to find the FireWire cable, the USB cable, the serial cable, or whatever. (Of course, in that entire carton, you have every type of cable except the one you need.)

With the advent of efficient and relatively inexpensive cordless connectivity, you can toss out those cables. Along with those cables, you can toss out the dust and dirt that always accumulate in that rat's nest of cables behind or under your desk. What people are relying on today is the ability of devices such as computers and printers to communicate directly over the Internet. When you can send a print job via email to your printer (or to a colleague's printer halfway around the world), cables do not matter that much any more. (Of course, things such as security do matter so that you do not receive junk print jobs from halfway around the world.)

This standardization on Internet communication protocols is a large part of what has made mobile devices possible. It takes some getting used to the idea of connecting a printer to a computer through the Internet, particularly when the devices are sitting right next to one another, but the gains in productivity can be tremendous.

Along with the standardization on Internet communication protocols, the world is moving more and more toward the use of cloud computing and data storage that, like your printer, might be located a continent away. You can use mobile devices the way you have always used computers, but experiment with the new technologies to see what you can do.

As you experiment with these technologies, you will come to see why some of the initial complaints about mobile devices such as iPhone and iPad are irrelevant. The clamor for USB ports, serial ports, and special connectors ignored the movement toward these standard protocols.

5

IN THIS CHAPTER

- Using Auto-Sizing to Handle Device Orientation
- Speed Up and Reduce Typing with Value Lists and Auto-Entry
- Save Space with Conditional Formatting

Preparing FileMaker for Mobile Use

Preparing your data for mobile use is more than just making it accessible to mobile devices running Bento, FileMaker Go, or browsers that can access your Instant Web Publishing (IWP) or Custom Web Publishing (CWP) databases. Sometimes, it is a matter of creating a brand-new database in FileMaker or Bento. That means you can think about the mobile issues from the very beginning. Other times, it is a matter of taking an existing database and making it usable with FileMaker Go or with CWP/IWP.

Whether you are starting from scratch or modifying an existing database, you'll find that life (and database design) is different in the mobile sphere. This chapter is about those differences and how you approach the challenges and opportunities that they present. If you are an experienced FileMaker database designer, you will probably notice that you already know many of the tools that are needed for successful mobile database design. In the mobile world, it is often a matter of reprioritizing your use of the tools. Some things that are so simple and obvious on a desktop computer (using the mouse or keyboard, for example) are not so easy on a mobile device. By the same token, some things that do not matter much or are just sort of nice to have on a desktop become critical on a mobile device.

This chapter focuses on what you can do to new or existing FileMaker databases to make them more mobile-friendly. In many (if not most) of these cases, the changes make your databases more user-friendly in many ways. (The changes certainly make your databases more up to date and consistent with current interface design best practices.)

The evolution of graphical user interfaces (including FileMaker) has been going on for more than two decades. Many of the most efficient ways of dealing with mobile devices use these evolved interfaces and developer tools. You already know or have seen the basics of FileMaker, and this chapter not only helps you prepare your databases for mobile devices, but it also helps you bring your code up to date.

The emphasis in this chapter is on changes you can make to your databases to make them more mobile-friendly whether you will be accessing them with FileMaker Go or through FileMaker web publishing technologies (CWP and IWP). There are more specific details in the chapters that deal with FileMaker Go and FileMaker web publishing. This is not a complete discussion of FileMaker features because it only addresses those features that you can use to help prepare your databases for mobile devices and the web.

Understanding the Mobile Difference

From the standpoint of databases, the mobile difference consists of three aspects:

- Working with a limited keyboard
- Managing your new look-and-feel
- Working around menu commands

Just about everything you do to make your database accessible on a mobile device is derived from one or more of these aspects.

Working with a Limited Keyboard

There is no regular keyboard on a mobile device. You might have the on-screen keyboards that are found on iPad and iPhone, or you might have a small keyboard on other types of smartphones. In either case, experienced and speedy typists will be slowed down in their data entry. If your database solution is input-intensive, this can be a serious problem. Fortunately, the solutions are simple.

Before jumping into the next section, sit back and think about whether your database solution is truly input-intensive. It is perfectly natural to think about the hard work and efforts that you and others have put into building your database. But there are many solutions that lend themselves very easily to use on mobile devices where the mobility is used by people who are accessing data rather than entering it. Because you can build solutions that use a shared database, you can let people with desktop-based computers use their keyboards while the mobile users access the data that has been typed in. This is a remarkably common paradigm.

Add a Wireless Keyboard

You can pair a Bluetooth keyboard with an iPad or iPhone. This is often the simplest solution. Apple's wireless keyboard sells for $69. Many people already use a wireless keyboard for either a desktop or laptop computer, so you might have one that you can use. (If you do, the first step is to un-pair the keyboard from your computer.) If you are pairing a keyboard with your iPhone or iPad, think about how you'll set up the keyboard and the device; most people find it very difficult to type without being able to glance at the text that they're typing. This normally means putting your iPad or iPhone in a vertical position so that you can see the screen. You can accomplish this with a dock (usually less than $30) or by propping the device up against a stack of books. The dock is easier.

The combination of a wireless keyboard and iPad is common enough that you can find bags and wallets designed to store them both. Waterfield Designs has one of them at http://sfbags.com/products/ipad-cases/wallet-ipad.php.

Make Your Input Tappable

Whether you use an external keyboard or not, you can streamline input so that entry uses a different paradigm: the taps that users are accustomed to on mobile devices. In FileMaker (and Bento) terms, this often means using tappable interface objects such as buttons, checkboxes, radio buttons, and pop-up menus to avoid the need for keyboard entry. This makes for a very different look to your database interface.

On a desktop-based computer, input is done with a mouse and the keyboard. For text-intensive input, commands that require a user to switch to the mouse can be a distraction. (Even switching to a trackpad on a laptop can slow people down even though it is very close to the keyboard.)

On mobile devices, which many people hold in their non-dominant hand (the left hand for right-handed people), using their dominant hand for taps and gestures is simple and does not slow them down in the way that switching from a keyboard to a mouse can do. What that means is that you might choose to arrange your interface using every possible tool that can be tapped instead of data entry field. This requires a new way of thinking for many people, but FileMaker's value lists can be a big help here (see "Working with Value Lists" later in this chapter).

Managing Your New Look-and-Feel

Mobile devices are new types of computing devices, so you and your users need to adapt to the new paradigm. Beyond the issues of the keyboard (which might be onscreen) and the menu commands (which are missing), you have some general look-and-feel issues to deal with. There are two major look-and-feel issues to deal with:

- Do you want your database to look mobile or match a desktop model?
- How do you handle rotation (between portrait and landscape orientations)?

Choose Your Database's Look-and-Feel Paradigm

When it comes to iPhone, your FileMaker database is going to look more or less like an iPhone app. There really is little choice because of the size of the screen. On the other hand, iPad poses some questions and opportunities. If you are converting from a traditional desktop-based FileMaker solution, you can move it to iPad without much modification in many cases. Even if you do some rearranging of screen elements, you can take as a design goal that it should look and behave as much as possible like the existing desktop-based solution. That means using similar graphics and a similar type of data organization. This strategy is most effective if the same people who use the desktop-based solution will sometimes use the mobile solution. Everything that they know about the desktop solution still applies to the mobile solution.

In fact, you might have confronted this issue before if you have moved your database to the web. Many people want a browser-based version of their database to look and behave as much as possible the same way as a desktop-based version. (That is one of the reasons for using IWP. With relatively little effort, you get the same look-and-feel on the web and on the desktop.)

But just as you might have discovered when you put up a browser-based interface to your database, sometimes you have to rethink and even redesign the solution. The most common reason for this is that the people using FileMaker Pro to access the database on desktop-based computers might be heavier users than the browser-based users. For example, they might be your in-house customer support people, while many of the browser-based users might be customers themselves. For that reason, you might have adopted web practices and interfaces that require less training and support for the customer users. Your in-house customer support staff might sneer that they can work much faster with the FileMaker commands and their knowledge of FileMaker Pro, but that is just a reflection of the difference between the two groups.

Nothing has changed, particularly when you are looking at iPad. The question is do you want it to look like a native iOS app or do you want it to look like a FileMaker solution? Of course, as you think about answering this question, you might have to confront the fact that you are in the process of developing several interfaces to your database: one for browsers, one for iPhone, one for iPad, and another for FileMaker Pro users. (And, yes, you can add another version for people who are accessing your FileMaker database through SQL.)

The more distinct the people using each interface are, the less complication you have because each group of users only needs to learn one interface. Of course, if you are supporting everything, you need to start becoming an expert in multiple technologies. That is likely to be an issue, particularly as you deploy a new interface. The first people who are using an iOS interface to your database might need some basic training in iPhone or iPad because the users have never encountered those devices before. Fortunately, that is likely to be a temporary situation.

Use Autosizing for Layout Objects

Chapter 2, "Introducing the FileMaker Architecture," introduced you to autosizing of objects on layouts. The rotation of a mobile device results in the automatic resizing of its window. If you have set autosizing, your interface elements will be sized and located appropriately.

Get in the habit of setting autosizing controls so that you are ready for iOS devices. The default behavior provided by autosizing might turn out to need some tweaks, but at the cost of a few mouse clicks as you are working on layouts, you might find that you will have created default behavior that works well in FileMaker Go.

 How to Use Autosizing to Change an Object's Size

Here is how to change an object's size with autosizing:

1. Create a new shape or other interface object in a layout, as shown in Figure 5.1

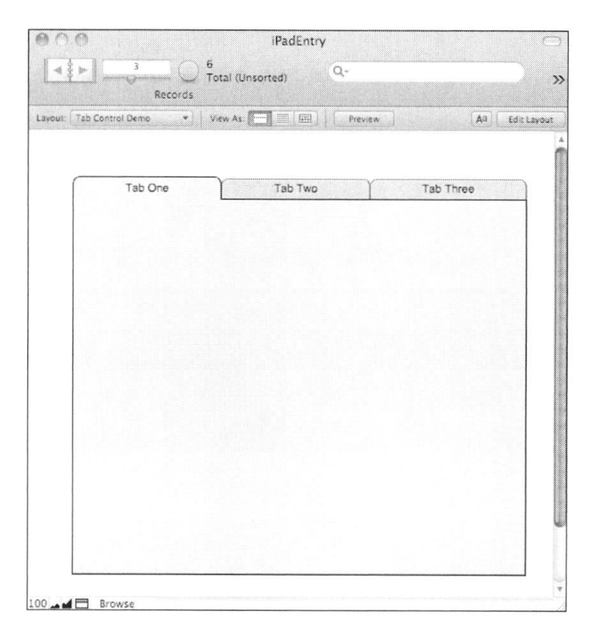

Figure 5.1
Create a new object on a layout.

2. While you are in Layout mode, open the inspector as shown in Figure 5.2. It might open automatically; if not, choose View, Inspector or ⌘-I (Mac) or Control+I (Windows).

   Objects start out being anchored to the top and left of their window. This keeps them in the same place relative to the upper-left corner of the window; it is the behavior that people expect in most cases.

3. Click the lines to the right and bottom of the autosizing display. Note that the padlocks are no longer dimmed out; this means that those edges of the object are pinned not only to the top and left but also to the bottom and right of the container it is in (in this case, that's the window). This means that the object will be resized with the window. Figure 5.3 shows the Inspector with these settings.

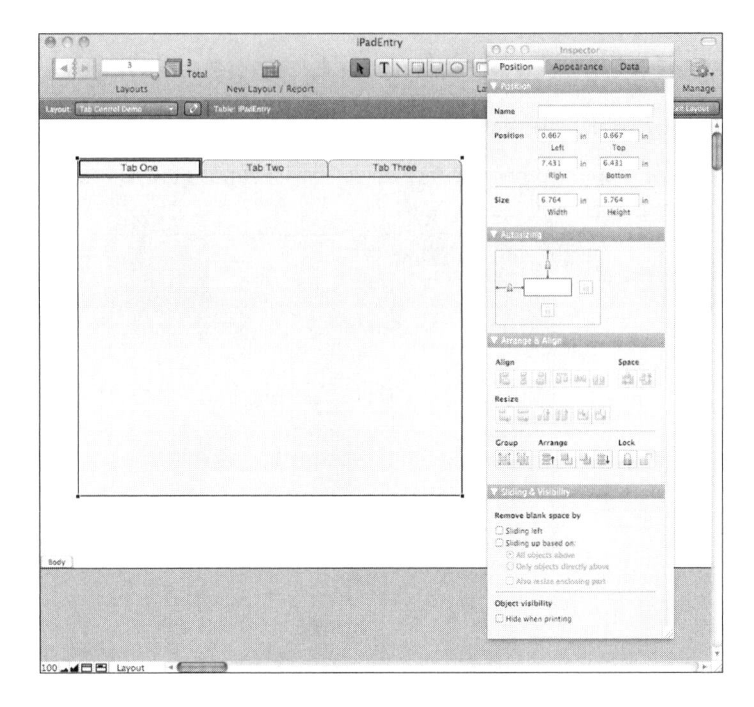

Figure 5.2
Open the Inspector.

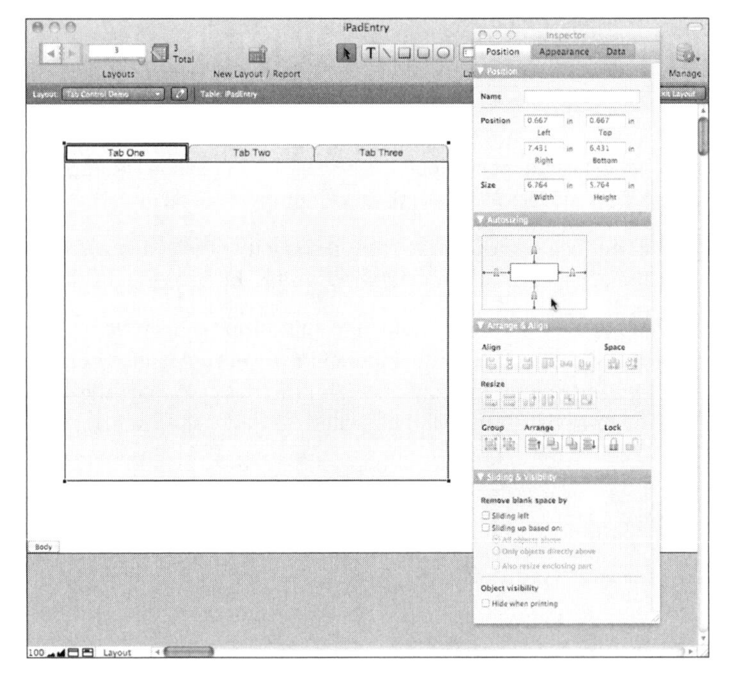

Figure 5.3
Anchor the object to all four sides.

4. Go into Layout mode and resize the window, as shown in Figure 5.4. The object resizes proportionately.

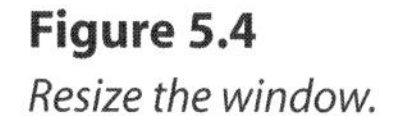

Figure 5.4
Resize the window.

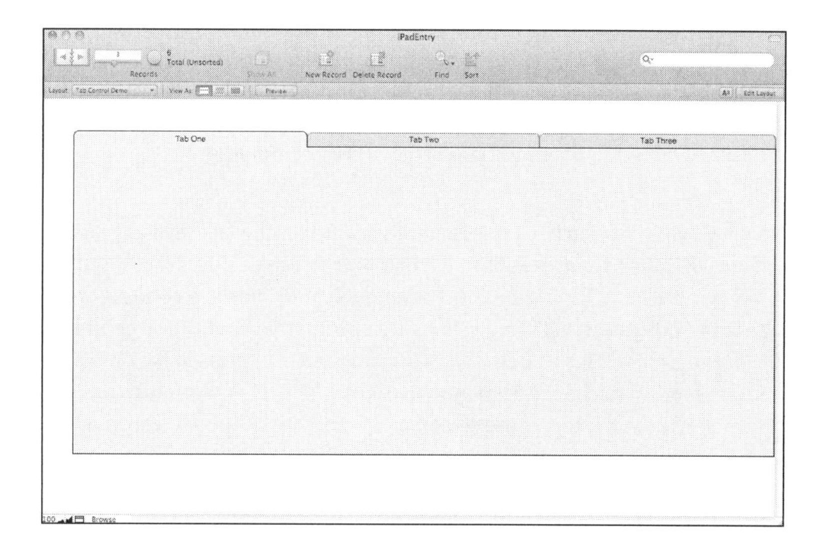

In general, when you have a data entry field that can contain variable-length text or variable-sized graphics, you want to lock it to all four sides of its container so that it grows and shrinks in the way that users expect. When you have an object with a size that you do not expect to change, it should be anchored either to the top and left or to the bottom and right.

An example of an object that should not resize is a button. Depending on your layout design, you might want to place it in different places, but it should always be the same size. If you are using a button in the lower right, such as an OK or Submit button, anchor it to the right and bottom but not to the top and left.

If you are creating a set of buttons or other elements at the top of a window so that the user has a toolbar-like set of controls in the database window, these controls are normally anchored to the top and left.

Use the Align, Group command to group items together; you can then anchor the grouped items. This can avoid some bizarre autosizing behavior. For example, it is a good idea in many cases to group a field label and the field itself so that they move together. In other cases, you want to anchor the label to the top and side and the field to the top, side, and right. In that case, the label remains in one place as the device rotates and the field lengthens or shortens depending on the orientation of the device. As you proceed, get a basic idea of how your autosizing works by resizing the window. The actual effects are different when you are running in FileMaker Go and actually rotating the device, but pre-flighting the autosizing in a resized window in FileMaker Pro uncovers many issues early in the design process.

Working Around Menu Commands

There is one critical point to realize about menus on a mobile device: There are none. Everything that you and the people who use your databases have learned to do with menus over the years now no longer applies on a mobile device. Fortunately, that is not such a dramatic change as you might think.

With FileMaker Go, some menu commands are visible to users, and you can use custom menus to modify the few commands that are presented to them. However, if it is possible to expose those commands directly to the user with buttons and other controls rather than requiring a tap to reveal commands, it might be more efficient.

One reason why it is not such a big change is because for the last several years, menu commands have been deemphasized. That arises in part from the fact that screens on desktop-based computers are bigger than ever; there is more room for buttons and other dedicated interface controls on these larger screens. Menu commands themselves have started to migrate onto the body of the screen. Shortcut menus (sometimes called *contextual menus*) let you click while a modifier key is pressed (⌘ on Mac OS X) or the right button on a multi-button mouse) to choose from among the commands that apply to the selected object. For these reasons, the menu bar at the top of the screen or window has gradually become less important. The fact that it is totally gone on the iOS devices (and most other mobile devices) is just the next step in the evolution of user interfaces.

Which button is which and what specific actions they control is generally configurable by the user. This accomodates such basic issues as handedness as well as personal preferences.

When you are accessing a FileMaker database over the web, you do have a menu bar, but it is of no use to you; that menu is the browser's menu bar.

Start Adding Non-Menu Functionality Now

The trend has been toward deemphasizing menu commands, but when you use a mobile device to access your FileMaker databases, you find that the trend has taken over. The solution and work-around is simple; use buttons and other interface items instead of menu commands. Of course, that is easier said than done if you are suddenly blind-sided by it.

You can adjust to this new type of architecture gradually over time. Even if you are not thinking of putting your databases on a mobile device now, start by recognizing that the trend is away from menu commands. As you implement new features in existing databases, start looking for alternatives to menu commands. If you have an installed base, there is no need to antagonize them by removing the menu bar, but you can add alternative ways of accessing data and functionality.

Review Your Menu Commands

Even if you are not adding new features to your existing database, spend some time (whenever you have it) thinking about how your current database could function without menu commands. Even if you are not adding new features, part of routine maintenance can consist of looking for places where you do not need menu commands.

There is only one way to really understand the difference between traditional desktop-based computers and mobile devices: Use the device yourself. It is possible to build or modify a database for deployment with FileMaker Go and never use it on a mobile device yourself, but chances are that your interface will not be as good as it can be. "Using the device" does not mean entering one record and putting the device down again. It means using it repeatedly over a period of time with a variety of apps so that you get the hang of it—and learn what works and what doesn't.

Data Entry Without Typing

If you need to provide for intensive data entry on a device with a limited keyboard, you have a wide range of tools available to you with FileMaker. You might not have used them much until now because the keyboard is such an easy tool for people to use and—until now—you could always assume that it would be right in front of a user at all times. Times have changed.

FileMaker provides four sets of tools for data entry that provide automated data entry or that are particularly well adapted to mobile devices where you might want to explore keyboard alternatives. They are discussed in this section. The tools are as follows:

- **Auto-enter options.** You set these when you design the database so that data is automatically filled in for your users. You can allow users to subsequently change the data if you want. This means that there is less data to be typed in. Common examples of auto-entered data are the current date (or a variation of it such as the date of the Friday of the week following today), the name of the user, and the value for the field in question that was entered in the previous record. No keyboard is needed unless you want to change the data.
- **Value lists.** These drive pop-up menus, radio buttons, and checkboxes. You can set them up with constant values or you can allow users to modify them in various ways. Pop-up menus, radio buttons, and checkboxes are terrific on mobile devices. In fact, some people find them easier to use with their fingertip than with a mouse.
- **Custom modal dialogs.** You can let people enter data from a modal dialog that you create. In this way, you can create a dialog in which people choose (with the tap of a finger) from several data values. This lets you create the equivalent of pop-up menus that contain images instead of text or that contain more text than a menu, radio button, or checkbox can easily accommodate.
- **Autocompletion.** FileMaker can remember other values that have been entered in a field and use them to provide autocompletion tips for users. (This option is available in FileMaker Pro, but not in FileMaker Go.)

In all of these cases, you wind up entering data by merely creating a new record, tapping a choice, or typing only a few characters instead of a word. They might not have been designed for devices with small keyboards, but they certainly work spectacularly well on them.

Using Auto-Enter Options

Many people who use FileMaker have never explored the options available when you create a new field in a table. You set them from the Options button at the right of the bottom of the Manage Database dialog you see in Figure 5.5. You open the Manage Database dialog from File, Manage, Database or with ⌘-Shift-D (Mac OS X) or Ctrl+Shift+D (Windows). Options apply only for a specific field, so you must have selected a field in the Fields tab of the Manage Database dialog before the Options button is enabled. (This means that for a new field, you must name the field and click Create before you can set its options.)

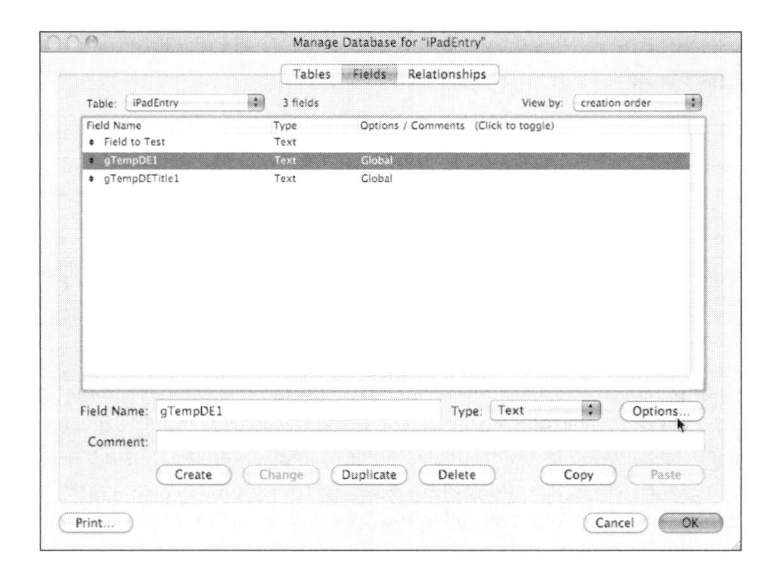

Figure 5.5
Set data entry options.

With the Manage Database dialog open, click the Options button to open the dialog shown in Figure 5.6. Click the Auto-Enter tab. There are seven checkbox options for auto-entry. Although all of them can be useful, the first three are particularly helpful for managing databases and monitoring updates; they let you automatically enter timestamps or user names for creation and modification of a record. You also can automatically assign a unique serial number to each record.

But for improving data entry on a mobile device, it is the four remaining options that can help you out. They can totally automate data entry in many cases; in other cases, they automate part of the process (you might have to enter just part of the field's data). When you are dealing with a small keyboard, any automation is welcome.

Note the checkbox at the bottom of Figure 5.6: Prohibit modification of value during data entry. If that checkbox is checked, auto-enter puts the appropriate value in the field, and the user will be unable to modify it. That is the correct behavior for data such as the name of the person who created the record or a unique serial number. The data is the data, and you want to rely on its integrity. However, when you want to modify the data—which is often the case with keyboard-saving auto-enter data—you want that checkbox off.

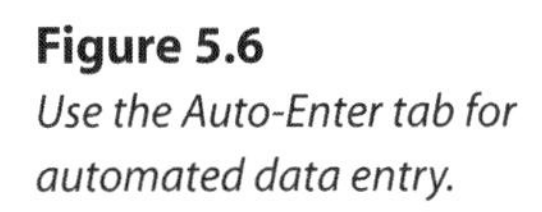

Figure 5.6

Use the Auto-Enter tab for automated data entry.

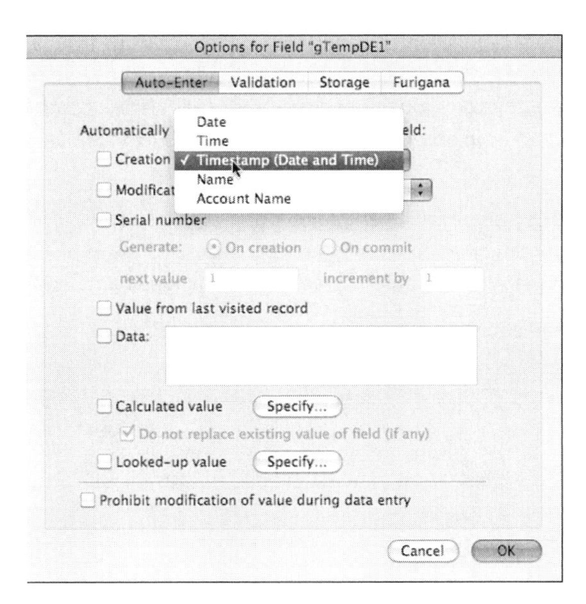

These are the auto-entry options that you can choose from:

- **Value from last visited record.** This repeats the value for this field from the last record you visited. This option is particularly useful if you are entering data that is sorted. For example, if you are entering sales data from a report or from transaction records, you might want to use this option for the transaction date and arrange the data entry in chronological order. You manually can type in the date of the first sale and, after that, the date will be used for each successive record. When you get to the next day's data (remember, this assumes sorted input data), you just change the date and then continue entering data for that new date. Thus, you only enter the date once for each day. If this is what you are doing, remember to uncheck the Prohibit Modification checkbox at the bottom.
- **Data.** This option lets you specify data that is automatically entered into the field. It is commonly used for a default value such as No Data Entered. Some FileMaker database solutions have a data integrity field (it might have radio button values such as Unchecked, Verification in Progress, OK, and Failed). In that case, you might insert Unchecked automatically into the field and then only have to modify it as circumstances change. If you are using one of these scenarios, you also have to uncheck the Prohibit modification checkbox at the bottom.
- **Calculated value.** You saw an overview of calculations in Chapter 2. You can use that interface to perform any calculation and to automatically insert the result value in the field. Remember that calculations can use variables and functions so that you can incorporate the date, time, user name, operating system, and much more into your calculation. Remember to specify the result type of the calculation in the lower left of the Specify Calculation dialog (by default, it is a number, but that is wrong for text calculations).

  Depending on exactly how you use the calculation, you might or might not want to prohibit modification of the calculated value. Note the checkbox option to not replace existing values in the field. If this is checked, any data that is already entered is preserved; the calculation does not overwrite it. What that means in effect is that the calculation is the backup procedure in cases where no data is entered.

- **Looked-up value.** The final auto-enter option lets you fill in a value from a related record. If you create a relationship between two tables, you can reference a field from the related table in the first one. For example, in an inventory system, you might store product IDs (stock control units or SKUs) that uniquely identify each product. You can enter the name of a product and its SKU in a table listing all of your supplies.

  With the relationship in place, you can enter the stock levels of a specific SKU and the related supply name value can be printed on a report. All you need to enter is the SKU. If the product name is changed, you change it only in the supplies table; thereafter, the changed name appears on invoices.

  Figure 5.7 shows this relationship in place.

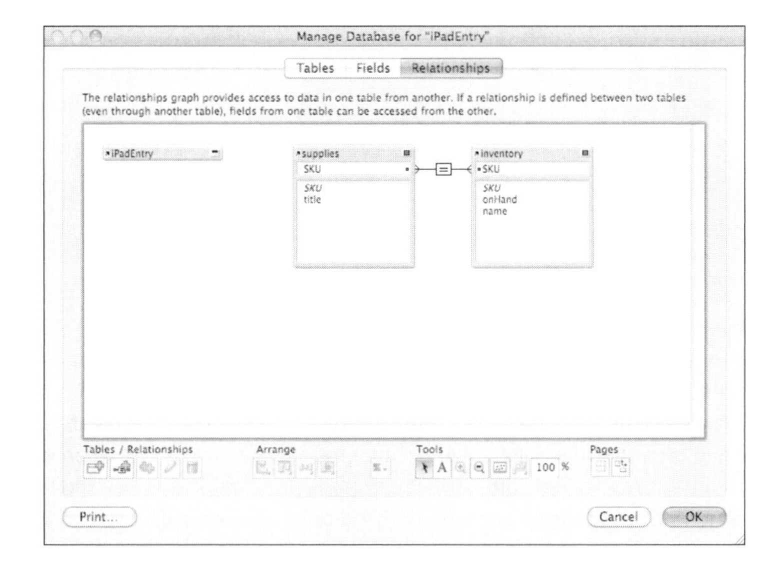

Figure 5.7
Create a relationship between supplies and inventory levels.

  The whole point of auto-entering data is to save typing, so that seems to be a good choice—and in many cases, it is. But sometimes you do want to be able to modify the looked-up value. That is where this auto-enter option comes into play. Instead of using the related value from the purchases table, auto-entry copies it into a field where it is now separated from the related value. Either can now be changed without affecting the other.

  With the relationship from Figure 5.7 in place, Figure 5.8 shows how you can set a lookup. You specify the related table you want to use and pick the field with the value that you want to look up. That value is automatically entered into the field for which you are setting the Looked-up Value option. You can choose not to import an empty value. That means that if there isn't a value, whatever is already in the field is preserved.

All of these techniques eliminate or reduce the need for the use of a keyboard, and that is the goal on mobile devices.

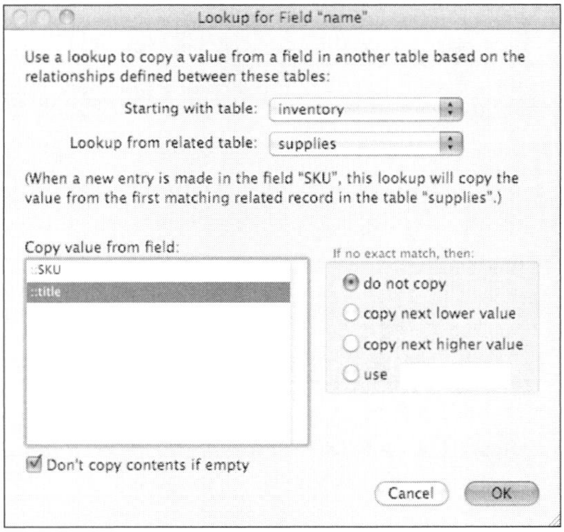

Figure 5.8
Implement an auto-enter lookup.

Working with Value Lists

Value lists serve two primary purposes in FileMaker:

- Values for checkboxes and radio buttons
- Sort orders

It is the first purpose that you look at when you are trying to save keystrokes in FileMaker Pro. You set up value lists with values such as Yes and No, colors for products, and any other values that you might want to see in a set of radio buttons or checkboxes. A value list is set up independently of a database field, but you can connect a value list to any field. This enables you to do things such as setting up a value list consisting of Yes and No and then using it in many contexts.

The values in a value list can be specified when you design the database; you can also allow users to modify the values as they are entering data. This is an important distinction both in terms of the management of the database, but also because specifying value lists and their values is done by using the File, Manage, Value Lists command—and the Manage commands affect the database structure, so they are the primary FileMaker Pro commands that are not supported in FileMaker Go. The commands that enable users to modify value lists are supported in FileMaker Go because they are part of the data entry system rather than the commands in File, Manage. Thus, they are available on FileMaker Go in most cases.

Here is an overview of value lists that you can use with FileMaker Go to save data entry time and improve efficiency.

Creating and Editing a Value List

All of a database's value lists are stored together. Create a new one by choosing File, Manage, Value lists to open the dialog shown in Figure 5.9. Click New to create a new value list or Edit to edit an existing list.

The fact that value lists are stored in a database means that in a multi-database solution, you need to be sensitive to which database a value list is located in. Fortunately, as you will see, it is easy to arrange to share them.

Figure 5.9

Create or edit a value list.

Whether you click New or Edit, you see the value list edit dialog shown in Figure 5.10. The edit dialog has three major sections that determine the content of the value list, as follows:

Figure 5.10

Specify the values for the value list.

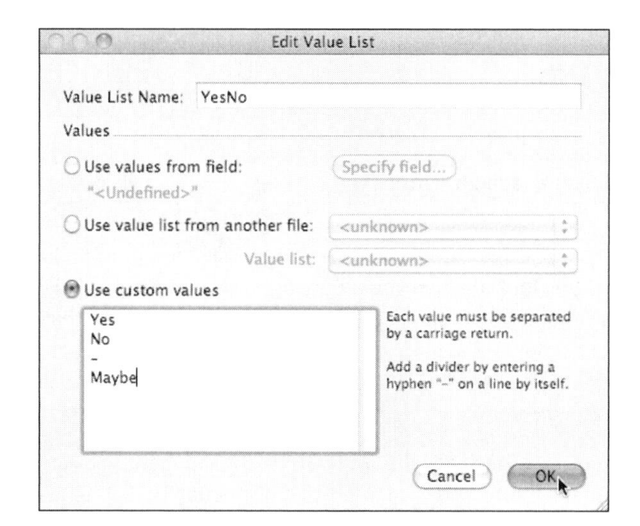

- **Field values.** You can specify a field in the database that will contain the values for the list. This is an easy way to convert user-inputted values into a pop-up or drop-down menu (you can also use a value list to power radio buttons and checkboxes). One of the advantages of creating a value list from field values is that when users select a value, they select it from an interface element such as a menu that does not allow users to modify the data. If they were copying values from an edit field, they might accidentally modify the existing value. This field may be in its own separate table. The field values can be reused—five values can be used over and over again in a table of several hundred different records.
- **Value list from another file.** If you choose a value list from another file, you simply select the file and then the value list you want to use. (You specify the other file as an external data source, as described in Chapter 4, "Working with Mobile Devices.")
- **Values for the value list.** At the bottom of the dialog, you can simply type in the values for the list. Note that a hyphen can be used as a separator.

Using a Value List

The value list is separate from the fields it works with. You connect them in Layout mode. To use a value list, create a field that will store the data. It should be a text field, as shown in Figure 5.11.

Figure 5.11

Create a text field to store the value.

In Layout mode, add the field to the layout, as shown in Figure 5.12. In the Data tab of the inspector, choose the field type you want to use. As you can see, you can choose from pop-up or drop-down menus, radio buttons, and checkboxes. (Edit fields allow typing data directly into them, so they are not used with value lists, particularly because in this context you are trying to avoid typing on a mobile device.) Depending on which type of field you are using, you might have additional options, as follows:

- **Other values.** This enables people to type in values that are not in the value list and have them stored in the field. This option is available for pop-up menus, checkboxes, and radio buttons.
- **Editing of value list.** This lets users modify the value list itself. After they have done so, they might actually choose to use a different value. However, this option provides the only mechanism by which the value list itself can be modified with FileMaker Go, and, as such, it can be very important. (Note that you can use privilege sets in Manage, Security to limit the ability to modify value lists to specific users.)

Drop-down lists also have options to show an arrow to drop the list down as well as an option to auto-complete data entry based on the value list. Neither of these modifies the value list itself.

Figure 5.13 shows how FileMaker Go shows a value list for a field that allows both other values and editing of the value list.

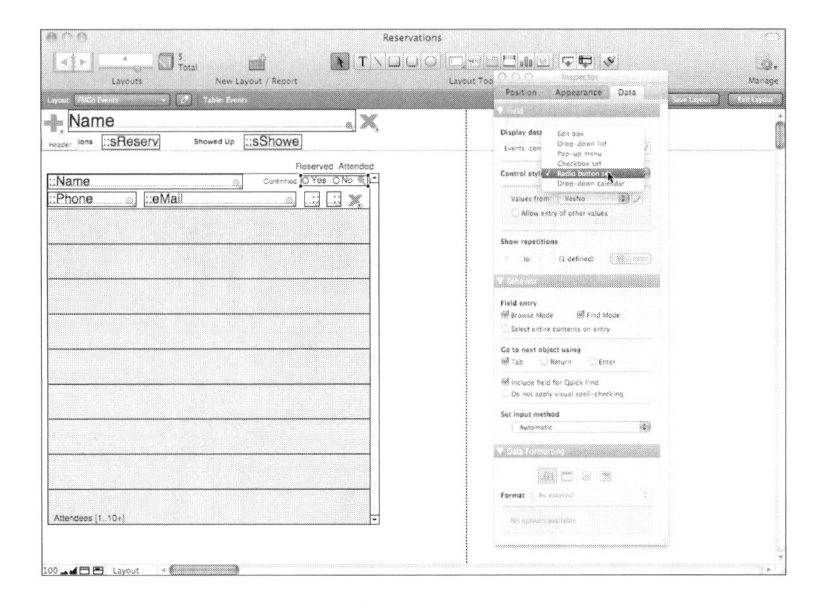

Figure 5.12
Connect the value list to the field.

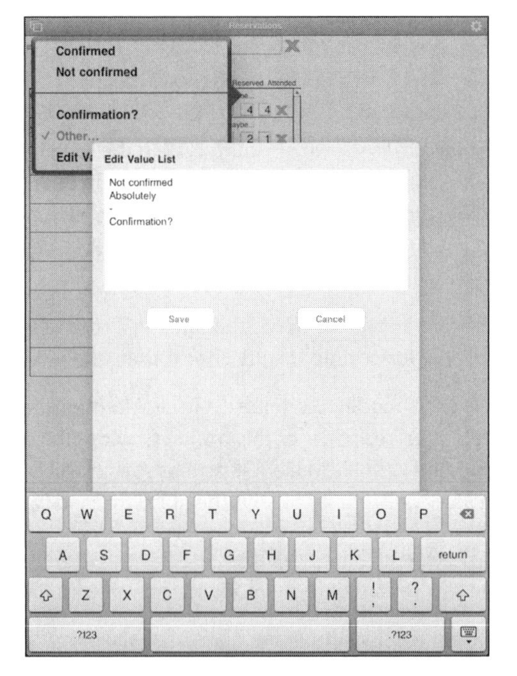

Figure 5.13
Allow users to select value, use other values, and add to the value list.

Value List Validation

Value lists are one of the tools that you can use to minimize keyboard input and thus make your solutions more mobile-friendly. However, there are a few issues to remember when you use value lists.

First of all, remember that the value list is a tool for speeding and automating data entry. Because you can use the option to let users add other values to the fields, you cannot be certain that only the value list values are actually in the field. And if you modify the value list, that modifies the data entry tool for users to use from that time forward; non-complying values that are already in the field will remain.

You can work around this to a certain extent by providing a validation option for a field. With the field selected in the Manage Database dialog, click Options. In the Validation tab, you can add a validation rule that requires that values be in a value list that you select. However, remember that this, too, only applies prospectively—that is, for future data entry. You can combine this option with a privilege set so that the validation is performed in all cases, but only certain users can override the error.

Saving Space with Value Lists

When you are running on a mobile device—whether on FileMaker Go or with FileMaker web publishing—space is a constant concern. Although a Yes/No value list can be an excellent interface tool in many cases, you should look at your particular case very carefully. The issue is clearly demonstrated in Figure 5.14, which uses a Yes/No value list.

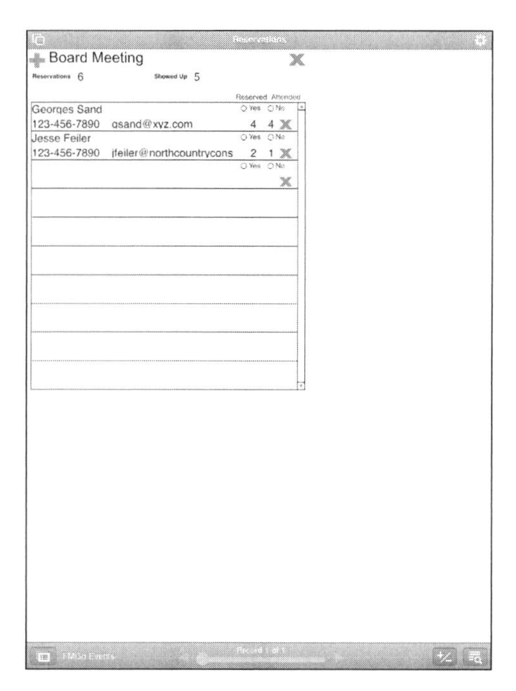

Figure 5.14
Using a Yes/No value list on a mobile device.

Here, a Yes/No value list is applied to a field in a portal, which means that the field appears in each row of the portal. Obviously, just having the two radio buttons can be a space saver, but this comes at the price of possible confusion: What is the question to which the answer is yes or no?

For layouts such as this, one of the standard solutions is to use self-explanatory radio button choices, as shown in Figure 5.15.

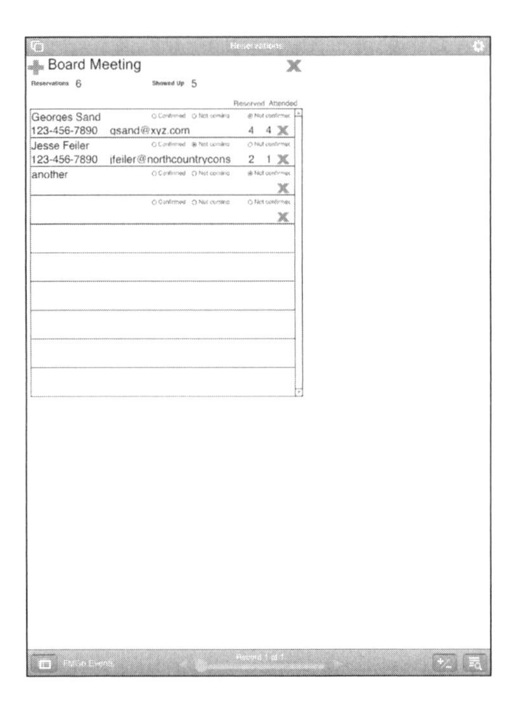

Figure 5.15
Make your value list values clear.

There is no one right answer for this in all cases. Basically, you have to consider whether the space you save by using short values such as yes and no is eaten up by a label that needs to identify them. And, by the same token, using longer values that remove any confusion gets you right back to the starting point: the space problem.

Fortunately, FileMaker gives you plenty of options, so it is just up to you to choose which you prefer.

Using Conditional Formatting

The general issue described in the previous section is that of how to save the space that is needed for a field label. The fact is that in many cases, when a data entry field contains data, it is fairly clear what the data should be, and you might be able to skip the label. This technique is used in many apps on iPhone where space is at a great premium, and, because the technique is now built into Apple's development tools, it is used in many other places as well. (This technique is used widely and is not only used by Apple and its development tools.)

The idea is to fill a data entry field with a description of the data that belongs in it (such as Client Name). To show users that this is a label and not data, the default text is shown in a distinctive font or color (often it is presented in gray).

Figure 5.16 shows three data entry fields that use this technique. The first one (Ambrose Barker) is a traditional field. The field begins as empty and has a label at the left (sometimes the label is above the field).

The second field contains data that a user has entered, Mabel Halifax; note that the text is black. In the third, no data has yet been entered, so the field contains the data description (Enter Client Name), and it is shown in gray.

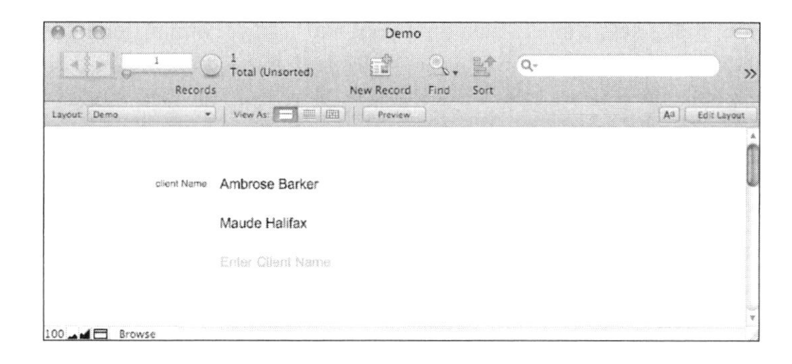

Figure 5.16
Make your fields self-labeling to save space.

All of this is done in two steps when you set up the database and the layout. Having done that, you can reclaim the space that might otherwise have been used for labels. (Of course, certain types of data do need labels; if you have a pair of date fields that represent starting and stopping dates, you need to identify them. But in many cases, you can use this technique to save space.)

Using Self-Labeling Fields

To make fields self-labeling (that is, with the description of the data inside the field), here are the steps:

1. Create the database field and the field on the layout as you normally would.
2. Select the database field and click the Options button at the lower right of the Manage Database dialog to open the Options dialog.
3. Use the Auto-Enter tab to set an initial value (such as Enter Client Name) for the field, as shown in Figure 5.17.

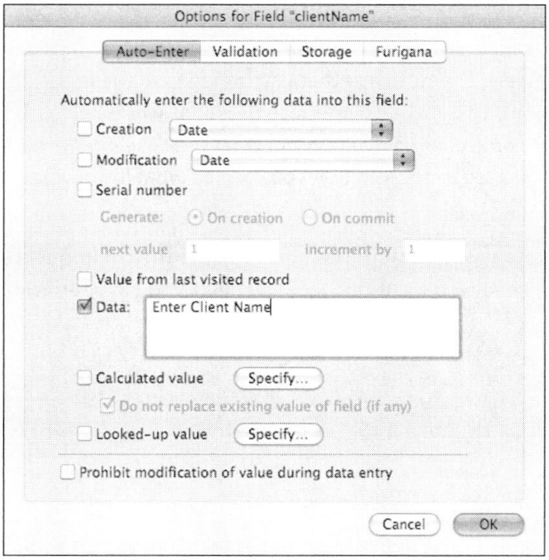

Figure 5.17
Auto-enter the field label.

4. In Layout mode, select the field on the layout and choose Format, Conditional to open the Conditional Formatting dialog, as shown in Figure 5.18. (Note that in Figure 5.18, all three fields used in the demonstration of Figure 5.18 are selected at the same time. Normally you would format each field separately.)

   The condition is set equal to the field label, and the formatting is a dark gray. This is enough to let users know what belongs in the field without having to use up space on the layout.

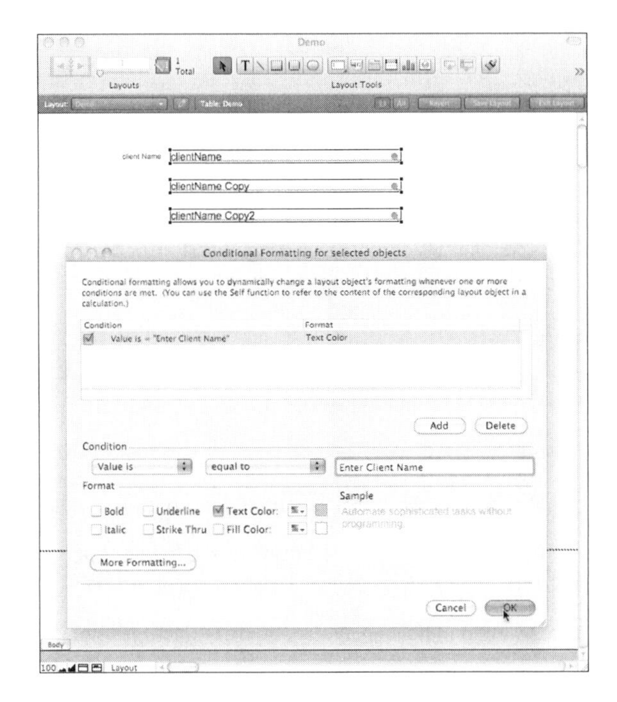

Figure 5.18
Add conditional formatting to the data entry field.

Further Steps

In addition to using mobile devices yourself so that you understand the issues, take some time to explore discussion groups and chat rooms that focus on interface design as well as on mobile apps. Read the reviews in the App Store to see what people complain about and what they like. Try downloading some apps that are a bit different from the ones you are used to so that you can see what interface problems and solutions arise in different subject areas.

As you explore discussion groups and chat rooms, you soon see the common threads that appear, and you can use these discussions to help you focus on what you need to do in your own mobile interfaces.

6

IN THIS CHAPTER

- Installing and Deploying FileMaker Server
- Running FileMaker Server
- Using FileMaker Server Admin Console

Introducing FileMaker Server

FileMaker Server is a critical component of many FileMaker Go-based solutions. Even for relatively small organizations, the ability to have mobile devices connected to a central database at all times quickly becomes an essential part of the operation. Immediate access to whatever matters—be it client records, photos of a storm drain needing repair, or the comments of colleagues—is part of the new way of operating. Thinking of FileMaker Server as a way to connect desktop computers in an office is all well and good, but your FileMaker Server installation can become the hub of your mobile operation. Even if yours is a one-person business, FileMaker Server shines in the mobile world. You might access your database from a desktop or laptop computer in your office, and the only mobile user might be you. Nevertheless, your access can be from everywhere and can be immediate. Arguably, FileMaker Server is an essential element of a FileMaker mobile solution.

About FileMaker Server

You use FileMaker Server to make your FileMaker Pro databases available to many users at once across a network (which can be the Internet). On its own, the FileMaker Pro software can host files for networked access from up to ten users at a time, in what's called a *peer-to-peer* configuration. The stability, security, and management features of FileMaker Server make it a preferred solution even for many environments that could manage with peer-to-peer solutions. However, small shops (particularly those with part-time FileMaker use) can function quite effectively with peer-to-peer configurations.

That configuration handles the needs of many small businesses. Because it is ten users at a time, many small businesses with as many as a dozen regular users of FileMaker Pro can manage with the restriction of ten simultaneous users; small businesses with much larger staffs, not all of whom use FileMaker, can also handle this restriction. Using FileMaker Pro enables you to use Instant Web Publishing (IWP) for up to five users.

However, for larger environments, serious web publishing, or environments where automated management of databases and backups are needed, FileMaker Server is what you need. In addition, for use by mobile devices, you need 24/7 access to your database, and FileMaker Server is the tool you need. Yes, you can run FileMaker Pro on a dedicated computer in your office, but the scheduling and backup features of FileMaker Server are really necessary if you will be storing mobile updates.

The FileMaker Server Product Line

Two products are available under the name FileMaker Server, as follows:

- **FileMaker Server.** FileMaker Server is used to provide concurrent access to as many as 250 networked users running FileMaker Pro client software; it also provides Custom Web Publishing (but not Instant Web Publishing) for 100 users.
- **FileMaker Server Advanced.** This product supports 250 or more FileMaker Pro clients, or 50 ODBC/JDBC clients, as well as an additional 100 web clients. It also supports IWP and ODBC.

➔ Complete technical specifications as well as hardware requirements are available at http://www.filemaker.com/products/filemaker-server/server-11-specifications.html. You can see the actual limits for each configuration as well as theoretical limits. This matters particularly for FileMaker Pro hosting on FileMaker Server Advanced. This deployment has been tested on 250 users but there is no built-in limit. As always with computers, actual performance depends on the hardware and networking configuration as well as on exactly what work is being done with the data. A comparison of all the FileMaker products is available at http://www.filemaker.com/products/compare/.

To upgrade from FileMaker Server to FileMaker Server Advanced, you simply purchase a new license key that unlocks the FileMaker Server Advanced functionality; there is no installation to be done.

WHAT IS A SERVER? WHAT IS A NETWORK?

Server has at least three meanings in this chapter; as a result, it is usually qualified as a FileMaker Server, file server, or network server.

FileMaker Server is the simplest term: It is the software product from FileMaker that lets you share databases for more than ten users at a time. It runs on a computer referred to as the FileMaker server (lowercase *s*) or, more awkwardly but precisely, "the computer running FileMaker Server."

A *file server* is a computer on a local area network on which people share files. It can be used for shared storage of corporate documents, as a backup location for individuals, or any other purpose. A network can have more than one file server.

A *network server* is a computer that manages a local area network. It might have shared network applications on it (such as mail). Examples of network server products are Mac OS X Server and Windows Server 2003/2008.

In a small environment, all three servers (FileMaker, file, and network servers) can be the same computer. It can even be a computer on which someone also does ordinary work by running word processing or other applications. This configuration really stresses the hardware, but it is a common situation in development environments where it is important to have all the features of all the computers available, but there is not a great deal of processing going on.

Any other configuration is possible. In fact, if you are using FileMaker Server Advanced's web publishing features, you can run that set of tools on a separate computer from the FileMaker Server computer.

The best performance of FileMaker (and of networking) comes when FileMaker Server and the network software are the only applications running on their respective computers. The reason is that both FileMaker and many networking tasks are large in number but relatively brief in duration: querying a database, sending an email, downloading a web page, and so forth. Most of the users of these computers want action as soon as they click or press the Enter key, and then they disappear from the server computer's environment while they type or read.

If you use some combination of computers and servers, make every possible effort to turn off file sharing on the computer running FileMaker Server. It provides its own sharing, and there can be corruption of databases if ordinary file sharing is turned on.

As for the second question, "What Is a Network?," the answer is this: a local area network, a wide area network, or the Internet. FileMaker can communicate over any TCP/IP network. You can even use peer-to-peer database hosting over the Internet; if you have a broadband connection, performance can be quite satisfactory. Dialup will drive you to drink.

FileMaker Server Versus Peer-to-Peer Database Hosting

With peer-to-peer sharing, no more than 10 database files can be served to no more than ten clients at a time. The peer-to-peer method uses a regular copy of FileMaker Pro or FileMaker Pro Advanced as the database host, so a deployment of this type also forgoes important features of FileMaker Server, especially the capability to make regular, scheduled backups of the databases. Although such schedules could be created with operating system-level scripting technologies, it's much simpler to use FileMaker Server's built-in tools.

If you do choose to begin with a peer-to-peer configuration for database sharing, it is recommended that you still treat this situation as a server-type deployment as far as possible. Give the database host its own dedicated machine on which to run—one that people won't casually use for other daily tasks; make sure that you have a reliable solution for regular backups. Make sure that the machine at least meets the minimum specifications for the FileMaker Pro client software and add a bit more RAM if you possibly can.

BACKING UP OPEN FILES

If you're backing up hosted FileMaker files by hand, be aware that you should never make a copy of a FileMaker file while it is open—even if it's not hosted and is in use by only a single user. FileMaker can guarantee that a database file is in a fully consistent state on disk only if the file has been closed properly by the server process. Otherwise, there might be database transactions that exist only in RAM that have not yet been committed to disk.

As you'll read in a later section, FileMaker Server's built-in backup capability handles the details of closing the files before backing them up. If you're working in a peer-to-peer setting, you don't have that luxury. You'll need to make sure that any automated solution you put into place takes into account the need to close each database file before backing it up.

FileMaker Server Capabilities

You've seen some of the features that set the FileMaker Server product line apart as a hosting solution: much greater scalability than the plain FileMaker Pro software and the capability to perform automated tasks such as backups. There are quite a number of other distinguishing features as well. Here are some of the most important:

- **Admin Console.** FileMaker Server comes with Admin Console, a Java application that can be used to administer one or several instances of FileMaker Server, potentially all running on different machines. A separate copy of Server Admin is created for each FileMaker Server instance to be managed.
- **Scheduling.** You can create schedules to run automatically to back up or verify databases, send messages, or run scripts.
- **Server-Side Scripting.** FileMaker Server can run FileMaker scripts as well as operating system scripts. You can schedule these within FileMaker Server. A particularly common use of these is to periodically run an export script that puts files into a known location (perhaps using `Get (DocumentsPath)`), where they can be retrieved by other applications.
- **SMTP Email.** FileMaker Server can send email on its own directly from an SMTP server rather than through a mail client that in turn accesses the SMTP server.
- **Email Notifications.** You can configure FileMaker Server 11 to provide email notifications of conditions and status to one or more email addresses. Thus, instead of having to check the server status, the server status will come to you.
- **Plug-In Management.** FileMaker Server can be configured to download plug-ins to FileMaker Pro clients in response to programmed requests from the clients, ensuring that clients will always have the latest versions of plug-ins installed on their own machines.
- **External Authentication.** FileMaker Server can be configured to check user credentials against a networked authentication source, such as a Windows Active Directory server or a Mac OS X Open Directory server.
- **Secure Transfer of Data.** When FileMaker Pro clients are used in conjunction with FileMaker Server, the transfer of data can be encrypted with Secure Sockets Layer (SSL).

In addition to these features, FileMaker Server offers a large number of other important functions, such as automatic consistency checking as files are opened, the capability to send messages to guests, to disconnect idle guests, to limit the visibility of database files based on user privileges, to be run in a scripted fashion from the command line, and to capture a variety of usage statistics and server event information for logging and analysis. All these features are discussed in the sections to come.

FileMaker Server Requirements

Like any piece of server software, FileMaker Server has certain minimum hardware and software requirements. You'll achieve the best results with a dedicated server; as with any piece of server software, it's best if FileMaker Server is the only significant server process running on a given machine. Forcing FileMaker Server to compete with other significant processes, such as mail services or domain controller services, is likely to hurt Server's performance.

The server machine, in addition to being dedicated as far as possible to FileMaker Server and having the minimum amount of file sharing enabled (preferably none), also needs the items discussed in the following sections.

Static IP Addresses

Clients must be able to connect to the server computer. It must either have a static IP address or a domain name that is set to a static IP address. If you are running a local area network, it is quite possible that a single Internet connection to your router is shared among all the computers. The router will have an IP address visible from the outside. This might or might not be static (in the case of cable connections, it is frequently renewed once a day with the same or a different address). The computers on the local network share that one changing IP address, but they might have a static IP address beginning with 192.168 or 10.0. This is determined by the configuration of the network server. It is under your control, not the control of your ISP. If the only access to your server computer is internal, you can provide it with a static internal IP address. However, if it has to be accessed from the outside, you must work with your ISP to provide it with a static IP address.

FileMaker Server is capable of *multihoming*, meaning that it can take full advantage of multiple physical network interfaces, each with its own IP address. FileMaker Server listens on all available network interfaces. As far as we know, it's not possible to configure FileMaker Server to ignore one or more of the available interfaces; if the interface is available, FileMaker Server tries to bind to port 5003 on that interface and begins listening for FileMaker traffic. The FileMaker client/server port number, 5003, is also not configurable.

Fast Hard Drive

Like any database, FileMaker Server is capable of being extremely disk-intensive. For some database operations, particularly those involving access to many records—such as a large update or a report—the speed of the server's hard disk might be the limiting factor. Redundant Array of Inexpensive Disks (RAID) technologies (whereby multiple physical disks are combined into a single *disk array*, for greater speed, greater recoverability, or both) are becoming ever cheaper, and some sort of RAID array might well be the right answer for you. When it comes to FileMaker Server performance, buy the biggest, fastest disk you can.

Fast Processors

A fast processor is a fairly obvious requirement for a server machine. But it's worth noting that FileMaker Server can take full advantage of multiple processors.

Lots of RAM

Again an obvious requirement: FileMaker Server is capable of using up to 800MB of RAM for its cache. Maximum cache RAM is determined as a fraction of installed RAM: The formula is roughly (physical RAM – 128) × 0.25. This means that to be able to use 800MB of cache memory, you'll need 4GB of RAM installed. This limitation on cache RAM was introduced in FileMaker Server 7.0v3. Because it is a 32-bit application, FileMaker Server can use only 2GB of RAM directly. Larger amounts of RAM will increase the available cache and are desirable if you're running components of FileMaker Server Advanced on the same machine as Server itself.

Turn Off Unnecessary Software

You do not want the computer running FileMaker Server to sleep, hibernate, or go into standby mode. A screensaver is also unnecessary—most of the time, your server computer will not even need a monitor, and the simplest screen saver of all—turning off the monitor's power switch—is the best.

Indexing Service (Windows) and Spotlight (Mac OS X) are great tools to help you find information on your computer, but they use resources in the background— processor power and disks, both of which are needed for FileMaker Server. If the computer running FileMaker Server does not need Indexing Service or Spotlight, turn off those options. These settings apply to all types of servers.

You also should disable antivirus software on the folder where the database files are stored (but only that folder).

Fast Network Connection

FileMaker is a client/server application, which means that FileMaker Pro clients remain in constant contact with a database host such as FileMaker Server. FileMaker Server constantly *polls* (attempts to contact) any connected clients to determine what they're doing and whether they're still connected. In addition, although Server is capable of handling a few more tasks than its predecessors, it still has to send quite a lot of data to the client for processing in certain kinds of operations. All this means that FileMaker is an extremely network-intensive platform that benefits greatly from increased network speed. A switched gigabit Ethernet network will provide good results.

Supported Operating Systems

FileMaker Server supports current operating systems on Mac OS X and Windows. For the latest technical specifications, see http://www.filemaker.com/products/filemaker-server/server-11-specifications.html.

Java Runtime Environment

You must have at least version 5 of the Java Runtime Environment installed. If you do not have it installed, you will be prompted to allow it to be automatically installed. You do not need any extra discs.

Data Center Environment

Although not strictly a requirement for running FileMaker Server, proper care and housing of server equipment is a necessity, one that's often overlooked, especially in the small- and medium-sized business sectors, some areas of education, and among nonprofit groups. These are all key groups of FileMaker users—ones that do not always have sufficient resources to build and maintain anything like a data center. Ideally, a server of any kind should be housed in a physically secure and isolated area, with appropriate cooling and ventilation, with technical staff on hand 24 hours a day to troubleshoot any issues that arise, and with automated monitoring software that periodically checks key functions on the server and notifies technical personnel by email or pager if any services are interrupted. Some organizations are fortunate enough to be able to house their FileMaker servers in such an environment. But even if you can't provide all those amenities, you can see to the key areas. The server should minimally be up off the floor, well ventilated, and under lock and key if possible. In addition, some sort of monitoring software is nice and need not break the bank; Nagios (http://www.nagios.org) is a popular and powerful open source monitoring package.

Nagios runs on UNIX but can monitor servers running on almost any platform. Many server monitoring packages exist for Windows deployment as well.

External Data Centers

A number of companies provide FileMaker hosting. Search the FileMaker website for "FileMaker hosting" to see a list. You can use a shared copy of FileMaker Server running at a remote site to run your databases and support your web publishing. The vendor will provide you with the tools to upload your databases and open them. Users connect using your domain name or the IP address of the FileMaker host. You do not have to purchase FileMaker Server because your monthly payment reimburses the hosting company for its purchase of the software. Your monthly payment also covers its data center environment, backups, and monitoring.

Web Server

If you are using FileMaker Server's Web Publishing features, FileMaker Server requires a web server. On Windows, this is IIS; on Mac OS X, it is Apache. You might have to install and configure IIS before beginning the FileMaker Server installation process. On Mac OS X, Apache is part of the standard installation; you might have to start it if you have not enabled it before. And if it has been removed, of course, you need to reinstall it.

If you are using Custom Web Publishing with PHP, the web server needs PHP; you can use a version of PHP on your own web server, or you can have it installed as part of the FileMaker Server installation process (it does not require any extra discs or licenses). Version 5.0 is the minimum required.

Installing and Deploying FileMaker Server

The process of installing and deploying FileMaker Server is different from installing software such as a word processing application or even FileMaker Pro. FileMaker Server runs in the background and has no user interface; you interact with it using Admin Console, a Java application that runs on the server computer or any other computer that has network access to the server computer.

Step 1 in the installation process is to uninstall any previous version of FileMaker Server. The easiest way to do this is to use the original distribution discs or disk images from which you did the installation. Follow the instructions in the Getting Started guide and through the links it contains to the FileMaker website. You will need to stop FileMaker Server itself and the Web Publishing Engine (if installed and running), run the uninstall process, and restart the server computer. If the server computer is running a previous version of FileMaker Server or any other applications, you should know that a restart might be required as part of the uninstall process. Unless you are starting an install on a computer that is currently not running any networked applications, you are probably going to find yourself doing this at night, on a weekend, or on a holiday. Doing this when the production environment can be stopped is very helpful. It might only take you an hour to install, deploy, and configure FileMaker Server, but that hour will go much faster if users are not poking their heads into the room or sending you text messages asking, "How much longer?"

The Installation Process

On Windows, all installed files are installed in a directory called Program Files\FileMaker\FileMaker Server. On the Mac OS, the FileMaker Server components are installed in /Library/FileMaker Server. The default install location can be changed on Windows, but not on the Mac OS.

After FileMaker Server is installed and deployed, you can manage it using Admin Console on any computer that has network access to the FileMaker Server computer. For the initial installation and deployment, however, you need hands-on or direct access to the FileMaker Server computer. If it normally runs without a keyboard and monitor, attach them before beginning the process. When everything is complete, you can turn off the monitor or even detach it.

Installation is handled by an automated process. All that you have to do is to enter your name and license code. Note that the license code company name must exactly match the name on the license you have received from FileMaker. If there are any misspellings or mistakes in the name, you must either correct them with FileMaker or grit your teeth and enter them in their incorrect version during the install process.

FileMaker Server Configurations

FileMaker Server can run in several configurations. There are three components to deal with, as follows:

- **Database server.** This is the FileMaker database engine. It controls access to the databases. In FileMaker Server terminology, whatever computer is running the database server is the *master* computer. All others (if any) are *worker* computers.
- **Web Publishing Engine.** This is the FileMaker software that enables Custom Web Publishing and IWP. IWP is available only for FileMaker Server Advanced, but Custom Web Publishing is available in both products.

- **Web server.** This is IIS (Windows) or Apache (Mac OS X). The web server must be configured and set up before you begin the FileMaker Server installation. This is the Web Server module that lets the Web Publishing Engine talk to the web server. Also, as part of the FileMaker Server installation process, PHP might be installed. If it is already installed on your web server, you can use that version as long as it is at least 4.3.

You can install these components on one, two, or three computers. Because the computers communicate using standard protocols, it does not matter which operating system is used on which computer in a multi-computer environment. Here are the configurations you can use:

- **Single computer.** The database server and the Web Publishing Engine on a single computer along with your web server. This is obviously the simplest installation, and it might be the best one to start with. Certainly, if you are creating a test environment to explore FileMaker Server, it is a logical place to start. This configuration has been successfully tested with up to 250 FileMaker Pro clients.
- **Two computers: web/database.** With two computers, you can put all the web components on one computer, leaving the database server alone on the other computer. This is generally the highest-performing configuration.
- **Two computers: FileMaker/web server.** Another configuration places the main FileMaker components (Web Publishing Engine and database server) together on one computer. The only software that must be installed on the web server is the Web Server module. This configuration might be advisable or even required where the web server is used to provide other services in addition to FileMaker.
- **Three computers.** Each of the components is on its own computer. This is suitable for very heavy loads because the web server does not have to compete for resources with anything else. Tuning the web server's performance might be easier in this environment. Note that because there is intercomputer communication between the web server and the Web Publishing Engine, this configuration can in some cases provide slightly poorer performance than a two-machine configuration.

Except in the very unusual situation in which everything is installed on a single computer and the only clients accessing FileMaker Server are on that same computer (an environment usable only for testing), there has to be communications among computers managing FileMaker Server, Admin Console, the FileMaker Pro clients, and the clients of the web server. These communications should be protected by firewalls on the various computers. Table 6.1 provides a list of the ports, their purposes, and which computers need access to them.

Firewalls

Whenever you are dealing with a network connection, you are usually dealing with firewalls and ports (if you are not, you are running a major security risk). Table 6.1 provides the official FileMaker Server port list. If you open these ports before you install FileMaker Server, things will go faster. The users of the ports are the Web Server (WS), Database Server (DS), Web Publishing Engine (WPE), Admin Console (AC), and FileMaker Pro clients (FPC).

Table 6.1 FILEMAKER SERVER FIREWALL PORTS

Port	Used For	WS	DS	WPE	AC	FPC
80	HTTP	X				X
5003	FileMaker sharing		X			X
16000	HTTP	X				X
16001	HTTPS for Admin Console		X		X	
16004	Admin Console	X	X	X	X	X
16006	FileMaker Server			X		
16008	FileMaker Server			X		
16010	Custom Web Publishing			X		
16012	FileMaker Server			X		
16014	FileMaker Server			X		
16016	Apache Jakarta Protocol			X		
16018	Apache Jakarta Protocol			X		
50003	FileMaker Server service/daemon		X			
50006	FileMaker Server service/daemon		X			

Selecting the Configuration

When you begin the installation process, the installer will present the window shown in Figure 6.1 to begin configuration. Choose whether this will be a single-machine or multiple-machine installation.

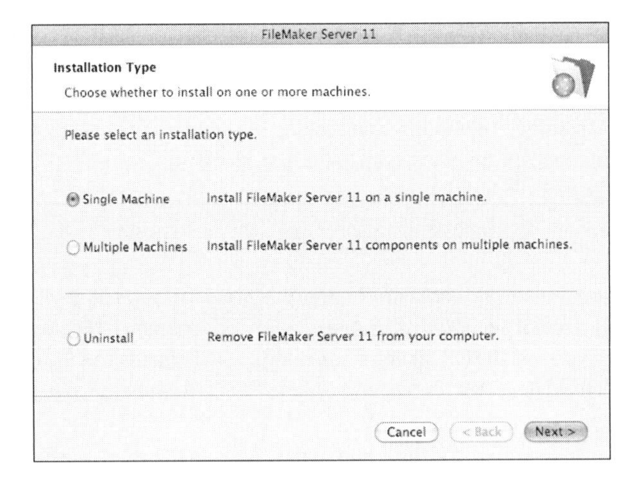

Figure 6.1
Choose a single- or multiple-machine configuration.

If you choose a multiple-machine configuration, you need to install FileMaker Server on each computer. The window shown in Figure 6.2 lets you specify for each computer whether it is a master computer (with the FileMaker Database Server installed on it) or a worker.

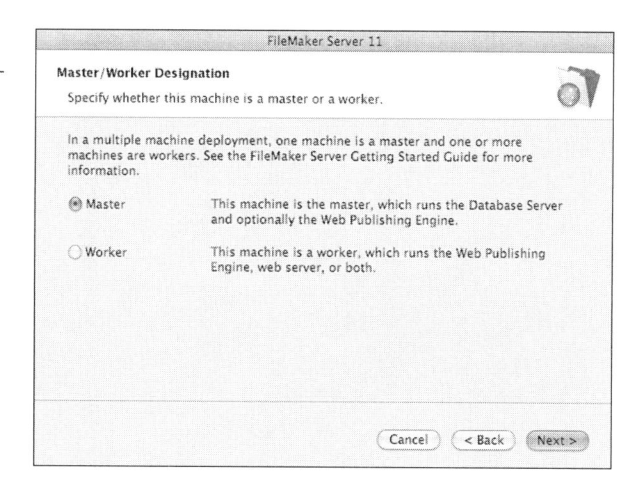

Figure 6.2
For multiple-machine configurations, install FileMaker Server on each computer.

The Deployment Process

After you have completed the basic installation process, you will be prompted to continue on to the Deployment assistant. If you choose not to do so, you can pick up at this point later by choosing Start, Programs, FileMaker Server, FMS 11 Start Page (Windows) or double-clicking the FMS 11 Start Page shortcut that was installed on the desktop. The FileMaker Server Start Page will appear followed shortly by the Admin Console start page shown in Figure 6.3.

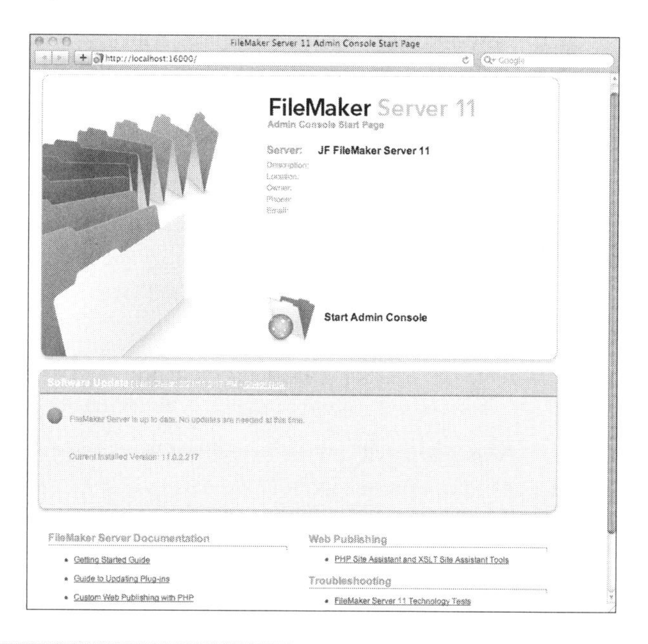

Figure 6.3
Admin Console start page.

In part because the Deployment assistant and Admin Console are Java applications and you might not have run Java applications on your computer before, there could be a little fiddling to get them to run. The Getting Started guide that is installed as part of the FileMaker Server documentation is an invaluable resource. After the installation and deployment process is done, you should not have to worry about these issues again.

In the installation process, you next go to the Deployment Assistant as shown in Figure 6.4.

Figure 6.4

The Deployment assistant guides you.

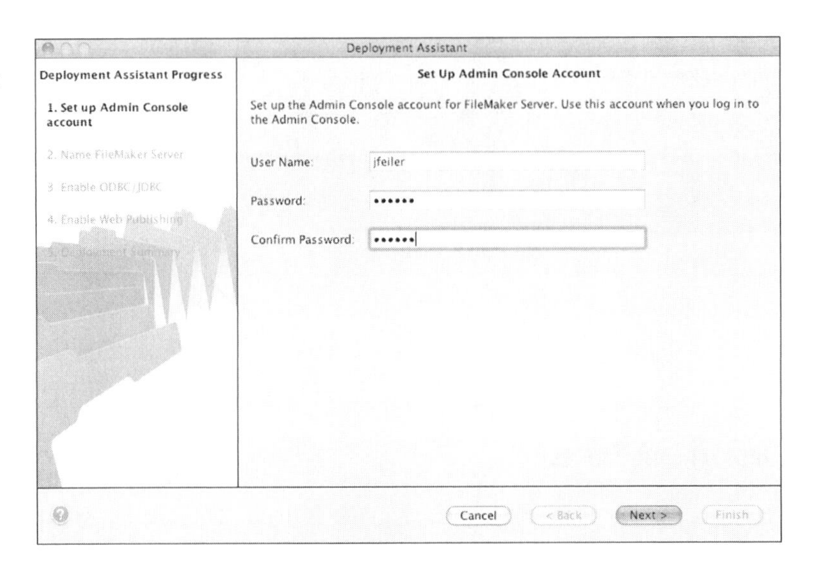

Along the way, you might see messages asking you if you want to allow access to specific resources such as the Java application. If you are confronted by messages such as these, remember that you have started a process of installation and deployment of a FileMaker product, so a message asking if FileMaker is a legitimate provider of software should be answered Yes or Always. If you launch an installation of FileMaker and see a message asking you to approve installation of software from another vendor, you might get suspicious, although Java, which is not a FileMaker product, is installed. If you get a warning of software from some other vendor or a name you do not recognize, you might want to contact FileMaker customer support.

There are five or six steps, depending on whether or not you are doing a single-machine configuration. All of these settings can be changed later:

- Set up the Admin Console account with an ID and password. This is the ID and password for full administration. See "Administrator Groups" later in this chapter for details on how you can customize and constrain passwords for specific functions.
- You name this particular FileMaker Server installation, providing a brief description and the name and contact information for the person responsible, as shown in Figure 6.5.
- You can enable ODBC/JDBC publishing if you are using FileMaker Server 10 Advanced. You do not need to enable ODBC/JDBC publishing to access ODBC/JDBC data published elsewhere and that you and your users want to consume. This interface is simple: Click Yes or No.
- You choose which web publishing technologies you want to use (XML, XSLT, Instant Web Publishing [IWP requires FileMaker Server 10 Advanced], and PHP), as shown in Figure 6.6.

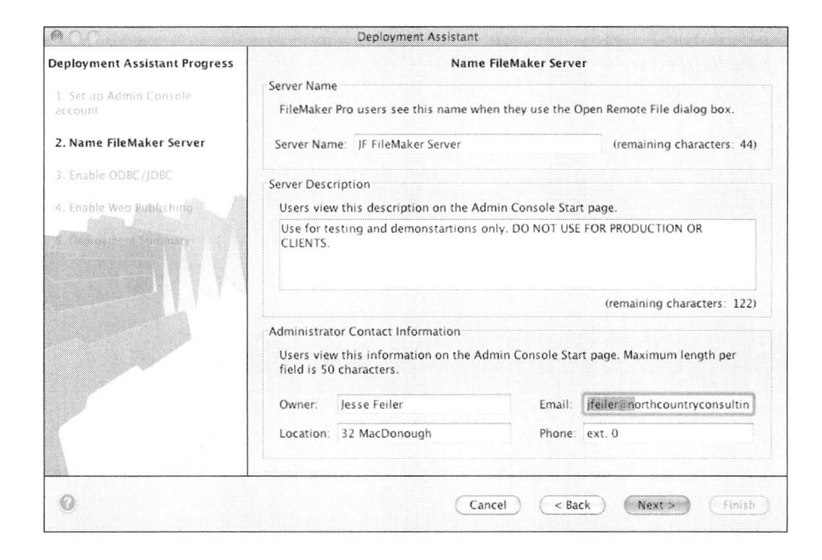

Figure 6.5

Name the FileMaker Server.

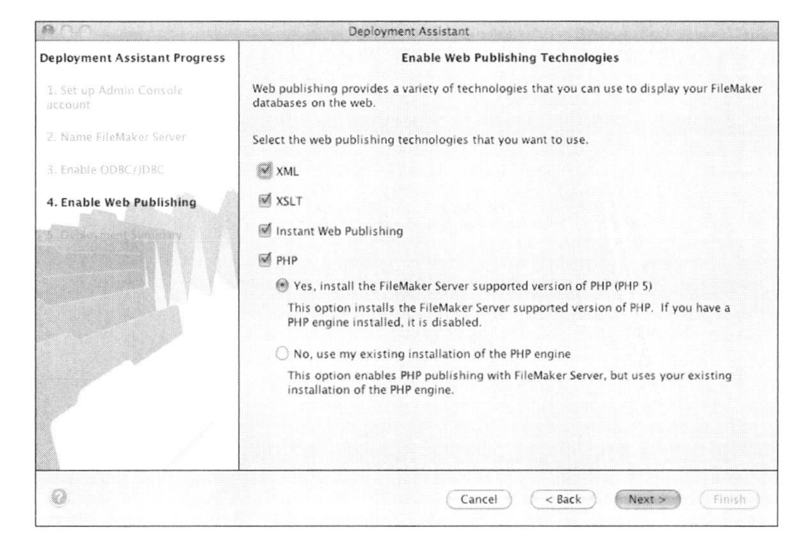

Figure 6.6

Choose your web publishing technologies.

- You can then choose your web server (or choose to select it later), as shown in Figure 6.7.
- A final summary screen reviews your choices, as shown in Figure 6.8. You can use the Back button to go back and change them.

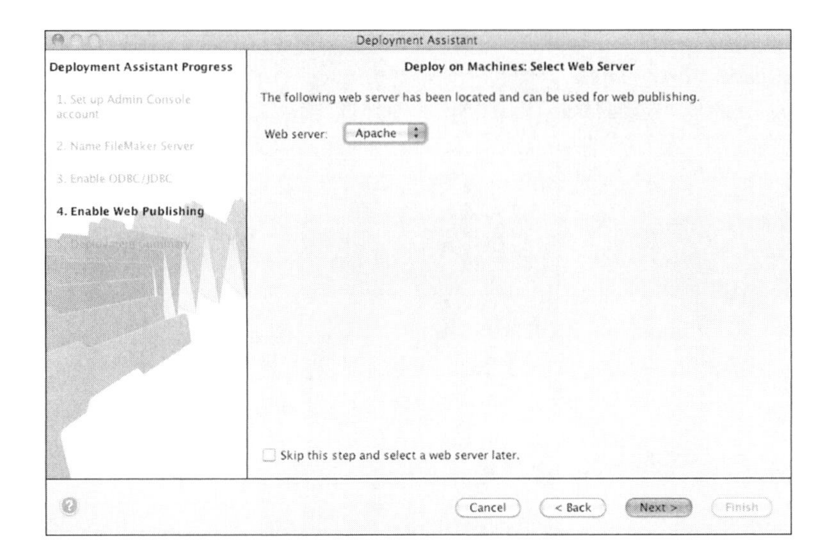

Figure 6.7
Select your web server.

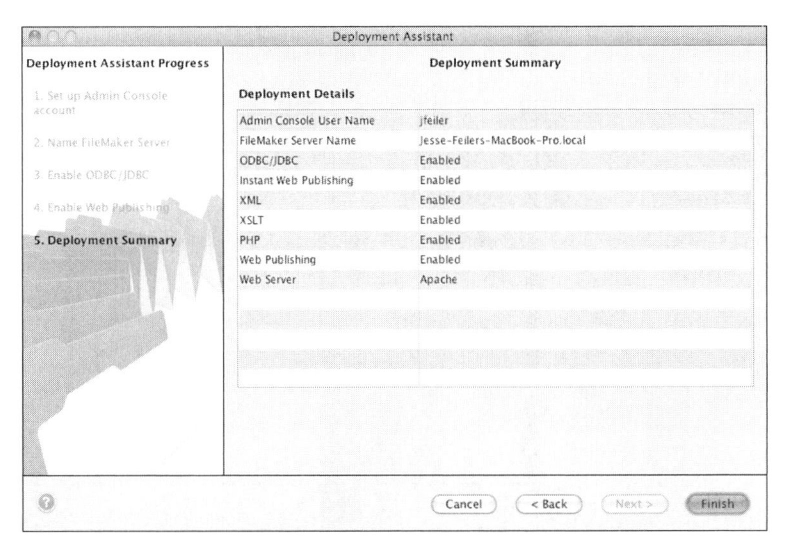

Figure 6.8
Review the installation summary.

You will be provided with information about the progress of the deployment. When it is complete, you will be invited to register (a good idea) and to run the technology tests (a critical step), as shown in Figure 6.9.

At this point, FileMaker Server is installed and deployed. You can use the Admin Console on any computer with network access to the FileMaker Server computer to administer it from now on.

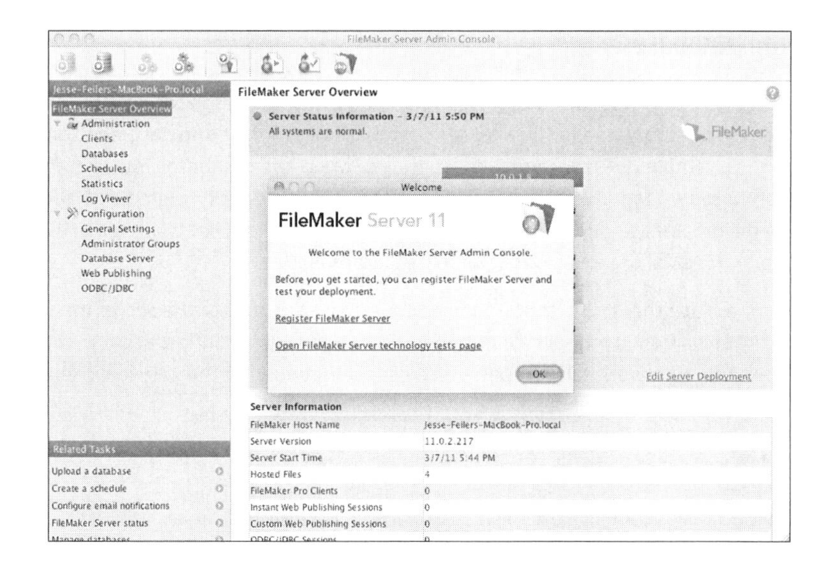

Figure 6.9
Run the tests.

4.11

If you will not be using a monitor or keyboard on the FileMaker Server computer, it is a good idea to take two additional steps before removing them. First, verify that any automated power equipment is working properly to avoid interruptions. In the case of an extended outage, the uninterruptible power supply (UPS) should shut down the server computers gracefully. Make certain that you can power them back on again successfully and that the databases open properly. If there are any problems, check them out before disconnecting the monitor and keyboard. Likewise, make certain that you can access the computer using Admin Console from another computer before disconnecting the monitor and keyboard.

Running FileMaker Server

Installing FileMaker Server installs two separate components, both of which run as services: FileMaker Server and the FileMaker Server Helper. They appear as two separate services (Windows) or processes (Mac OS). FileMaker Server does not function correctly without the FileMaker Server Helper service also running. Installing FileMaker Server Advanced will cause additional services to be added.

Starting and Stopping FileMaker Server

When you install FileMaker Server, you can choose whether to have these services start automatically, in which case they are started every time the server computer itself starts up, or manually, in which case you need to start the services by hand. The Admin Console lets you start and stop FileMaker Server manually.

Hosting Databases

FileMaker Server can host up to 250 FileMaker databases. When the server starts, it looks for files in the default database file directory, and in the alternate database directory if one has been specified. (We discuss how to specify the alternate directory later.) It also tries to open any databases found in the first directory level within either of those two top-level directories. Databases in more deeply nested directories are not opened. You can find the main database directory at c:\Program Files\FileMaker\FileMaker Server\Data\Databases (Windows) and /Library/FileMaker Server/Data/Databases (Mac OS X).

Take care to place these directories on hard drives that are local to the server machine. It's not at all a good idea to host files from a mapped or networked drive. In such a configuration, every database access has to be translated into a network call and passed across the network. At the very least, this approach is likely to cause significant loss of performance.

In the world of databases, it is common to speak of *starting* and *stopping* databases, which is basically the same as opening and closing them. FileMaker Server often uses the open/close verbs, but if you are working with people from other environments, make certain that you are both clear about what you mean by stopping a database—is that closing the database and leaving FileMaker Server running, or closing FileMaker Server and all its databases?

Using Admin Console

When you first install FileMaker Server, you might be prompted to install and open Admin Console. You have a variety of choices about when to do this, and you will have prompts and options to place shortcuts on the desktop. At that time or thereafter, you can open Admin Console manually. To do so, open a browser on any computer with network access to the computer where FileMaker Server is running (even the same computer). Enter the IP address of that computer and port number 16000. Here are three formats for that URL:

```
http://localhost:16000

http://10.0.1.2:16000

http://www.mydomain.com/rex:16000
```

The first is used if you are running the browser on the same computer as FileMaker Server. The second is used to address the server computer by its IP address (either locally or over the Internet). The third is used if you have a domain name and have configured a name for the computer running FileMaker Server. In all cases, you use port number 16000.

FileMaker Server responds by sending a small Java application to your computer. You will be asked to log in, and the application will open.

FileMaker Server Overview

Figure 6.10 shows FileMaker Server Overview. It is the default screen, and you can always return to it by clicking on FileMaker Server Overview at the left.

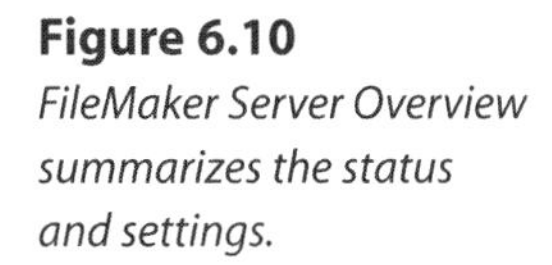

Figure 6.10
FileMaker Server Overview summarizes the status and settings.

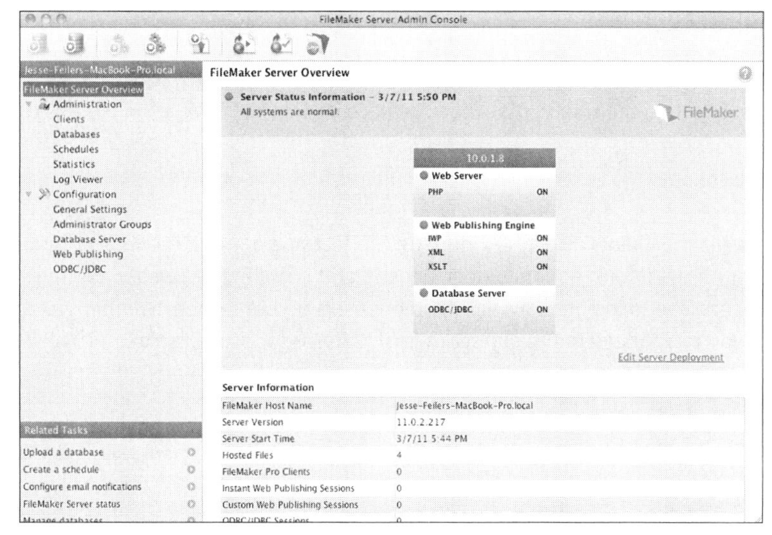

All the windows in Admin Console have a similar layout. The navigation pane at the left lets you view and change settings. The center and right of the window contains detailed information about whatever you are viewing; in the lower left, links let you go to related tasks and documentation. At the top of the window, in the toolbar, are eight icons:

- The first two let you start and stop the database engine (not an individual database).
- The third and fourth let you start and stop the Web Publishing Engine (if it is installed).
- The fifth starts the Upload Database assistant.
- The sixth lets you view the FileMaker Server Start page in a browser.
- The seventh opens the test page; you can run any needed tests again.
- The eighth launches the PHP Site Assistant.

The status overview at the right shows what is running on what machine. If you have a two- or three-computer configuration, there will be a slight space between the boxes representing the machines. In Figure 6.11, a single-computer configuration, the IP address of the single computer is shown. In a multiple-computer configuration, the IP address for each computer is shown.

You should know the IP address of the master computer, but if you do not, write it down the first time you see this display. It is the IP address people will need to connect with Admin Console (remember to add port 16000 to the IP address when you connect with a browser).

411

FileMaker Server provides the ability to configure and customize your installation whether it is large or small. If it is small, there are a few settings in Admin Console that you must know about; you can safely ignore the others unless something strange opens. Many people run FileMaker Server for years without changing the default settings. Note the IP address and the status of each of the FileMaker Server components.

Administration

The Administration section lets you manage clients, databases, schedules, and statistics. A summary screen lets you move among them, or you can click on each item in the navigation pane at the left.

Clients

You can see the Clients display in Figure 6.11. Select one or more clients and choose an action from the pop-up menu. Click Perform Action to do it. In general, it is not a good idea to disconnect clients; instead, send them a message to log off. However, if remote users do not respond (perhaps because they have gone out to lunch), you might need to disconnect them if you need to stop FileMaker Server.

Unless your FileMaker Server environment is small and in a confined area, you will probably use the Clients section to send messages to your users.

When you select a client, you can see details of that user as well as the database that the user has opened. This information is provided at the bottom of the screen.

Databases

The Databases section, shown in Figure 6.12, provides information for databases managed by FileMaker Server, whether or not they are open. A similar interface to that in Clients lets you send actions to the databases. Also, as with Clients, the information about the selected database is shown at the bottom. Thus, on Clients you can see the databases a specific client has open, and on Databases, you can see the clients that have a specific database open.

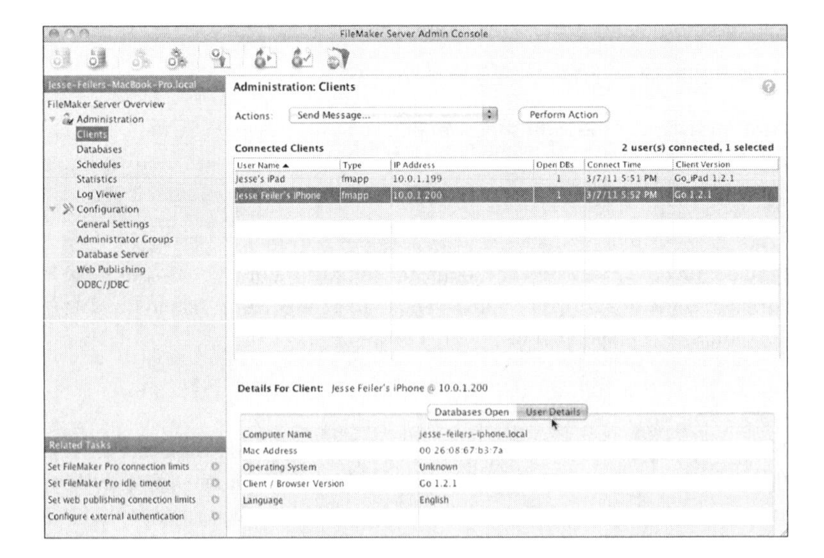

Figure 6.11
Manage clients.

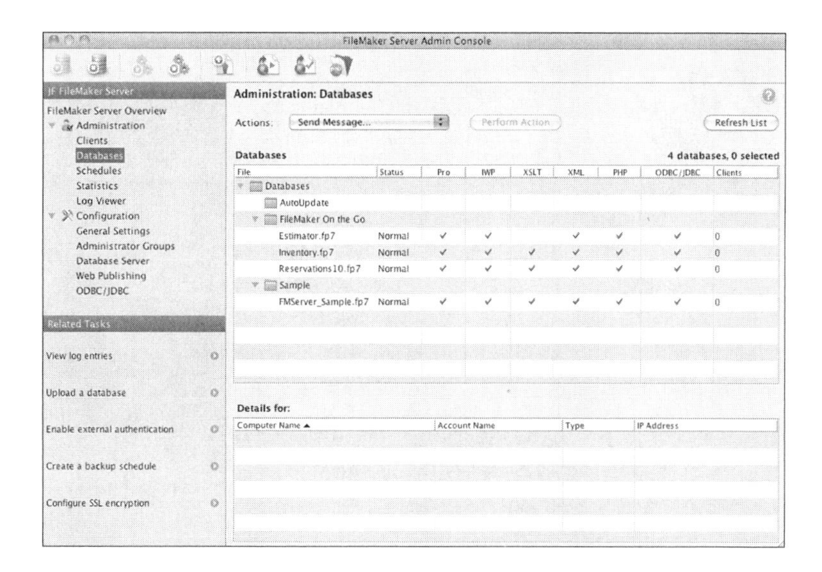

Figure 6.12
Manage databases.

Checkmarks indicate the features enabled in the databases. Note that these report the status; you set sharing and the other features in the databases themselves (in the File, Sharing submenu of FileMaker Pro). This display is useful when you have first added a database; you should check that the correct features are enabled.

You use the Databases section to open, close, and pause databases, as well as to send messages to database users without having to identify them on the Clients display. Pausing a database leaves it open, but flushes the cache and prevents reading or writing. It is most frequently used to create a copy of the database while it is technically open but not in the middle of processing. You can also right-click (Windows) or Ctrl-click (Mac OS X) on the database file for a contextual menu.

Schedules

FileMaker Server includes a powerful scheduling feature, as shown in Figure 6.13.

There are four types of events you can schedule: backups, verifications, execution of scripts and batch files, and messages. Using the scheduling feature, you can create, duplicate, delete, or edit schedules; you can also select them and execute them manually. Note that you can enable each schedule with the check box to the left of its name; you can also enable all individually enabled schedules with the check box at the top of the list. This capability lets you temporarily turn off all schedules without changing each one's status. You might want to do this for diagnostic purposes if you are experiencing slowness.

Figure 6.13

Schedule events.

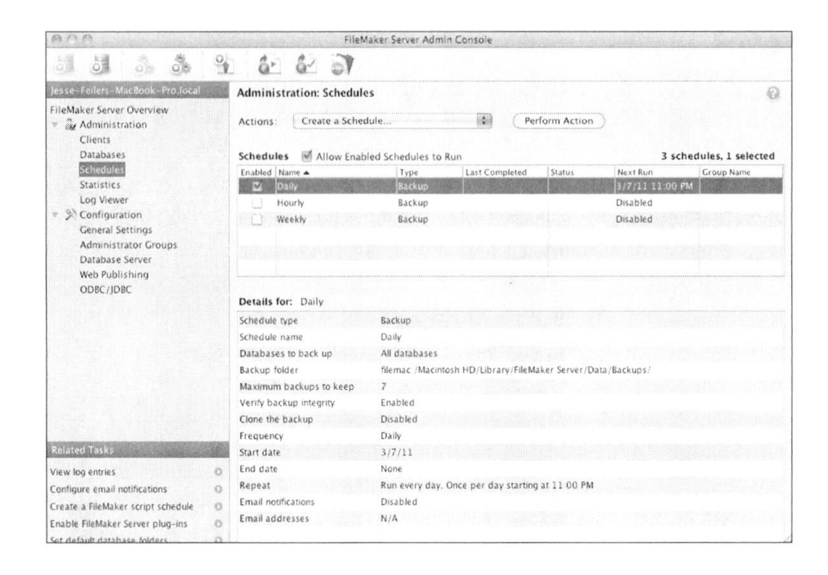

You will be walked through your choices when you create or edit a schedule. For a backup schedule, those choices are as follows:

- Select a task—backup, database verification, run a script, or send a message.
- Select from several predefined schedules (daily, weekly, and so forth).
- Select the databases to back up.
- Select the location to which they will be backed up. You also have a choice to verify the backup. This step takes more time, but it means that the backup copy is correct. If you can schedule backups to run overnight when there are few users of the system, this is a useful option. If you are automatically backing up databases, set your tape backup to back up the files from the backup folder, not the live databases. That way, you can avoid issues with backing up live databases that may temporarily be inconsistent.
- When you choose the predefined schedule, you are given the opportunity to refine or change the schedule giving it a start and stop date, selecting specific days of the week (such as omitting weekends), and specifying how often each day the schedule should be run. You can also specify the number of backups to keep.
- Name the schedule.
- You can specify an email notification to be sent. If you have asked to verify the backups, the verification status will be shown in the email. Note that you can have several recipients for the email notification. What this can mean is that instead of someone having the task of checking the backup schedule in the morning, all that has to happen is a check of email—not an extra step involving the database.
- The last step is a summary of the schedule. This information is also shown at the bottom of the Schedule window, shown previously in Figure 6.13.

Set up a backup schedule, ideally at least once a day with an email notification. Back up the databases to a known location, and then if you have an automated file backup, copy the backup files to another disk, tape, or whatever storage you are using. Make certain that the schedule of the file backup is set for a sufficiently long time after the database backup so that the files are created. For example, schedule the FileMaker database backup for 1 a.m. and the file backup for 5 a.m. Check periodically to see how long the FileMaker backup is taking (the email notification will help).

FileMaker Server now performs a live backup that requires significantly less time when the databases are unavailable. At the beginning of the backup process, FileMaker Server flushes the cache so that any data saved in memory is written to disk. Then it creates a dirty copy of the file. Users can still access and modify the original file while this copy is being made. After that's finished, the live database is paused and compared to the dirty copy; incremental changes are made to the copy so that it reflects the current state of the live file. The pause required for the incremental update is usually quite short and might not even be perceptible to users.

The Schedule assistant for scripts and email messages is quite similar. Note that you can run scripts from FileMaker or from your operating system, so the capabilities are quite large. In the case of FileMaker scripts, you select the database and provide the login information; in the case of system scripts, you choose the file.

Statistics

The Statistics window provides statistics, as shown in Figure 6.14.

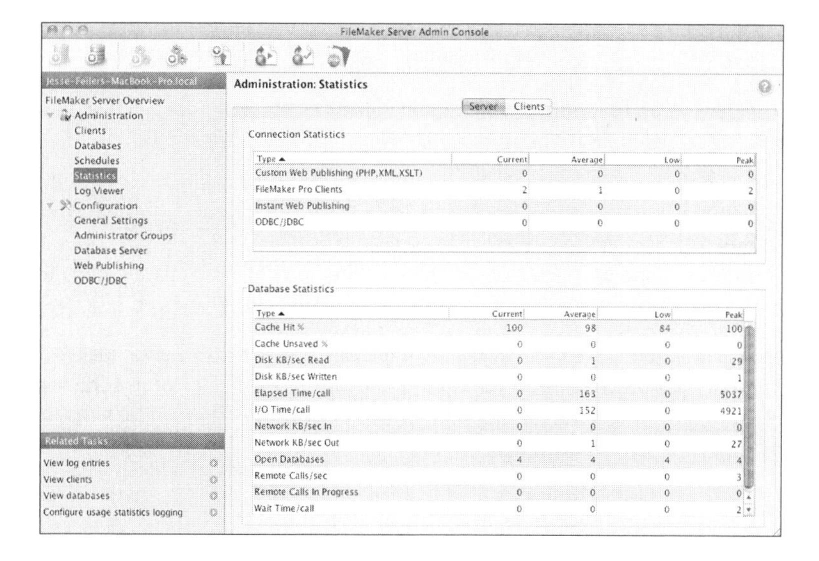

Figure 6.14

Review statistics periodically.

For each parameter, you can see the current, average, low, and peak values. Here's a list of some of what's monitored. Other items, such as times per call, are self-explanatory:

- **Clients.** This tells you the number of connected FileMaker Pro, Instant Web Publishing, Custom Web Publishing, and ODBC/JDBC clients.

- **Cache Hit %.** This number indicates how often FileMaker Server is finding the data it's looking for in the cache. Here, you want to see a number greater than 90%. Much less than that, and FileMaker is looking to the disk too often. In that case, it's a good idea to increase the size of the RAM cache on the Database tab of Database Server configuration. If the RAM cache is already as high as it can allowably go, you might want to consider adding more RAM to the machine, unless you've already reached the limit of 800MB of cache memory, which will be reached at 4GB of system RAM.
- **Cache Unsaved %.** Like many database servers, FileMaker Server sets aside an area of RAM (of a size configured by the administrator) to use as a cache. When a user makes a request for data, FileMaker Server checks first to see whether the data is in the cache, and if so fetches it from the cache more quickly than it could fetch it from disk. Over time, the contents of the cache are written out to disk. The period over which this occurs is governed by a setting on the Databases tab Database Server configuration. The setting is Cache Flush Distribution Interval. For example, if that value were set to one minute, FileMaker would attempt to write the whole cache out to disk over the course of a minute. The Cache Unsaved % should ideally be around 25% or less. If it's much greater than that, you might want to shorten the length of the cache flush period. Having too much unsaved data in the cache increases the odds of data corruption in the event of a crash.
- **Disk KB/Sec.** This gives you some idea of how much data is actually being written to disk over a given period. This is to some degree a measure of the extent to which the database files are being changed. If the files are being predominantly read from, the disk write activity should be low. If the files are constantly being written to, disk activity will be high. Keep an eye on this number if you expect that hard disk performance may be a bottleneck.
- **Network KB/Sec.** Average data transfer per second. This number tells you the extent to which the raw network bandwidth of the machine is being used up.

If you are having performance problems, this is the raw material that will help you to track down whether it is network, processor, memory, or other problems.

Log Viewer

The Log Viewer is shown in Figure 6.15. You can filter log entries by date and by the specific modules you want to view. Buttons are available to refresh the list as needed and to export data.

FileMaker Server logs have been available in the past, but the Log Viewer makes it easier than ever to view them and to track down issues. If something unusual happens, look at the Log Viewer before anything else. The columns are sortable by clicking their titles. In addition, clicking a single log entry can provide more details at the bottom of the Log Viewer.

Figure 6.15
Use the Log Viewer to troubleshoot issues.

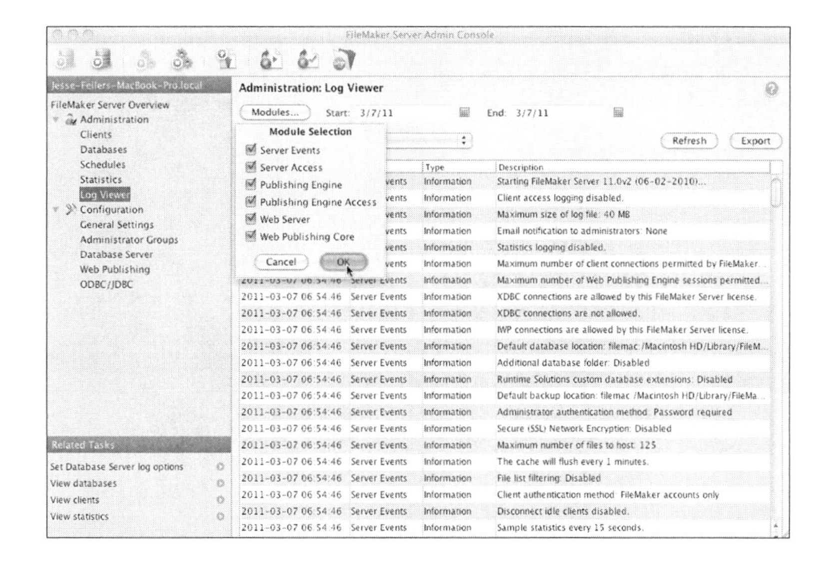

Configuration

The Configuration section of Admin Console lets you adjust settings as necessary. Most of the time, you will set and forget these settings: It is the Administrative settings that you use on a routine basis. Figure 6.16 shows the Configuration section. As you can see, there are four sections of the display, with links within each one. As you will see, those links take you directly to tabs.

Figure 6.16
Configure the databases.

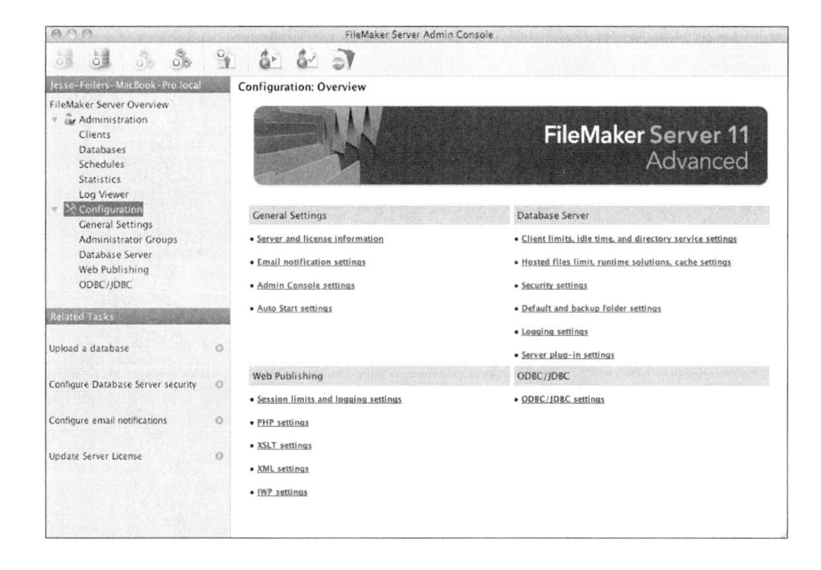

General Settings

The general settings are shown in Figure 6.17.

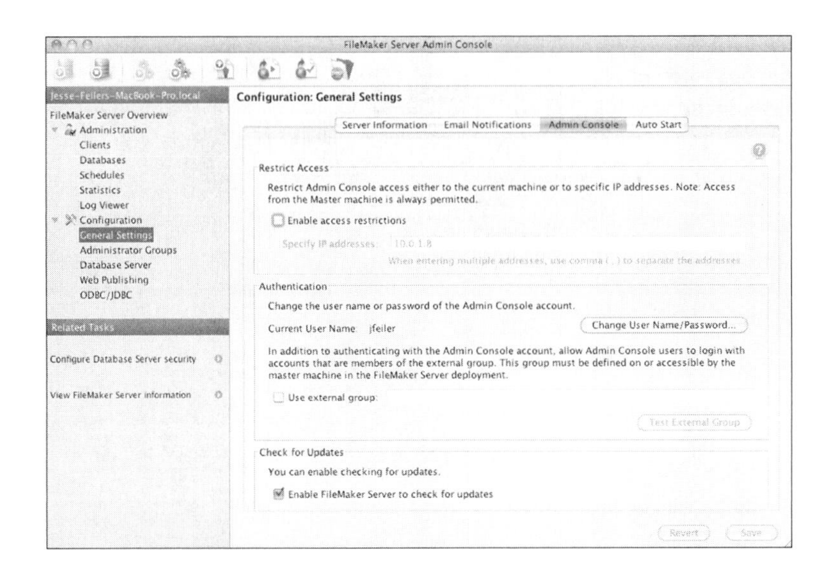

Figure 6.17
Control administrative settings.

Also in this area, you can rename the FileMaker Server or change the administrator's name and address; you can configure Admin Console to be able to be run only from certain IP addresses, and you can set the Auto Start options so that the Database Server and Web Publishing Engine start up when the computer starts. You can also change the Admin Console password and account name.

Administrator Groups

Beginning with FileMaker Server 11, you can create administrator groups to manage FileMaker Server. This enables you to delegate some—but not all—of your Admin Server management privileges. For the initial deployment, you provide an ID and password with all privileges. As you can see in Figure 6.18, you can then set up groups with specific privileges.

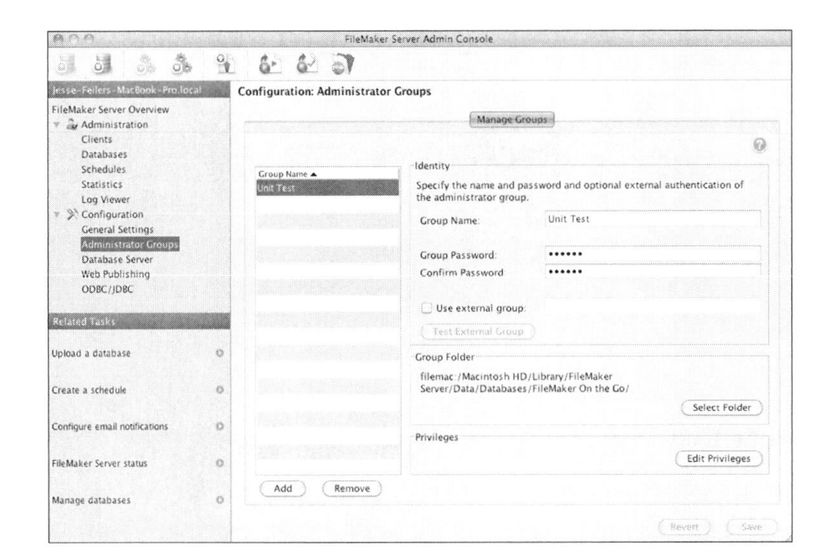

Figure 6.18
Set up administrator groups.

The interface for logging into Admin Server is as it always is: You provide an ID and a password. However, if you provide the ID and password for an administrator group, only specific functions are available as you see in Figure 6.19.

Figure 6.19
Specify privileges for an administrator group.

You can also select a folder for the group to use, as shown in Figure 6.20.

Figure 6.20
Choose a folder for the group in your databases folder.

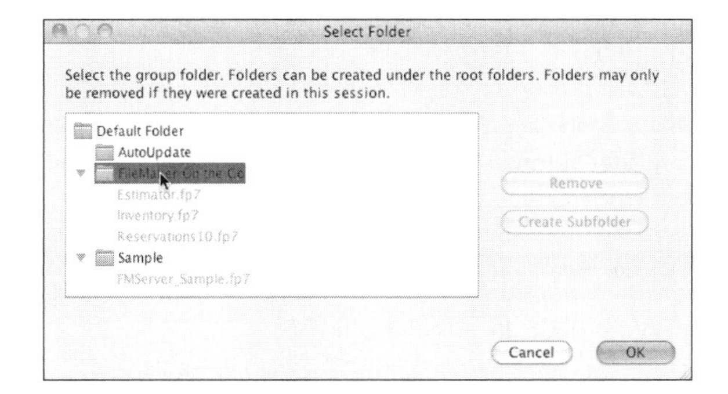

Database Server

The Database Server section provides the heart of your configuration settings. Here, you can set limits on the number of users, turn on secure communications, and set the default folders for backups and additional databases. Under FileMaker Pro Clients, you can use the Directory Assistant to select an external LDAP directory to be used.

You can specify the maximum number of FileMaker Pro clients that can connect at one time on the FileMaker Pro Clients tab, and on the Databases tab, you can control the maximum number of files that FileMaker Server will try to open. (You control the maximum number of web clients in the Web Publishing section.)

You might be able to improve performance by setting limits on the number of users that are lower than the defaults set in your FileMaker Server license. Doing so frees up resources, such as RAM, that FileMaker Server would otherwise need to keep in reserve for the possible higher loads. As a general rule, you should set these three numbers as low as you can.

You can also specify the amount of RAM to set aside for a database cache on the Databases tab. Admin Console lets you know what it thinks the maximum allowable cache size is, based on total available RAM. A good rule of thumb is to set the cache to half the allowable maximum to start and then raise it if your cache hit percentage dips too low (consistently less than 90% or so).

It's tempting to think you should just set the database cache to the largest possible size, but this isn't always the best option. Setting aside too large a cache can take RAM from other areas, such as the operating system, without necessarily being beneficial to FileMaker Server.

The last item in the Configuration section lets you turn on ODBC/JDBC sharing.

Automatically Updating Plug-Ins

Using plug-ins has become commonplace in FileMaker Pro solutions, both big and small. One of the perceived issues with plug-in use has traditionally been the difficulty of distributing them to client machines. Even with the advent of server-side plug-ins in FileMaker Pro 7, every client machine that needs to make use of plug-in functionality must have the plug-in installed and enabled.

FileMaker Server has a feature called Auto Update that simplifies the distribution of plug-ins to client machines. The concept is simple. Place your plug-ins in a designated folder on the server. When a user makes a client connection to a file hosted by FileMaker Server, you can have a script execute that checks the user's machine to see whether she has version such-and-such of such-and-such plug-in. If she doesn't, the script can automatically download the plug-in from the server. The plug-in is placed in the appropriate directory and enabled. The server thus provides automatic updates to client machines that request them, obviating the need to manually distribute plug-ins.

There's a bit of setup and scripting you have to do to make use of this feature, but it's certainly not more than an hour's work per solution. The time you'll save not having to run around to all the machines on your network updating plug-ins is certainly worth the investment of an hour.

There are essentially three tasks that you need to perform to use the Auto Update feature. These are as follows:

- Prepare FileMaker Server.
- Prepare FileMaker Pro.
- Add scripts to your solution files to perform the auto-update.

These tasks are covered in detail in the following sections.

Preparing FileMaker Server

To prepare FileMaker Server to provide automatic downloads of plug-ins, you must put the plug-ins in the appropriate folder on the server.

Inside the Data directory, you should have a folder called AutoUpdate. That's where your plug-ins go. You create a folder within the AutoUpdate directory for each plug-in that you want to be downloaded to client machines. Name the folder the same as the plug-in itself, sans extension.

Note that plug-in files often have different names on Mac OS X and Windows. If this is the case, each plug-in file needs to be treated as its own plug-in, with its own directory tree within the AutoUpdate folder. See the example using the UPLOADit plug-in later in this chapter for further details.

Within that folder, create a folder for each version of the plug-in that you want to make available. You can name the folders anything you like, but it's recommended that you simply use the version number of the plug-in. For example, within the MyPlugin folder, you might have a folder called 1.0 and another named 1.1.

Finally, place the actual plug-ins within the appropriate version folder. If you have both Mac and Windows users, you need to place both the Mac OS X and Windows versions of the plug-in in this same version-specific directory. If you are using a Windows version of FileMaker Server and need to allow Mac OS X clients to download plug-ins, be aware that you must compress the Mac version of the plug-in as a `.tar` archive. (See the "Mac OS X Plug-Ins on a Windows Server" sidebar that follows to learn how to do this.)

MAC OS X PLUG-INS ON A WINDOWS SERVER

If you have to make Macintosh versions of your plug-ins available from a Windows version of FileMaker Server, you need to bundle the Mac plug-in as a `.tar` archive and place the archive on the server. This ensures that Macintosh-specific file information is not lost during the transition of the file from one platform to another.

Tar, which stands for *tape archiver*, was originally developed to create tape backups on UNIX systems. It's now commonly used for bundling files for all sorts of purposes. You can use the Terminal application on Mac OS X to create a `.tar` archive that contains your plug-in.

From the command-line prompt in the Terminal application, navigate to the directory where the plug-in is located on your machine by using the `cd` command. You can learn more about changing directories by typing in **`man cd`** at the command prompt.

Say that the plug-in you are working with is called `foo.fmplugin`. You want to turn this into an archive called `foo.fmplugin.tar`. To do this, you would type the following at the command line:

```
tar -cf foo.fmplugin.tar foo.fmplugin
```

Take the `.tar` file and place that in the appropriate directory on your Windows server. The archive is automatically unbundled when Mac clients download the archive.

If you are using a Mac OS X version of FileMaker Server, you must make sure that any plug-ins you place on the server are owned by the fmsadmin group and have group read permissions.

A sample plug-in is installed with FileMaker Server so that you can see the directory and naming structures that you need to follow. There's also a sample FileMaker Pro database that contains scripts to download the sample plug-in. These are both valuable resources the first time you go about setting up an auto-upload routine.

To give you an additional, more real-world example, we walk through the steps you would take to build an auto-update routine for a different plug-in. The plug-in we've chosen as our guinea pig is UPLOADit, from Comm-Unity Networking Systems (www.cnsplug-ins.com). There's nothing special about this choice; we merely wanted to use something other than the sample plug-ins that ship with FileMaker. You follow the same steps for any plug-in that you use.

When you download UPLOADit, you'll get a folder full of demo files, instructions, and of course, the plug-ins themselves. The Mac version is called `UPLOADit_OSX.fmplugin`, and the Windows version is called `UPLOADit_Win.fmx`. The tasks you would have to undertake to prepare FileMaker Server to download these to client machines are as follows:

1. Create directories in the AutoUpdate folder on the server (\FileMaker Server\Data\Data\AutoUpdate\) called UPLOADit_OSX and UPLOADit_Win. You need to have both because the plug-ins have different names on the two platforms.
2. Create a directory within each of these folders called simply 1.0.
3. If FileMaker Server is running on Windows, bundle the Mac version of the plug-in into a `.tar` archive called `UPLOADit_OSX.fmplugin.tar` (see the previous sidebar titled "Mac OS X Plug-Ins on a Windows Server").
4. Copy the `.tar` archive and the Windows version of the plug-in to the appropriate 1.0 folder.

After those steps have been taken, and assuming the Auto Update feature of FileMaker Server has been enabled, client connections to the server can now begin requesting the UPLOADit plug-in. The actual download process is covered in the following sections.

Preparing FileMaker Pro

For a FileMaker Pro client to download plug-ins from FileMaker Server, the client needs to have the AutoUpdate plug-in installed and enabled. This plug-in is part of the typical installation of FileMaker Pro, so unless you've disabled the plug-in for some reason, chances are that the client application will be all prepared to download plug-ins.

As with all plug-ins, the AutoUpdate plug-in should be placed in the Extensions folder within the FileMaker Pro application directory. To confirm that the plug-in is enabled, go to the Plug-ins tab of the Preferences dialog.

As part of the routine for performing the actual download—which is described in detail in the next section—you'll write a script that checks that the AutoUpdate plug-in is installed and active. If it's not, you can show users a dialog telling them to call the database administrator or giving them instructions on how to obtain and enable the AutoUpdate plug-in.

Performing the Auto Update

The actual downloading of a plug-in from the server to the client machine is triggered by a script executed on the client machine. The AutoUpdate plug-in, which was discussed in the preceding section, has three functions, which all play a role in an auto-update routine. These three functions are as follows:

- **FMSAUC_Version (0).** This function returns a string containing the name and version number of the AutoUpdate plug-in itself.
- **FMSAUC_FindPlugin (*plug-in_name*).** This function returns a space-delimited list of the folder names on the server within the directory specified by the *plug-in name* parameter. The list, however, returns only folders that contain the specified plug-in. If there's no folder in the AutoUpdate directory on the server that's named the same as the specified parameter, this function returns a -1.
- **FMSAUC_UpdatePlugin (*plug-in_name_and_version*).** This function actually obtains the plug-in from the server. A string containing both the plug-in name and version should be used as the parameter. If the plug-in downloads with no error, the function returns a 0. Table 6.2 shows the other values that might be returned.

Table 6.2 ERROR CODES RETURNED BY FMSAUC UpdatePlugin

Error Code	Description
-1	The file to be downloaded is missing from the temporary folder.
-2	The Extensions\Saved folder to contain the backup of the outdated plug-in or support file couldn't be created.
-3	The file to be replaced on the client computer couldn't be deleted from the Extensions folder.
-4	The file to be replaced couldn't be moved to the Extensions\Saved folder.
-5	The downloaded file can't be copied to the Extensions folder.
-6	The downloaded file must be a plug-in file.
3	The AutoUpdate plug-in is disabled in the FileMaker Server Administration Client Connections Assistant, FileMaker Server Properties (Windows), or Configure, Clients (Mac OS).
5	The downloaded file can't be found in the AutoUpdate folder on the FileMaker Server computer.
6	An error occurred on the computer running FileMaker Server as the file was being downloaded.
100	The external function definition for FMSAUC_UpdatePlugIn contains an invalid or empty parameter.
101	The function call from the client computer to the computer running FileMaker Server failed. The server computer might be running a previous version of FileMaker Server.

To download a plug-in from the server, a user must first open a client session to a file that resides on the server. Plug-in downloads will not work from a peer-to-peer hosted file.

A typical auto-update routine consists of three tasks:

- Checking to see what version of the plug-in, if any, already resides on the client's workstation
- Checking whether the server has a more recent version
- If necessary, downloading the plug-in to the client workstation

If a certain plug-in is required for a file to operate as designed, you will want to have the auto-update routine be part of the file's startup script. That way, if for some reason the user isn't able to retrieve the plug-in, you can prevent her from entering the system. Whether you write the routine using just a single script or split it into three (or more) subscripts that are called from a master script is a matter of personal preference. In the example that follows, we use a single script because it's a bit easier to follow the logic. First, however, we briefly discuss each of the parts of the routine independently.

Check What's Already on the Workstation

Every plug-in should contain a function that returns the name and version number of the plug-in itself. By calling that identity function, you'll know not only whether the user's workstation already has the plug-in, but also what version of the plug-in it has (thereby possibly obviating the need to download the plug-in again). You need to manually install and enable the plug-in on a workstation so that you can find out what this function is supposed to return when everything is up to date.

In the case of the UPLOADit plug-in that is serving as our example, this function is called `Upld-Version`, and the version we're working with returns the string `UPLOADit v.1.0.0`. A quick call to this function at the beginning of your auto-update routine informs you whether the workstation already has everything it needs. If it returns nothing, or if it returns a different version number, the script needs to proceed with the update routine.

Of course, if the user's workstation doesn't have the AutoUpdate plug-in installed and enabled, there's no chance that a download can occur. You therefore need to check the version number of that plug-in as well; you do this with the function `FMSAUC_Version (0)`. As long as this function returns something—indeed, anything—the plug-in is active, and you can proceed. If not, you'll want to provide users with some feedback on what they need to do (such as calling the database administrator).

Check What's on the Server

You can check what version or versions of a plug-in are available for download from the server by using the function `FMSAUC_FindPlugin`. The parameter you pass should be the name of a folder you've set up on the server to contain plug-ins. If a folder with the specified name can't be found, the function returns a `-1`. If it is found, the function returns a string containing a space-delimited list of the version numbers of the plug-ins of that name that are available.

The version number string returned by this function contains the names of the folders you've created within the plug-in's directory; these might or might not correspond to the actual version numbers of the plug-in—that is, you can name the folder anything you want. As long as it's in the plug-in's directory and contains the specified plug-in, the folder name is included in the response generated by the `FMSAUC_FindPlugin` function. For the sample plug-in, the functions `FMSAUC_FindPlugin ("UPLOADit_OSX")` and `FMSAUC_FindPlugin ("UPLOADit_Win")` would both be expected to return `1.0`.

There's one other thing to know about the list of version numbers returned by the `FMSAUC_FindPlugin` function: It returns only the names of folders that actually contain a version of the plug-in that's appropriate for the client's operating system. That is, if you have a Mac version of the plug-in in a folder called 1.0.1, and a Windows version in a folder called 1.0.2, Mac clients see only the 1.0.1 directory and Windows clients the 1.0.2 directory. If both directories contain versions for both platforms, the function returns the string `1.0.1 1.0.2`.

Because the `FMSAUC_FindPlugin` function returns a space-delimited string, you must avoid using spaces in the names of the folders you create on the server. It is impossible to parse one folder name from another if they contain spaces.

After you determine the version numbers available on the server, you have to compare them to what the client already has to determine whether a new version should be downloaded. There are many ways you can go about comparing the local and remote version numbers, and there's no single right way that will work in all cases. You'll probably need to extract the numeric portion of the local version, using the `GetAsNumber` function or one of the text-parsing functions. Set up the name of the version folders on the server to facilitate easy comparison with what's actually returned.

It's rare that you'll ever need or want to have multiple versions of a plug-in available on the server. If you have only one version, you can simply check whether the local version equals the server version.

Download the Plug-In

If the workstation either doesn't have the plug-in or if your comparison of the local and server versions reveals that the local version needs to be updated, you'll use the `FMSAUC_UpdatePlugin` function to download the plug-in to the workstation. The parameter you pass to this function should contain both the plug-in name and the version number, separated by a space. For instance, to download the Mac version of the UPLOADit plug-in, you would use the following function:

```
FMSAUC_UpdatePlugin ("UPLOADit_OSX 1.0")
```

If desired, you can use FileMaker's string manipulation functions to dynamically build a string to pass as this parameter, using the results from the `FMSAUC_FindPlugin` function. If you know the name and version number you want, though, you can also hard-code it as has been done here.

If the user's machine already has a version of the plug-in, it is automatically moved to a directory named Saved (within the Extensions folder). The new plug-in is placed in the Extensions folder and is enabled for immediate use. There should be no user intervention necessary before, during, or after the download.

It's good practice to include a final check at the end of your update routine to ensure that the plug-in is indeed active. This would consist of another call to the version function of the particular plug-in. Assuming that all's well, your startup script can proceed with any other desired tasks.

Putting It All Together

The preceding sections discussed the tasks and principles involved in a typical auto-update routine. It should nonetheless be helpful to see a complete sample script from start to finish. The example again uses the UPLOADit plug-in and assumes a directory structure on the server, as described in the "Preparing FileMaker Server" section. Because the names of the Mac and Windows versions of the plug-ins are different, it's necessary to have some conditional logic that takes the client platform into consideration. Finally, in this script, we're simply interested in getting the version 1.0.0 plug-in on the user's machine. You could add more complex logic to automatically test for updates; this script would need to be edited slightly if an updated plug-in became available:

```
Set Variable [ $localVersion; Value: Upld-Version ]
If [ RightWords ($localVersion; 1) ≠ "1.0.0" ]
    If [IsEmpty (FMSAUC_Version (0))]
        Show Custom Dialog [ Title: "Warning"; Message: "You do not have the
        Auto Update plug-in installed on your workstation.  Please
        call Jasper, the database administrator, immediately";
```

```
        Buttons: "OK" ]
        Halt Script
    End If
    Set Variable [ $remoteVersion;
    Let ([paramName = Case ( Get (SystemPlatform) = -2;
             "UPLOADit_Win" ; "UPLOADit_OSX");
          versionString = FMSAUC_FindPlugin (paramName) ];
          RightWords (versionString; 1)) ]
    If [ IsEmpty ($remoteVersion) ]
        Show Custom Dialog [ Title: "Warning"; Message: "The UPLOADit plug-in
        could not be found on the server.  Please call Jasper, the
        database administrator, immediately"; Buttons: "OK" ]
        Halt Script
    End If
    Set Variable [ $error;
    Let ( [pluginName = Case (Get (SystemPlatform) = -2;
               "UPLOADit_Win"; "UPLOADit_OSX");
           version = $remoteVersion;
           paramName = pluginName & " " & version] ;
           FMSAUC_UpdatePlugin (paramName)) ]
    If [$error ≠ 0 ]
        Show Custom Dialog [ Title: "Error Downloading Required Plug-in";
        Message: "There was an error encountered during an attempt
        to download a plug-in required by this database.  ERR = " &
        AutoUpdatePlugin::gError; Buttons: "OK" ]
        Halt Script
    End If
End If
```

As you can see, this script has three error traps in it. You would want to change the error handling to be appropriate for your solution. We've just put Halt Script steps in here, but you might want to exit the application or take the user back to a main menu layout.

Further Steps

Try installing FileMaker Server with the default settings either on a dedicated computer or on your own computer. It comes with its own sample databases that are automatically opened for you. After the installation is done and you've run the tests, attempt to open the test databases from FileMaker Go on an iOS device.

Working with your iOS device and Admin Console on a desktop computer, examine the logs as well as the Clients and Databases sections as you work with the test databases. You can see what information is being logged. Becoming familiar with Admin Server's interface before there is a problem helps you deal with problems as they do occur.

IN THIS CHAPTER

- Opening and Closing Databases
- Setting Up Shared Databases on FileMaker Pro and FileMaker Go
- Adjusting Settings
- Getting Help
- Managing Open Databases
- Taking Photos with the Camera and Storing Them in the Database

In the first part of this book, you saw the basics of FileMaker and how you can make changes to your databases to work more effectively on mobile devices, whether you are using FileMaker Go, Bento, or FileMaker Web Publishing. This part of the book explores the specifics of FileMaker Go, and the following parts explore Bento and then FileMaker Web Publishing.

This chapter starts the FileMaker Go part by showing you how to use FileMaker Go—that is, how you (or anyone else) works on the database to browse or search data, add or sort records, or run scripts. You also see how to use the FileMaker Go interface on both iPad and iPhone.

→ In Chapter 8, "Optimizing FileMaker Databases for FileMaker Go," you see how to make changes to your databases specifically for FileMaker Go. These include some extended privileges that can improve security on a mobile device running FileMaker Go. By comparison, Chapter 5, "Preparing FileMaker for Mobile Use," explores adjustments and changes you should consider when getting your databases ready for mobile use with FileMaker Go, Bento, or FileMaker Web Publishing that is designed for a mobile device.

411

Remember that FileMaker Go does not support changes to the database structure or interface changes that you make in Layout mode or by creating scripts, setting up security, creating value lists (although you can update them), or implementing triggers, calculations, or conditional formatting. In practical terms, that generally means that you need access to a copy of FileMaker Pro or FileMaker Pro Advanced to do database development and maintence. FileMaker Go is perfect for use of an already-developed database.

The basic FileMaker Go interface is simple to use for the basic tasks you need to perform: opening and closing databases; browsing, editing, searching, and sorting records; running scripts; and exporting data with PDF files, FileMaker databases, or printing.

The FileMaker Go interface is only the beginning. Because you can write scripts in FileMaker Pro and attach them to interface buttons and conditional triggers, FileMaker Go can do many more things than the basic interface provides. Those added features are developed by the database designer (who might be you!).

➔ See Chapter 8 for more about the possibilities of functionality you can add for FileMaker Go.

About the Examples in This Chapter

This chapter uses two examples: Inventory and Reservations. Both of these examples work well on mobile devices. If you are keeping track of inventory, you might check items in as they are received; you also might periodically conduct an inventory in your offices (perhaps for stock-taking or insurance purposes). Whether you are tracking inventory on a loading dock or a shipping/receiving area, as well as if you are tracking down assets in your facility, being able to locate them, find and enter serial numbers if necessary, and photograph them if needed all are activities that work well on mobile devices.

With Reservations, it is often convenient to check in people with a mobile device. The process of entering reservations can be done on a mobile device or just as easily on a desktop-oriented computer. Similarly, entering inventory for retail operations often can be done easily at a desk, whereas tracking down assets and even entering sales can be done more easily with a mobile device.

Either of these database solutions can be used on a mobile device, with FileMaker Go accessing the database that is stored on the device. In some cases, you might want the database to be hosted on FileMaker Server so that it is shared among several devices. Even if it is not shared among several devices, there are many cases where it is more convenient to host it on a server rather than storing it on a mobile device. (FileMaker Server and its host computer are not accidentally left on a kitchen counter at home as frequently as an iPhone might be.)

In order to give you experience with both shared databases and local databases on a mobile device, this chapter describes deployment and use of Inventory with FileMaker Server or FileMaker Pro sharing and Reservations with deployment on iPad or iPhone. You can follow those procedures for either one or for other databases; the techniques are the same for all databases.

You can download Reservations from the author's website at northcountryconsulting.com. Inventory is one of the Starter Solutions included with FileMaker Pro.

Inventory

This is one of the FileMaker Starter Solutions. It lets you keep track of inventory items for sale; you can also use it for inventory items that you use in your business, including assets such as computers, chairs, and coffee makers.

It is used here in that second sense: You can take a photo to record a serial number of computers and assets of that kind.

Figure 7.1 shows a record from Inventory for a specific inventory item as it appears in FileMaker Go on iPad. Its name and description appear at the top left. At the left of the layout, you see the types of information you care about when you are managing inventory of your assets: serial number, when and from whom it was purchased, what it is, and where it is located.

In the center of the layout, you see the information you might use for tracking inventory of items for sale: the amount in stock, the reorder level, the cost (to you), and the price (to your customers), along with the weight and dimensions.

At the right of the layout is space for a picture of the item being inventoried.

This is a good example of the starter solutions. Depending on how you want to use it, you might want to remove the retail-oriented fields in the center or the asset management fields at the left. You might also want to rearrange fields and perhaps add other fields specific to your needs. Although these options have always been present, they might take on added importance as you start to think about FileMaker Go and mobile devices because space is at a premium on mobile devices.

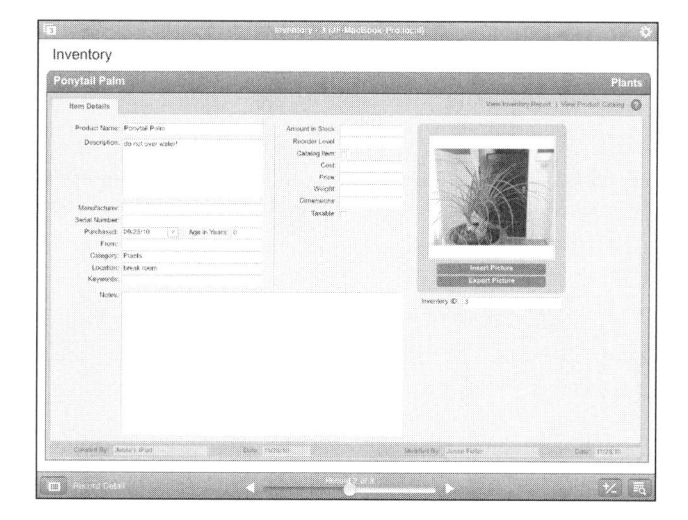

Figure 7.1
Use the Inventory Starter Solution with FileMaker Go on iPad.

Reservations

This example was introduced in Part I, "Data to Go." It lets you keep track of reservations for events, including plays, restaurant seatings, and meetings. After the reservations have been made, you can check people in as they arrive. Figure 7.2 shows Reservations running in FileMaker Go on iPhone.

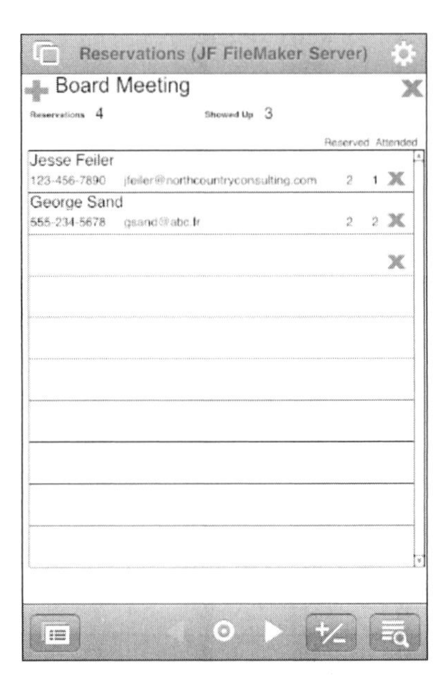

Figure 7.2
Use Reservations with FileMaker go on iPhone.

Hosting Reservations on a Mobile Device

If you want to use a FileMaker database on a mobile device with FileMaker Go, you can install it directly on the device.

→ For more on the database installation process, see Chapter 3, "Managing Data on the Move." The section "Putting Local Databases onto your iPad or iPhone" has the details, including several figures to guide you through the process.

Here are the three techniques you can use:

- **iTunes.** When an iOS device is connected to your computer, the Apps tab in iTunes gives you access to the files each app stores on the mobile device. Here are the steps to follow.

  1. Connect your mobile device directly to your computer (that is, not just with wireless).
  2. Open iTunes.
  3. Select the device at the left of the iTunes window.
  4. Click the Apps tab.
  5. Scroll down to the bottom of the window and select FileMaker Go.
  6. Drag the FileMaker database to the FileMaker Go section of the window.
  7. You might need to click Sync in the lower right of the window to transfer the database to the device. It does not hurt to click it twice, so if the transfer does occur and is so fast you do not see it (a common case), just click again.
  8. Disconnect the device after all syncing is done, and you are ready to go.

- **Email.** Email the FileMaker database to yourself (or have someone else do so). Read the email message on your iPad and then tap the enclosed FileMaker file to open it. You have an option to open it in FileMaker Go. If you choose that option, it will be opened and stored in the FileMaker Go directory for future use.
- **Web.** Place the FileMaker database on a website to which you have access (or have someone else do so). Click the file's link (or simply type in a URL that goes directly to the file). It is downloaded and opened in FileMaker Go. As with email databases, it is stored in the FileMaker Go directory for future use.

From now on, you can use the database on FileMaker Go.

Starting Out by Sharing Inventory

The version of Inventory that you see in this chapter is exactly the Starter Solution that comes with FileMaker Pro. The only change that has been made is that a single record has been added so that there is some data to display. (In the following chapter, you see how to make some FileMaker Go-specific changes to Inventory.) This part of the chapter shows you how to use Inventory when it is deployed on FileMaker Server; you can just as easily deploy Reservations or any other database to FileMaker Server. If you want to use Inventory with it deployed on your iPhone or iPad instead of on FileMaker Server, follow the steps in "Hosting and Using Reservations on a Mobile Device" later in this chapter.

You can share a database from FileMaker Pro without deploying it on FileMaker Server. You are limited to nine clients, and you must remember that the computer running FileMaker Pro must be available (and FileMaker Pro must be running) in order for mobile devices to access the shared database. From the mobile device, it does not matter if you are accessing the database from FileMaker Pro sharing or from FileMaker Server.

There are two steps to take before you can start using a database with FileMaker Go. If you are using a database in an existing environment, someone else will have done these steps for you:

- **Create or obtain the database.** This can be a database designed just for the current project or a long-standing FileMaker database that has been used for years.
- **Share the database.** Choose your hosting method and deploy the database.

As noted previously, this section shows you how to use the Inventory Starter Solution; both hosting methods are described.

Creating the Database

The first step is to get the database. In this case, you can start from scratch with a Starter Solution. In FileMaker Pro, start by choosing File, New from Starter Solution to open the dialog shown in Figure 7.3. Choose the Inventory Starter Solution.

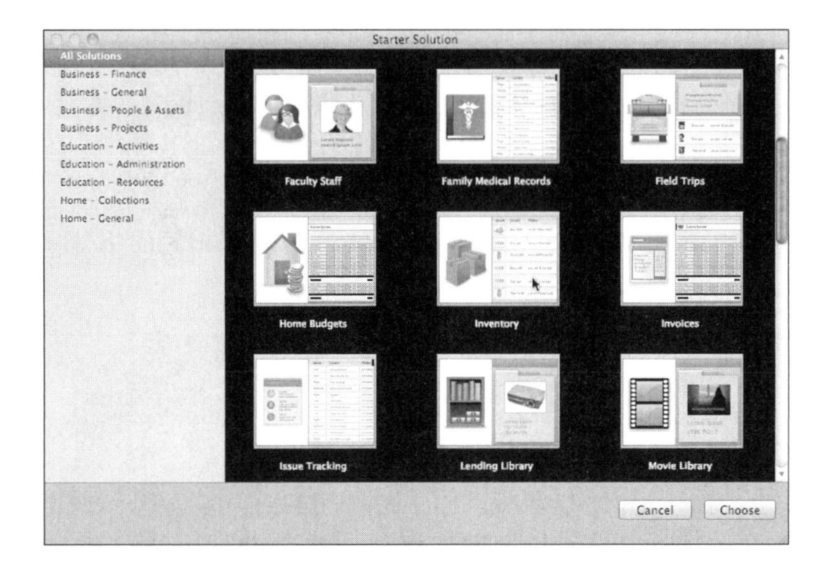

Figure 7.3
Choose the Inventory Starter Solution.

If you are planning to share a database, it is a good idea to allow access over FileMaker Network before you forget about it. Go to Manage, Security, choose Privilege Sets, and then choose the Privilege Set you want to allow to share the database. Click Edit to open the dialog shown in Figure 7.4.

➔ For more on privilege sets, see "Talking Security" in Chapter 2, "Introducing the FileMaker Architecture."

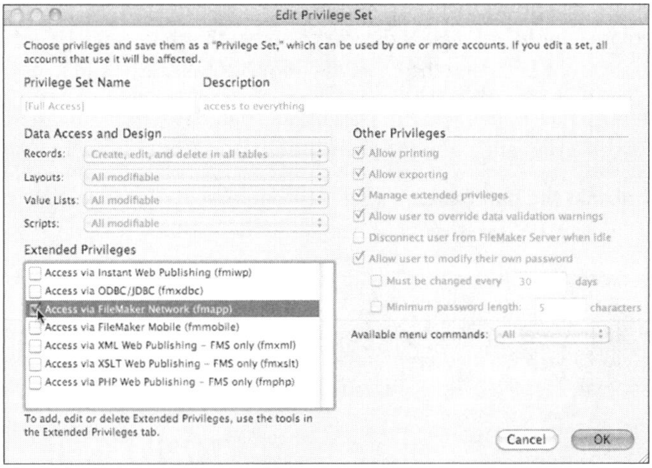

Figure 7.4
Edit a privilege set to allow sharing.

Choose Access via FileMaker Network under Extended Privileges in the lower left of the window, as shown in Figure 7.4. FileMaker Mobile refers to a no-longer supported product and technology. It has nothing to do with FileMaker Go.

You are now ready to share your database.

Sharing the Database with FileMaker Pro

If you are using FileMaker Pro to share the database, you first need to open it in FileMaker Pro. Then, choose File, Sharing, as shown in Figure 7.5.

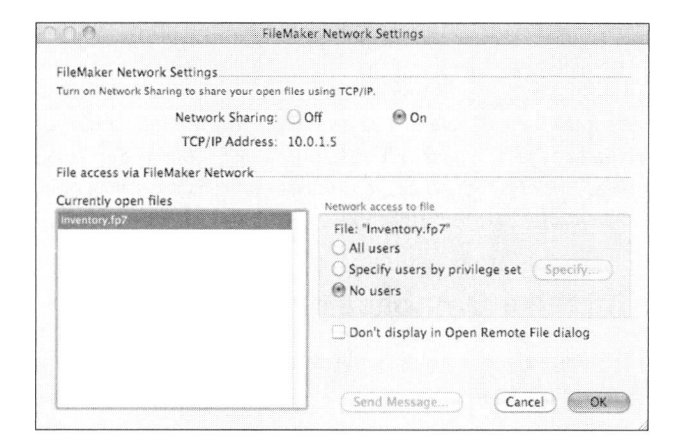

Figure 7.5
Set up sharing in FileMaker Pro.

Turn on Network Sharing at the top of the dialog. If you cannot turn it on, check FileMaker Pro Help for troubleshooting tips.

You should see all your open files at the left of the window; select the one you want to share. By default, the network access rules at the right may be set to No Users; change the access rules to All Users.

You can use privilege sets to limit network access, but it is easier to start with allowing all users so that you can test if the database is properly configured. Of course, be careful about doing this on a public server. If you are using FileMaker Pro to share databases, you are limited to nine users so the risks are lessened.

Sharing the Database with FileMaker Server

If you want to share your database with more than nine people, or if you want a more robust server experience than FileMaker Pro, FileMaker Server is for you. When you use FileMaker Pro to host a shared file, you have to remember to keep it running so that people can connect to it whenever they need to. You also have to remember to be careful about using it yourself. The best performance for shared databases happens when the server computer and the server software are dedicated only to sharing the necessary databases.

This is true with the smallest workgroup in an office to large enterprises with hundreds (and sometimes more) remote users accessing a shared database. When it comes to accessing databases with FileMaker Go, this is often a more relevant consideration. The portability of mobile devices means that people who are all over the place can access the single shared database as long as they have a network connection (and, with iPhone or iPad, that connection can be over the mobile network or over Wi-Fi networks).

➔ For more information on setting up FileMaker Server itself (rather than setting up a database on it), see Chapter 6, "Introducing FileMaker Server."

After you have set up and configured FileMaker Server as described in Chapter 6, add your database to the server and start it running. Download Reservations from the author's website, use the Inventory Starter Solution, or use any other FileMaker database you want. To install and configure the database, you need a web browser, and a network connection to the computer where FileMaker Server is running.

You can run your browser on the same computer if you want; in many environments, the computer that runs FileMaker Server is not available for any hands-on uses so that the performance of FileMaker Server is as good as possible. In those cases, you have to use another computer.

The FileMaker database must be available to you from the computer on which you are running the browser. Because the actual upload of the database will be handled by FileMaker Server Admin, you do not have to worry about tools such as FTP or direct access to the files on the computer running FileMaker Server.

How to Install a Database on FileMaker Server

With FileMaker Server running somewhere on a network to which you can connect, and with a copy of the database you want to install available to you, you are ready to install the database on FileMaker Server:

1. **Using your browser, connect to FileMaker Server Admin Console.** Use the URL of the server computer; if it is 10.0.1.5, the URL for Admin Console is http://10.0.1.5:16000. If FileMaker Server is running on the same computer as your browser, you can use http://localhost:16000. When prompted, enter your name and password.
2. **Check the database status.** Click Databases at the left of the Admin Console window, as shown in Figure 7.6. This display shows the status of currently installed database on your copy of FileMaker Server, so you might see different databases and statuses.

   If no databases are open or if no databases are present, you might want to refer to Chapter 6 and redo the installation just to make certain that FileMaker Server is running properly.

   Click Upload a Database in the lower left to start the Upload Database Assistant.
3. **(Optional) Select or create a subfolder to put your database in.** You can have one level of folders below the top-level folder, so you cannot build a complex file/folder hierarchy. If you want to use a subfolder, click Create Subfolder, as shown in Figure 7.7.
4. **(Optional) Name the new subfolder, as shown in Figure 7.8.**

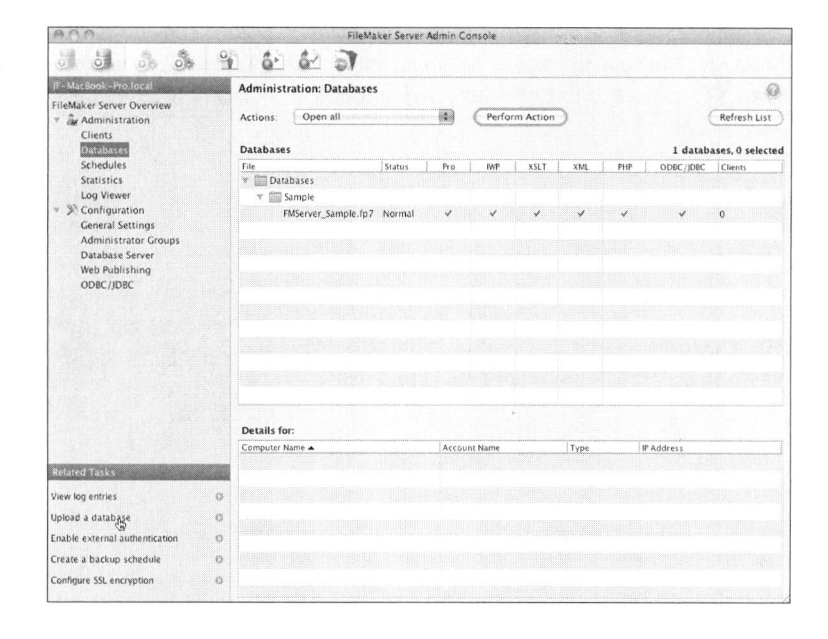

Figure 7.6
Check database statuses and start to upload a database.

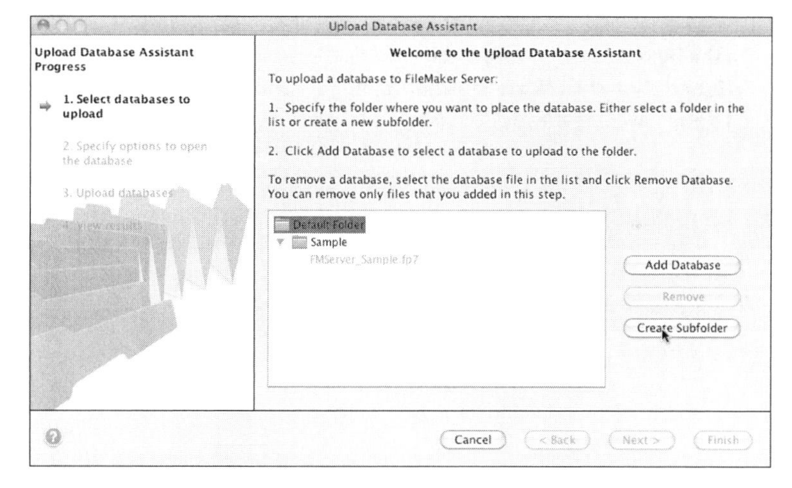

Figure 7.7
Create a new subfolder if necessary.

New Folder
Please enter a folder name
Name: FileMaker On the Go
Cancel OK

Figure 7.8
Name a new subfolder.

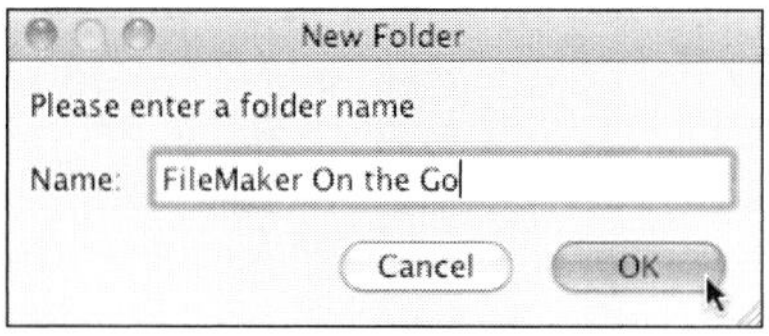

5. **Select the folder or subfolder to use, as shown in Figure 7.9, and then click Add Database.** A standard File Open dialog enables you to navigate to the database you want to upload. You must have access to it, but it can be on your own computer or on any networked computer or disk to which you have access.

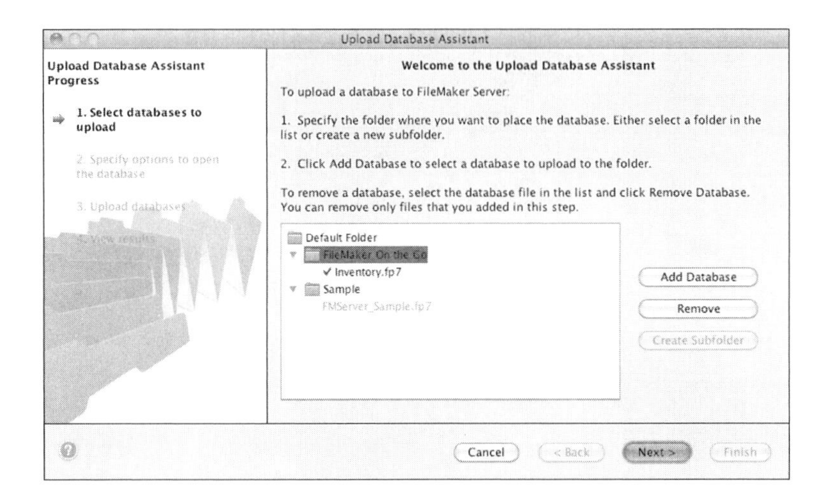

Figure 7.9
Add the database.

6. **On the next screen, choose to automatically open the database after upload, as shown in Figure 7.10, and then click Next.** This checks integrity and also updates the database privileges to allow sharing if you have not already done so.

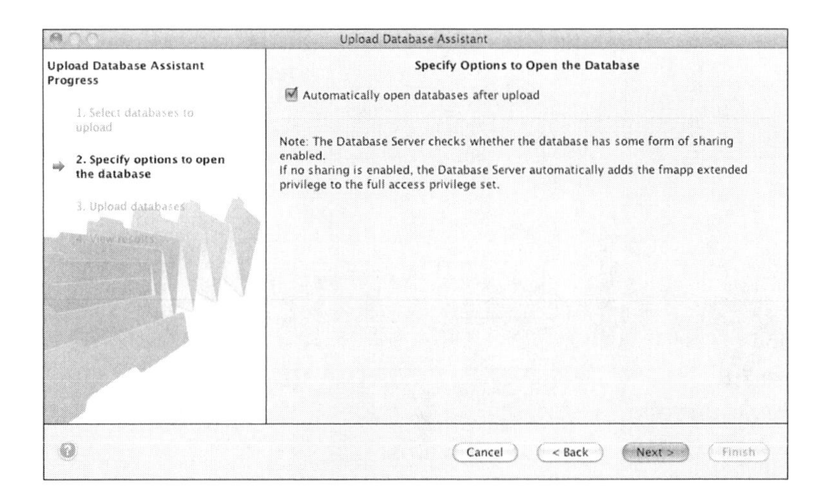

Figure 7.10
Check integrity after upload.

7. **The upload starts automatically after you click Next.** Depending on the size of the file and the network, this can take a while. When it is done, click Next (see Figure 7.11).

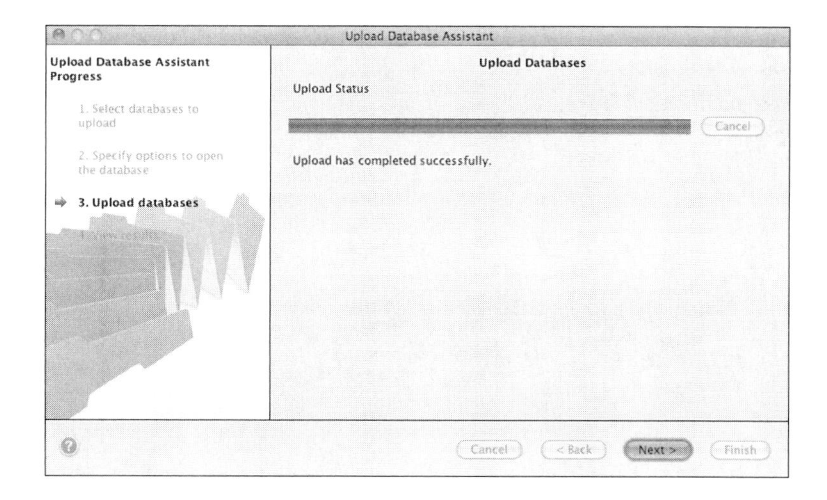

Figure 7.11
Wait for the upload to complete.

8. **Check the upload status, as shown in Figure 7.12.** If it is OK, the file is now accessible over the network. If you have configured FileMaker Server to start automatically (the default setting), that means that whenever the computer on which it is installed is restarted, FileMaker Server starts up, and all of its databases are available. This is an ideal configuration for a database that is shared by people using mobile devices who might be miles away from the server.

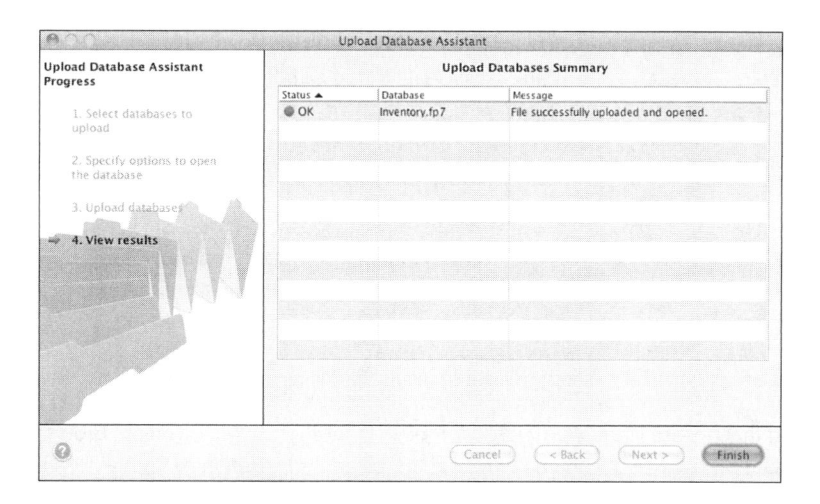

Figure 7.12
Check the upload status.

9. **Confirm the installation by using File, Open Remote from FileMaker Pro on any computer that has access to the server, as shown in Figure 7.13.** You can also use FileMaker Go on a mobile device, as described later in this chapter.

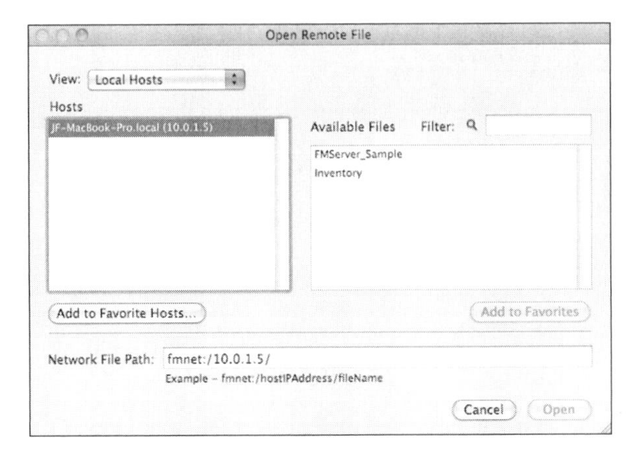

Figure 7.13

Confirm the installation from a computer running FileMaker Pro.

Opening and Closing Files

After your files are accessible on a server or installed on a mobile device (using iTunes, the web, or email), users can open them.

Opening Files on iPhone

Figure 7.14 shows the sequence on iPhone. When you launch FileMaker Go, the first screen (shown in Figure 7.14a) lets you choose recent files; you can also choose to open the File Browser. If you choose the File Browser, you see the screen shown in Figure 7.14b. This lets you choose from files on the local device, recent files (the same list you saw in Figure 7.14a), and local network hosts, as shown in Figure 7.14c. You can use Add Host at the bottom of the screen to add a host to your favorites so you do not have to remember the name or address.

If you are opening a file that is on FileMaker Server, you select the host from Favorite Hosts if it is there; otherwise, use Add Host to add it to your favorites. If you are opening a file that is on your device, you choose Files on Device.

When you have opened the list of files on your device or the list of files on a host, you view them as seen in Figure 7.15. When you are browsing files on a host (see Figure 7.15a), you can tap at the bottom of the screen to add that host to your favorites. As you see in Figure 7.15b, when you are browsing files on your local device, you can swipe across a filename to delete it. (You cannot delete files from a host when you are using FileMaker Go.)

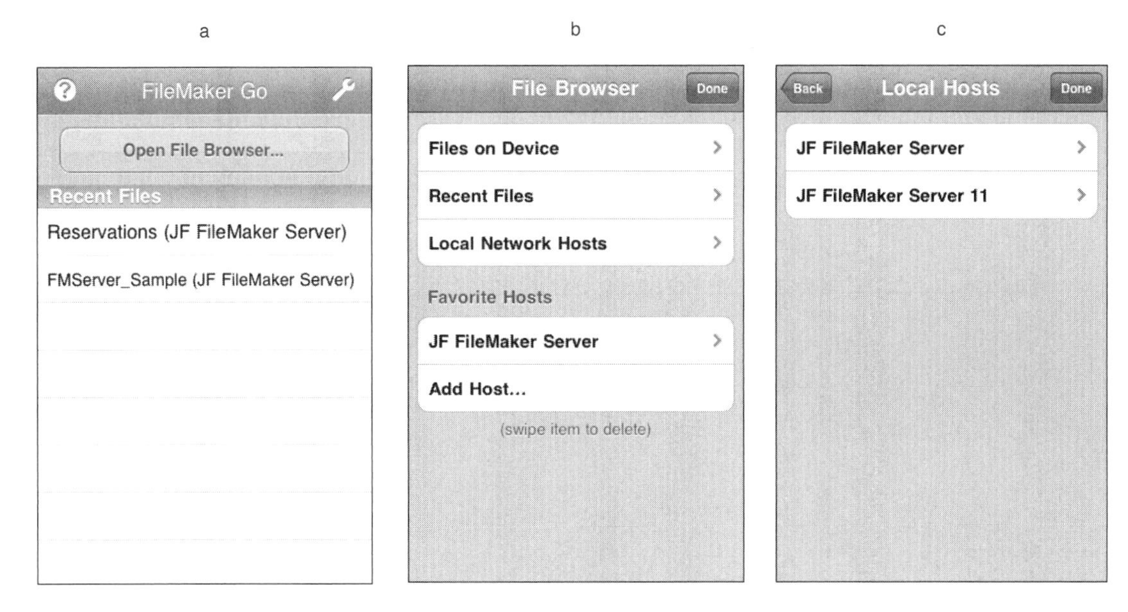

Figure 7.14
Find FileMaker files on iPhone using FileMaker Go.

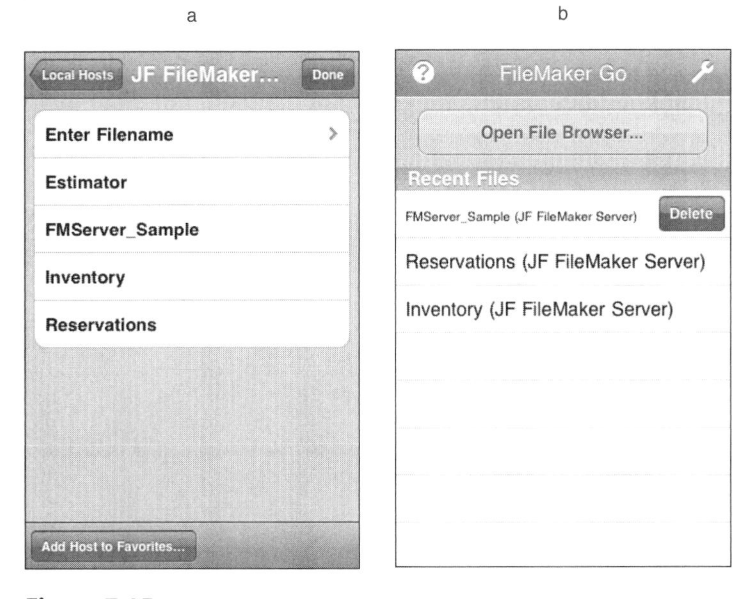

Figure 7.15
Open files on iPhone using FileMaker Go.

Opening Files on iPad

On iPad, there are fewer steps because the screen is bigger and more information can be shown at one time. When you launch FileMaker Go, you see the screen shown in Figure 7.16.

Figure 7.16
Open databases in FileMaker Go on iPad.

At the left, you see the databases installed on the device; the most recent non-local files are on the right. The name of the host is below each recent database name.

Favorite hosts are listed at the top of the right side of the screen. At the top right, you can browse for hosts on your local network (the icon with the magnifying glass), and you can add new hosts (the icon with the +) by typing in the name and address, as shown in Figure 7.17.

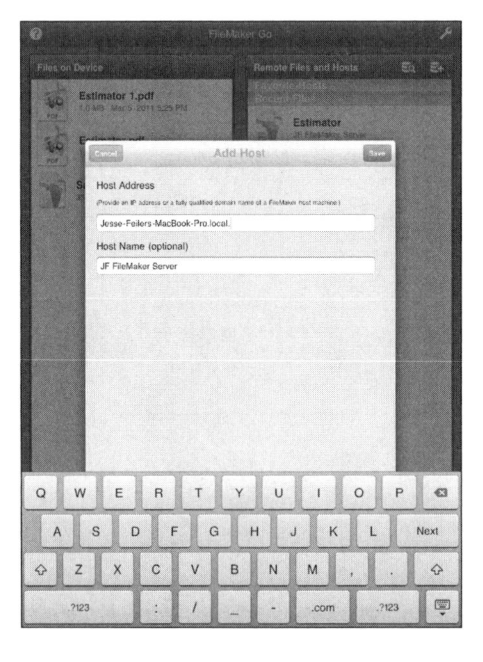

Figure 7.17
Type in the name and address of a host.

After you have located a host either from favorites or from hosts on your local network, you can choose from the open databases, as shown in Figure 7.18. (Compare this with Figure 7.15.)

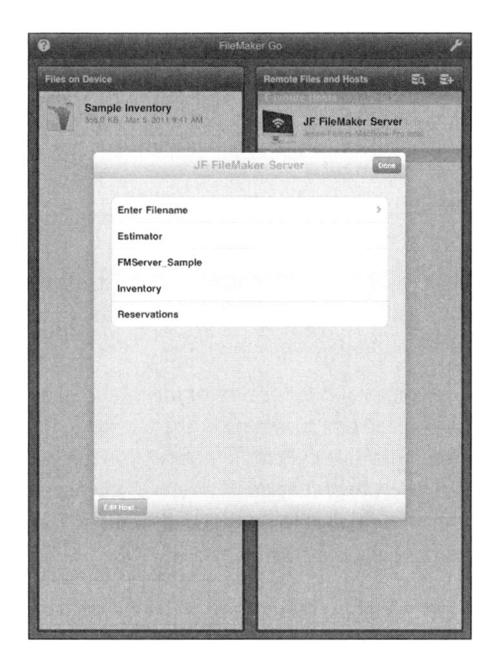

Figure 7.18
Open the databases on iPad using FileMaker Go.

Reviewing FileMaker Go Settings

The wrench at the top right of the first screen on both iPad and iPhone takes you to settings, as shown in Figure 7.19.

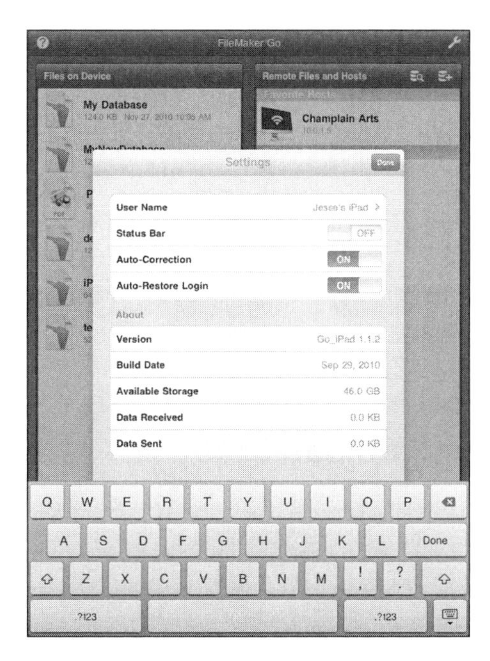

Figure 7.19
Review and change settings.

These are the settings you can change. They apply only to FileMaker Go; no other apps are affected by these settings:

- **User name.** By default, the user name for FileMaker Go is the name you have set for your device. You can change it here: Tap User Name, and a keyboard appears. You might want to change a user name such as Jesse Feiler's iPad to Jesse Feiler; alternatively, if you are worried about a log of transactions, you might want to change the name to Jesse Feiler's iPad so that you can keep things straight.
- **Status bar.** Turning the status bar on and off is a matter of taste. The status bar is the bar at the top of the screen with the time, battery status, and network status. Turning it off for FileMaker Go means that the status bar does not overlap the FileMaker Go interface.
- **Auto-Correction.** This lets you turn auto-correction on and off for the text that you type into FileMaker Go. If you are using FileMaker databases with FileMaker Go as well as with FileMaker Pro, you might want to make this setting the same on both environments.
- **Auto-Restore login.** This feature comes about because of the nature of app launching on iOS. Apps that are launched but not currently used can hibernate in the background; in addition, you might simply switch from one app to another. FileMaker Go remembers if you left with an opened database, and if this setting is off, it asks you if you want to go back to the record you were using. With this setting turned on, you are simply returned automatically to where you were in FileMaker Go. That behavior (returning to where you were) is the behavior of most apps on iOS.

→ There is more on Auto-Restore and ways you can control the launch experience in Chapter 8.

Other settings shown in Figure 7.19 provide information about your copy of FileMaker. The last two items—the amount of data received and sent—are particularly important for users with limited data plans. This typically is the case for users of iPhone or iPad who send data over the mobile phone network as opposed to using a Wi-Fi connection. It is the traffic over mobile phone networks that can sometimes cause extra charges on your bill.

Using FileMaker Go Help

The question mark at the upper left of the first FileMaker Go screen takes you to the help information. It is provided over the web, so if you do not have a web connection (perhaps you are using Airplane Mode), you cannot get to the help.

The URL for the website is not publicized, so it might change in the future. For now, though, it is www.filemaker.com/products/filemaker-go/for-ipad/help/1/.

Working with a FileMaker Go Database

After you have chosen a FileMaker database to open, you're in the main FileMaker Go screen, as shown in Figure 7.20.

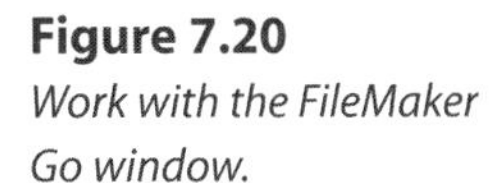

Figure 7.20
Work with the FileMaker Go window.

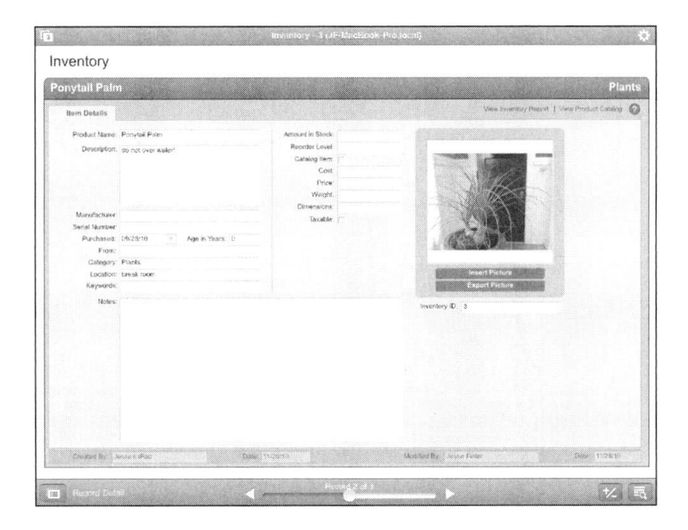

The center of the screen can look very much like a FileMaker Pro window (and that is what lies at the bottom of it). As you see in Chapter 8, you can create layouts that faithfully reproduce desktop-oriented layouts; you can also create layouts that look almost as if they were native iOS apps. What is important to note about the screen shown in Figure 7.20 is that each of the four corners has icons for special controls.

Working with Views

In the top left, the icon representing several screens lets you switch among different views; they might be showing different databases or they might be showing different views of the same one. Just as in FileMaker Pro, if you open two windows on the same database, you can look at different records (or the same record in different layouts). FileMaker makes certain that you do not edit the same record in two places at the same time.

The icon in the upper left shows you the number of open views on the database. You can switch from one view to another by tapping this icon. Figure 7.21 shows the result of such a tap.

Swipe left and right to move from one view to another. When you find the one you want to work with, just tap it, and it fills the screen, as shown previously in Figure 7.20.

You can close a database view by tapping the X in the upper left. If there are multiple views on the same database, the database itself is left open.

The + in the lower right lets you create a new view on the same database. It starts as being just a copy of the view, but after you have created the new view, you can navigate in it or the original to other locations.

Notice in Figure 7.21 that on iPad, a control in the center of the top of the view lets you choose between views and files. If you choose files, you are then back in the screen shown previously in Figure 7.16, where you can select a local or hosted file. (If you look closely, the entire screen shown in Figure 7.16 appears below the bar shown at the top of Figure 7.21, so that you can switch back to that view if you want.)

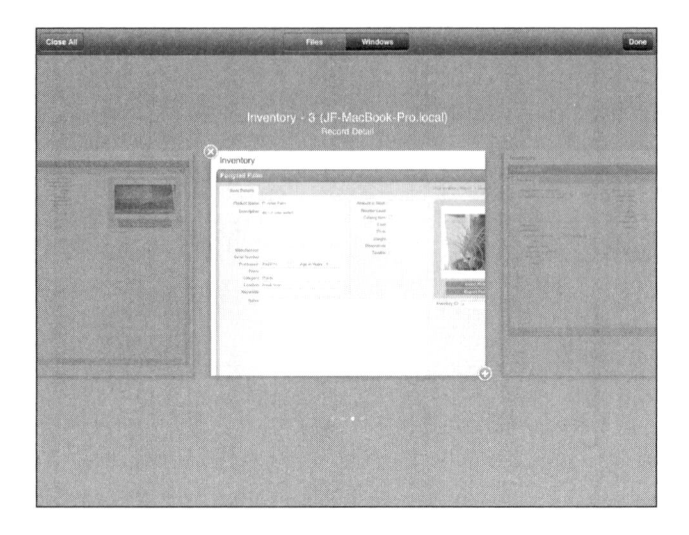

Figure 7.21
Switch database views.

If you are used to FileMaker Pro, you might notice that this functionality is pretty much the same as is provided in the Windows menu.

Working with Additional Commands

In the top right of the screen, the gear icon lets you choose from additional commands, as shown in Figure 7.22. (It is part of the standard Mac OS X and iOS interface that such an icon displays additional commands.)

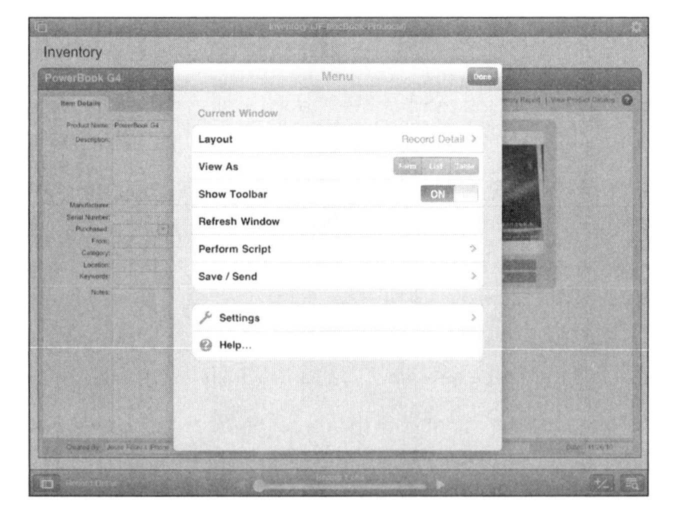

Figure 7.22
Use the gear icon to display additional commands.

Many of these commands exist in FileMaker Pro—generally with the same names and terminology. In FileMaker Pro, they are often menu commands, so they are implemented differently on FileMaker Go where there is no menu bar.

Choosing Layouts

When you look at a FileMaker window, what you are seeing is a *layout*—a combination of data fields, graphic objects, and buttons that let you navigate through a database and modify the data in it. Creating and modifying layouts is a task that is done with FileMaker Pro; in Chapter 8, you find some tips and techniques for creating and modifying layouts specifically for FileMaker Go.

You can have multiple layouts for a database. Typically (and hopefully) they are given meaningful names. Although creating a layout is a task for a designer, choosing the layout to use is usually a task for the end-user, so the names are important. Layout at the top of Figure 7.22 shows you the name of the current layout. You can tap that row to choose another layout from a list. The icon in the lower left of the FileMaker Go window also brings up this list of layouts for you to select. Furthermore, the name of the current layout is displayed next to the icon in the lower left of the window. Tapping it brings up the same list of layouts you can get to by tapping Layout in the additional commands shown in Figure 7.22. The list of layouts is shown in Figure 7.23.

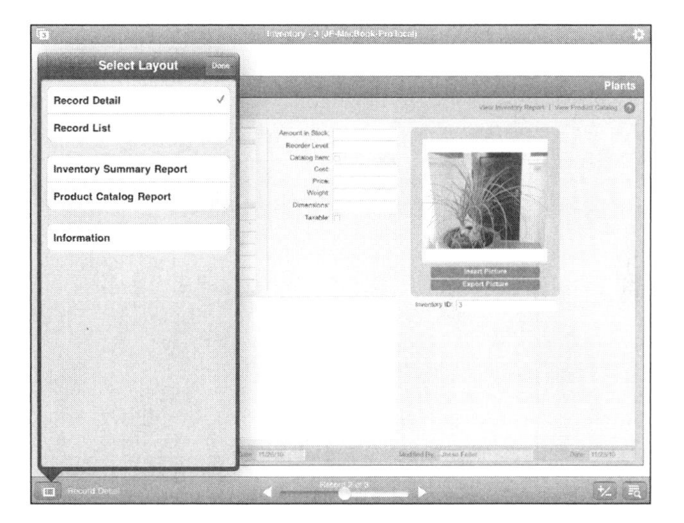

Figure 7.23
Choose a layout.

Using View Types

Layouts can be displayed in three different types of views:

- **Form.** In Form view, the layout showing the current record fills the screen.
- **List.** In List view, records are shown in a scrolling list using the current form. Often layouts are specifically designed for List mode. In those cases, the layouts are designed to be smaller (at least vertically) so that they can easily be seen in the list. Figure 7.24 shows the same layout shown previously in this chapter in List mode.

Figure 7.24

View records in List view.

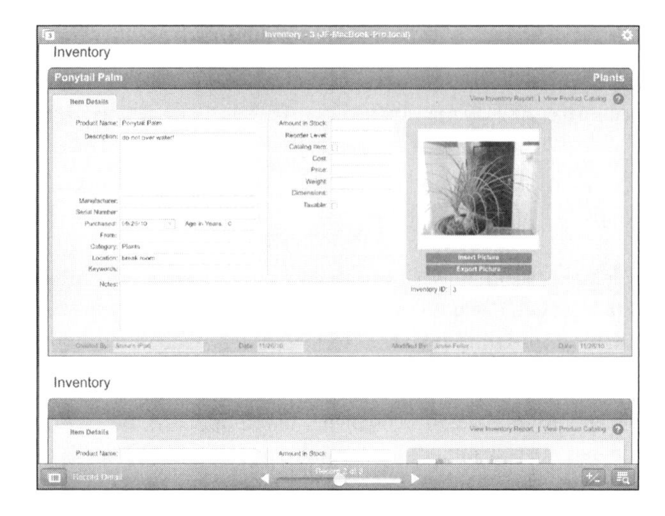

- **Table.** In Table view, the graphics of the current layout are not shown: FileMaker Go just takes the data and displays it in a spreadsheet-like table, as shown in Figure 7.25. You can sort any of the columns by tapping the header; tap the column header again, and the sort is reversed.

Figure 7.25

View records in Table view.

You can use the View As buttons in the additional menus to choose the mode you want to use to display the current layout.

The layout and the view type work together. For that reason, when you change a view type, FileMaker Go asks you if you want to save that choice along with the layout, as you can see in Figure 7.26.

Showing the Toolbar

The toolbar at the bottom of the window has the layout icon at the left, the slider and arrows for moving forward and back through the records in the database, and icons in the lower right. You can turn it on or off. Figure 7.27 shows the toolbar on (top) and off (bottom).

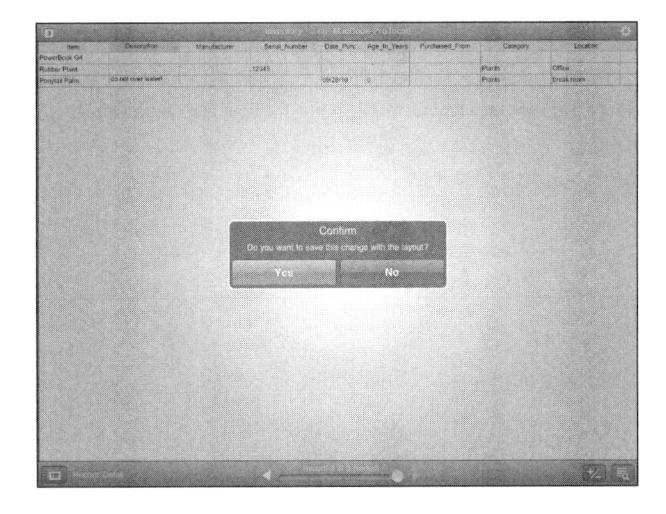

Figure 7.26
You can save your view selection with the layout.

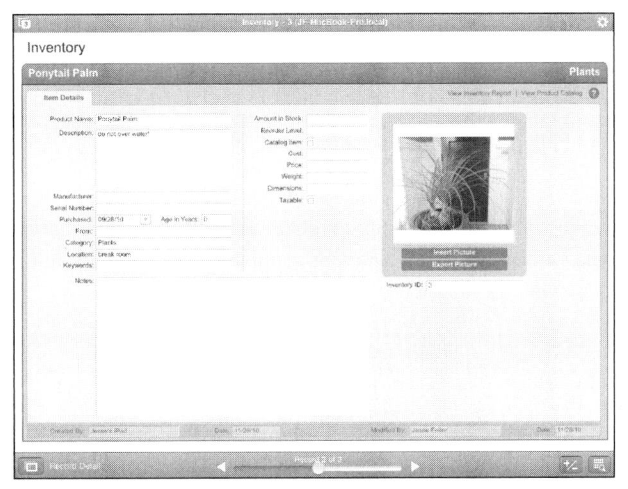

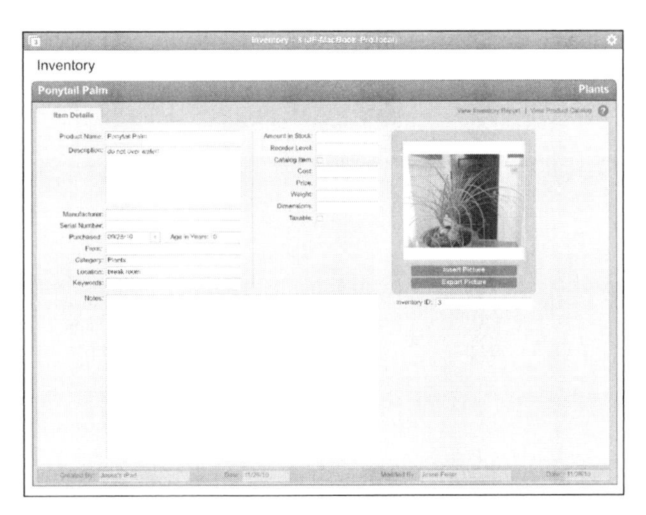

Figure 7.27
Show or hide the toolbar.

Do not confuse the toolbar with the status bar. The toolbar is part of FileMaker Go, and it appears at the bottom of the screen. The status bar is part of iOS (it shows the Wi-Fi or mobile status, the time, and the battery level).

Hiding the toolbar might seem to make no sense because people cannot navigate through your database. However, it can make a great deal of sense if you (or the database designer) have incorporated navigation tools into the layout itself.

Refreshing the Window

You can refresh the window from the additional commands.

Using Help, Settings, and Performing Scripts

Additional commands also gives you access to FileMaker Go help (discussed earlier in this chapter in "Using FileMaker Go Help") and settings (discussed in "Reviewing FileMaker Go Settings").

In addition, you can perform scripts. The scripts you can perform are those written by the developer. Chapter 8 discusses scripts in more detail.

Using Save/Send

There are two Save/Send commands available, and they all are enabled (or disabled) based on security settings configured for the database. (Once again, the options for designers are in Chapter 8.) They are shown in Figure 7.28.

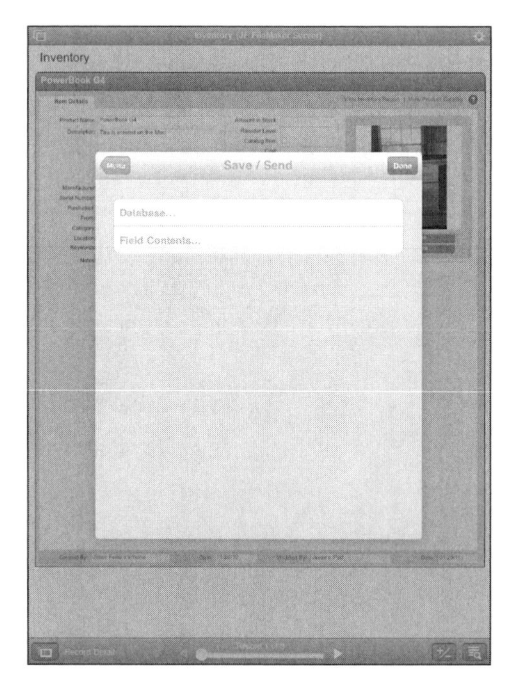

Figure 7.28
Use Save/Send commands.

The following are the Save/Send options that are available (subject to security constraints):

- **Database.** You can send the database via email, save it, save and re-open it, or cancel the request, as you see in Figure 7.29.

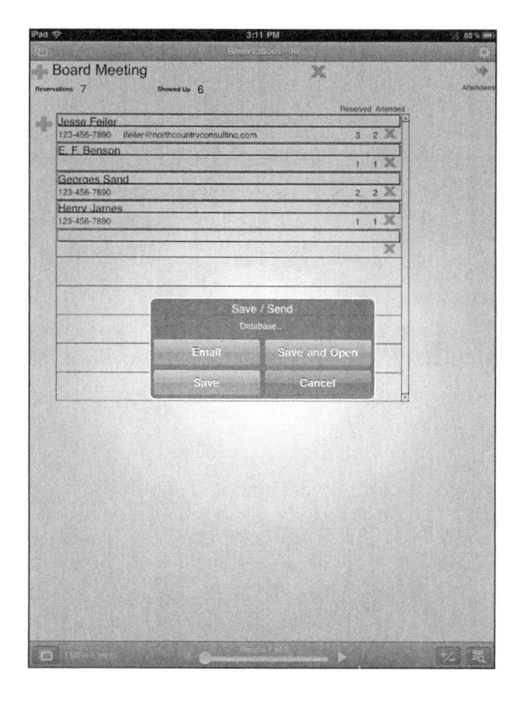

Figure 7.29
Save or send the database.

If you save the database, you can retrieve it from the FileMaker Go area of iTunes, as described previously in Chapter 4, "Working with Mobile Devices," in the "Moving FileMaker Databases to and from Your Mobile Device" section.

- **Field Contents.** If you have tapped in a field to select it, you can then export its contents. The interface is the same as for the database, but your choices are Email, View, Save, or Cancel. If you choose to view the content of a container field, and the content is an image, it is enlarged to fill the screen if possible, as shown in Figure 7.30. When field content is emailed, it is sent as an attachment—txt, jpeg, and so forth.

➔ With FileMaker Go 1.2, you also have a variety of PDF export options that are available through the printing interface. See Chapter 10, "Using Printing and Charting with FileMaker Go."

Figure 7.30
Graphics are enlarged on iPad as possible.

Working with Records

This section shows you how to work with records by navigating through them and editing them. The following section shows you how to search through records.

Navigating Through the Database

In FileMaker Pro, you have the option to create a new database from scratch or from a starter solution, but remember that FileMaker Go enables you to work only with existing databases. (That is not a limitation. It is what FileMaker Go does.)

In keeping with the needs of mobile devices, the window focuses on data to a greater extent than is possible on desktop-oriented software that runs on devices with larger screens. There are no menus, scrollbars, or window management tools, such as close buttons; the focus is on the data. The top and bottom of the screen are prime areas of interest for the user interface, and the corners are particularly important.

Figure 7.31a shows FileMaker Go running on iPad. This screen is almost identical on iPhone except, of course, that it is a bit smaller. That smaller size is responsible for the one difference in the iPhone version, and it is a very small difference indeed.

Figure 7.31 shows the iPhone version of the FileMaker Go window.

At the bottom, you have arrows to go to the next and previous record, just as you do on iPad. However, you do not have the slider that lets you slide from record to record. Tap the dot in the middle of the next/previous display, and the slider appears, as you see in Figure 7.32.

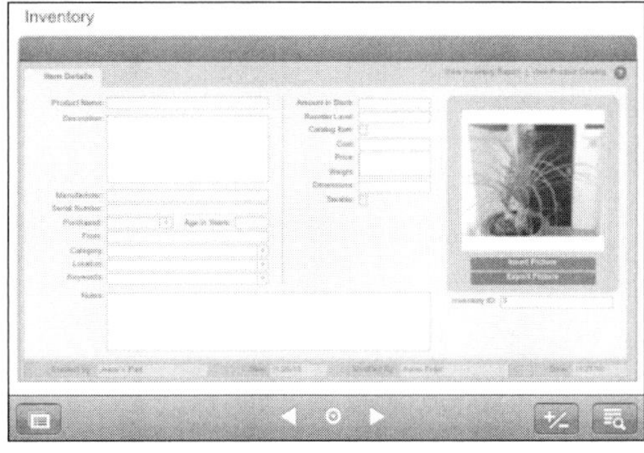

Figure 7.31
FileMaker Go on iPhone.

Figure 7.32
Slide from record to record on iPhone.

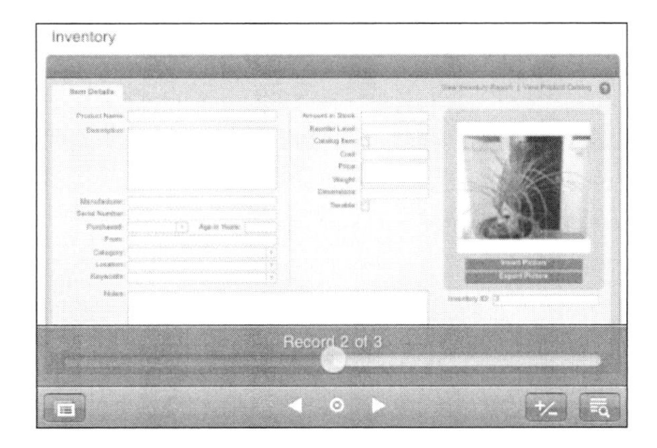

The slider interface requires a bit more space so that a person can use it with a fingertip; thus, it is available only when you need it. Other than that, the iPad and iPhone interfaces are almost identical. Most of the figures in this chapter use iPad because they are clearer when reproduced in the book.

Managing and Editing Records

At the lower right of the window is an icon that lets you work at the record level. As you see in Figure 7.33, you can add or delete records, duplicate the current record, or delete records that you have found in a search.

At the bottom of the menu shown in Figure 7.33, Exit Record lets you stop editing but doesn't navigate away from the record. This is important when you are working with a portal of related records. Exit Record releases the lock on the related record in the portal that you are editing. After you have started editing a main record, Exit Record changes to Save Record and Revert Record. Saving or reverting the record performs the necessary database update if needed and then allows other people to edit the record.

Figure 7.33
Work with records.

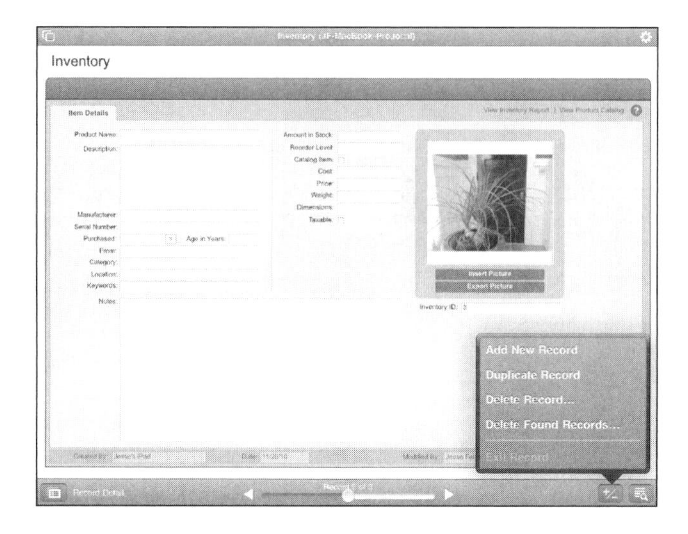

FileMaker Go and iOS provide data entry tools to help you add your data easily to the database on a mobile device. These include an onscreen keyboard, a camera, a date picker, and value lists. FileMaker Go presents the appropriate input interface depending on the type of data the field holds. (That is yet another reason for database designers to identify their fields specifically. A text field can hold a date easily, and FileMaker can make the necessary conversions. However, if you want a date picker to be presented when someone taps in the field, you have to identify the field as a date rather than a generic text field.)

The major input devices are described in the following section.

Using the Keyboard

When you tap into an editable text field, the keyboard slides up, as you see in Figure 7.34. In addition to the keyboard, the layout itself might adjust itself so that the field that you are editing is not hidden by the keyboard.

Figure 7.34
Edit text with the onscreen keyboard.

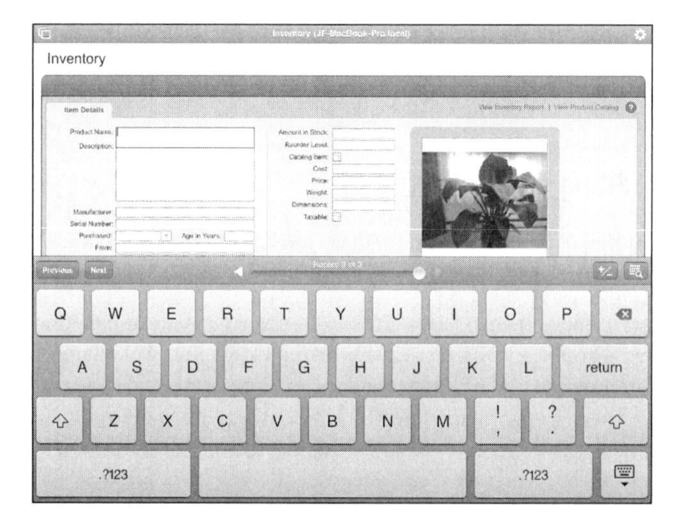

As the keyboard slides up, it appears to push up the navigation controls. Notice that just above the keyboard, you have a slider to move from record to record, as well as Next and Previous icons to move from record to record. In addition, the two icons from the bottom of the screen slide up so that you can access them. Hide the keyboard with the lower-right button (the keyboard with a down-pointing arrow).

The keyboard shown in Figure 7.34 is the keyboard that automatically slides up when you tap in a text field. If you tap in a number field, the keyboard you see is a numeric keyboard. There is more on this in Chapter 8.

On iPhone, the behavior is almost identical, but as you can see in Figure 7.35, the slider is not provided on iPhone.

Figure 7.35
Use the iPhone keyboard.

Using the Camera

When you tap into a container field, you can add a photo to it (if your mobile device has a camera). You can also add a photo from the photo library on your computer. The ability to take a photo and automatically add it to the database is one of the extraordinary features of FileMaker Go. If you are using a shared database, the moment you finish the process of adding the photo, everyone else with access to the database can see it.

Adding a Photo to a Database

Here is how to add a new photo to a container field in FileMaker Go:

1. **Tap the field where the photo will go.**
2. **As Figure 7.36 shows, you have a choice of adding a photo from your library or taking a new one. Tap Take Photo.**
3. **Take the photo.**
4. **You can preview the image, as shown in Figure 7.37.** Tap Use or Retake depending on whether you want to use the photo.

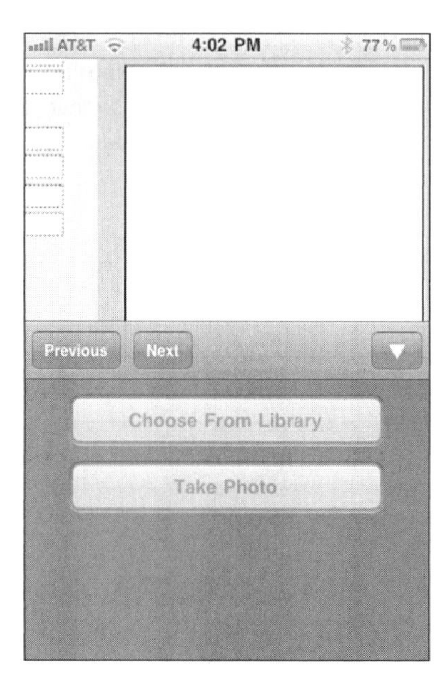

Figure 7.36
Choose the library or take a new photo.

Figure 7.37
You can preview the photo and decide whether or not to use it.

5. **You can decide the resolution to use: full, small, medium, or large.** Smaller photos take less space and can download faster to remote users. Just tap the indicator at the top of the screen to make your choice.

6. **If you tap Use, the photo is stored in the database and is available to all users.**

If you already have a photo in the field, tapping the field gives you additional options, as you can see in Figure 7.38. You can choose a new photo or take a photo, but you can also open the existing photo or email it to someone.

Figure 7.38
Work with an existing photo.

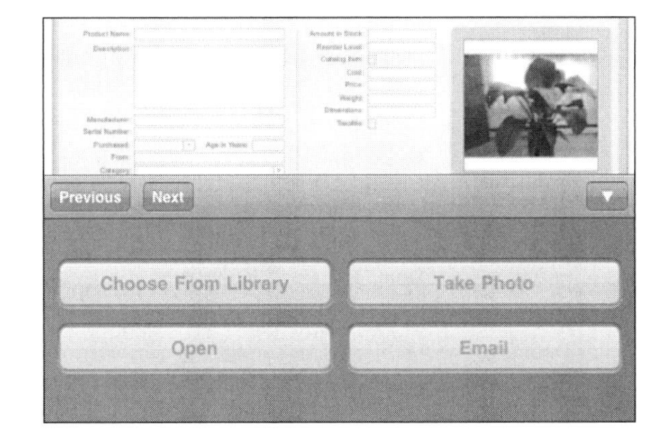

Using the Date Picker

When you tap in a date field, you bring up a date picker rather than a keyboard, as shown in Figure 7.39. (There is a comparable time picker, as well as a composite date/time picker.) This is a standard iOS interface element that is used in many contexts. Note that in accordance with iOS interface standards, a small arrow points to the field that is being edited.

Figure 7.39
Use a date picker.

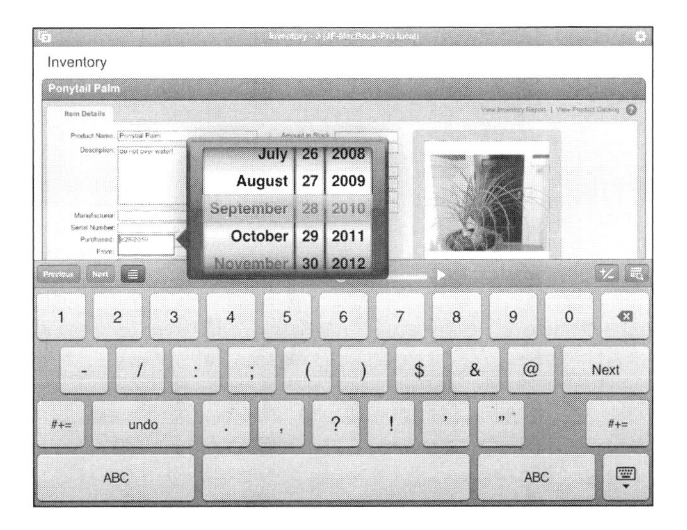

Using Value Lists

FileMaker enables developers to create *value lists* with multiple choice values for checkboxes, radio buttons, or pop-up and drop-down menus. Those menus are implemented on iOS as pickers, such as the one you see in Figure 7.40. They are used throughout iOS, so users do not need to learn how to use them just for FileMaker Go.

Figure 7.40

Use a value list.

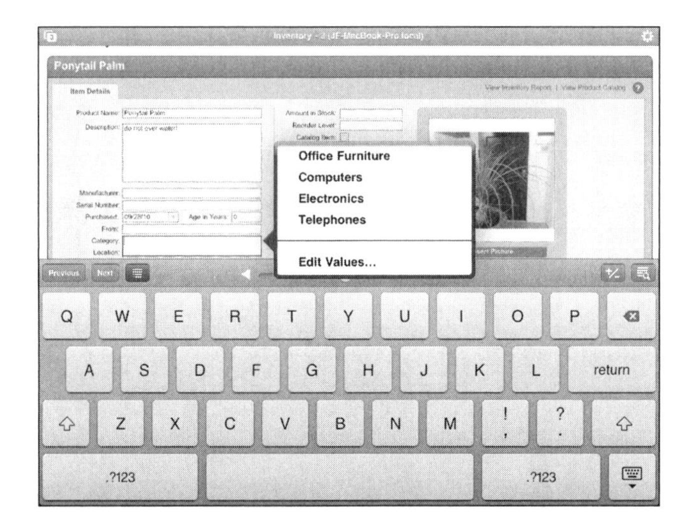

Note in Figure 7.40 that you have the option to edit the values in the value list. This is an option that the FileMaker database designer can choose to set for each list. If it is set, you can edit the list, as shown in Figure 7.41. These edits take effect starting with your changes, but they do not affect existing data. If you delete an item in the value list, those values that are already in the database remain.

Figure 7.41

You can edit a value list.

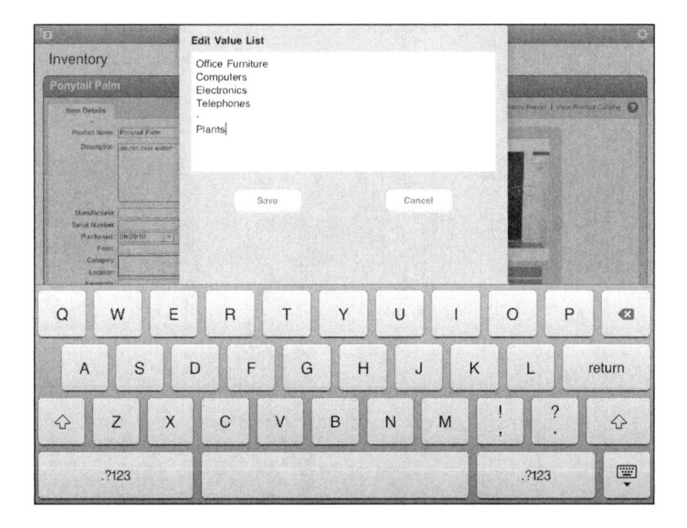

4 1 1

Creating and editing value lists is something that has to be done with FileMaker Pro. However, after a value list is set up, the designer can allow users of FileMaker Go to edit it, as seen here.

Searching for Data

Storing data in a database is all well and good, but being able to find the data you are looking for is at least as important, if not more so. FileMaker Pro provides four modes to let you accomplish your database tasks:

- **Browse** mode lets you enter and view data. Most of this chapter has demonstrated Browse mode.
- **Find** mode lets you search for data. That process is described in this section.
- **Layout** mode in FileMaker Pro lets you design layouts. Along with the commands to modify the database structure, Layout mode is not provided in FileMaker Go.
- **Preview** mode shows what your layouts will look like when they are printed. Not only are they formatted for paper rather than the scrollable expanse of a display, but summary and sub-summary fields are calculated in Preview mode. Until FileMaker Pro 10, summary and sub-summary fields were *only* calculated in Preview mode. Now, however, they are available in Browse mode, and many people believe that the necessity for Preview mode is significantly decreased, except for compatibility with older databases. Preview mode is not supported in FileMaker Go.

Because neither Layout nor Preview mode is supported in FileMaker Go, that leaves only two modes: Browse and Find. That makes it possible to further simplify the interface. On FileMaker Go, there is a separate Find mode that you can enter and leave as you wish. When you leave Find mode on FileMaker Go, you are in Browse mode, but there is not an explicit command to go to Browse mode; you simply exit from Find mode. For people whose first acquaintance with FileMaker is FileMaker Go, this does not even register; it is simple and obvious. Experienced FileMaker users might wonder how to get back to the Browse mode that they know and love; the answer is to just leave Find mode.

Using Quick Find

There is a further simplification to Find mode on FileMaker Go. It is not unique to FileMaker Go; it was introduced in FileMaker Pro 11. This is Quick Find. In some ways, Quick Find can be difficult to explain because it is so simple. Many experienced database users do not see how it can be useful—however, within a few minutes, they are converted. The idea behind Quick Find is that it searches the database for a word or phrase. That much seems simple and intuitive, but here is the issue: If you search a database for a word such as Adams, you get several people (including two presidents of the United States), several cities, many streets and avenues, and a multitude of other matches. The traditional database way of searching is to search for a specific word or phrase in a specific field—street, state, last name, and so forth. At first glance, the Quick Find broad search seemed to get too much undifferentiated data (at least to experienced database users).

Maybe it was the advent of search engines such as Alta Vista, Google, and now Bing, but after a while, it became obvious that many people prefer the trade-off of a quick search versus a more precise one. FileMaker now provides both: the traditional search mechanism, in which you can search for a word or phrase in a specific field, and the new Quick Find, which does a search-engine-like search.

To use Quick Find, tap Quick Find mode from the icon on the bottom right of the FileMaker Go window, as you see in Figure 7.42.

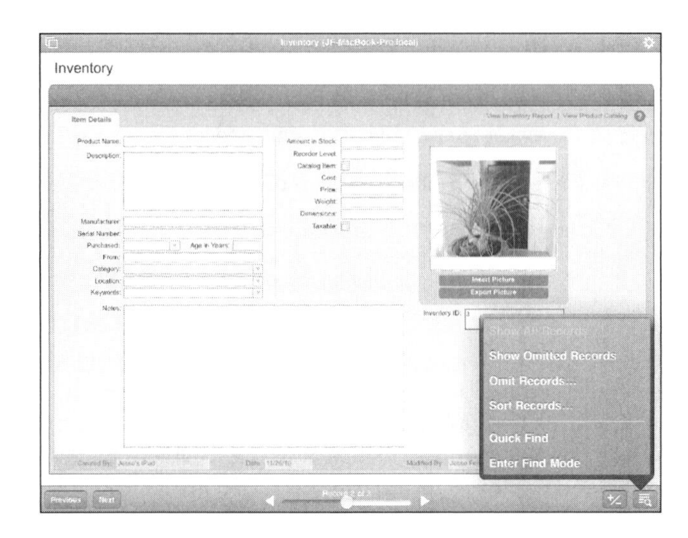

Figure 7.42
Enter Quick Find.

This brings up the keyboard; you can type in the word or phrase for which you want to search, as you see in Figure 7.43.

Figure 7.43
Enter your search term.

Tap Search to start the search. FileMaker Go searches all fields for the text, and it displays the results, as shown in Figure 7.44.

Tapping OK dismisses the keyboard, and you can now browse the results, as shown in Figure 7.45. Notice the center of the toolbar at the bottom: It gives you the search results. 1 of 2 / 3 means that it is displaying records one of two found records from a total of three records.

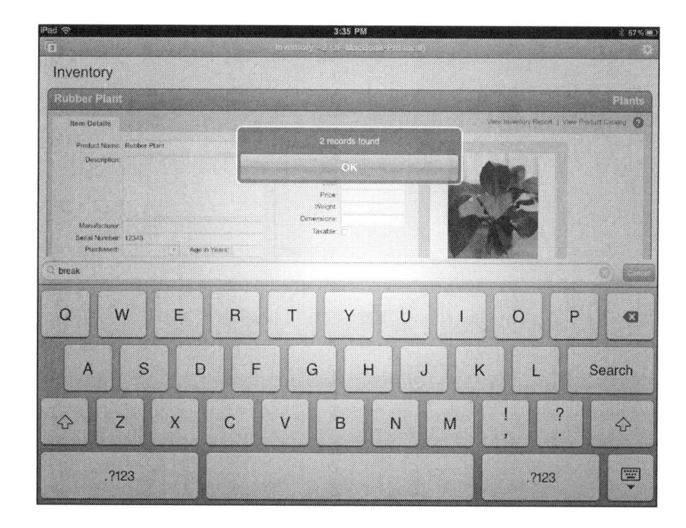

Figure 7.44
FileMaker Go displays the results.

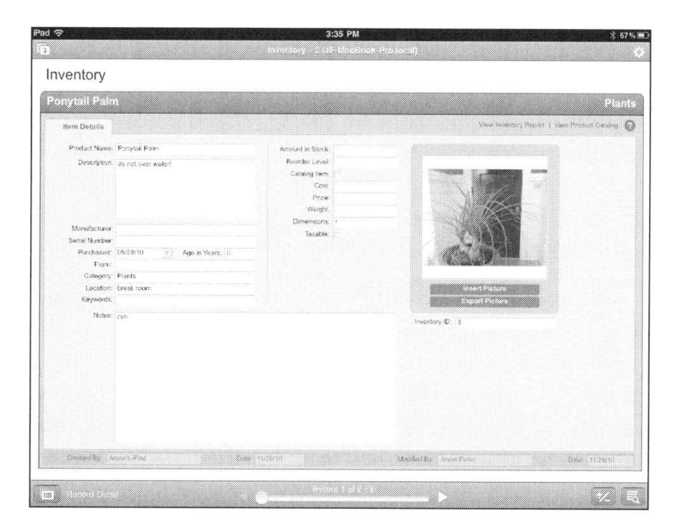

Figure 7.45
You can browse the records found.

The records that are found are called the *found set*. You operate on the found set until you choose Show All Records from the same menu shown previously in Figure 7.42 or until you perform another find and create a new found set. If you have not performed a find, Show All Records is not available. That menu also offers the option to reverse the operation and show only the omitted (not the found records).

Notice that when you have a found set, the +/- icon at the bottom right now has the Delete Found Records command enabled. This means that you can find all the records you do not want and either omit them from the records you are browsing or just delete them from the database.

Using Find Mode

All of this has been done with Quick Find. You can go into Find mode, as shown in Figure 7.42, where you have much more power. Find mode is similar in the sense that you perform a find and then obtain a found set, which you can then omit from the display or delete from the database.

Searching by Fields

Where Find mode is different is in the ways in which you can search. Instead of searching for a word or phrase anywhere in the database, you can search for words or phrases in specific fields of the database. You can enter words or phrases in several fields; when you perform the find, FileMaker Go searches for those words or phrases in only those fields. If you enter Adams in the last name field and Massachusetts in the state field, you get every Adams in the database who lives in Massachusetts. (In database parlance, this is called a *logical and.*)

Using Find Requests

You also can create *find requests.* The combination of Adams/last name and Massachusetts/state is a find request that finds all records with both conditions being true. You can create a separate find request for each part: find request one might be Adams/last name, and find request two might be Massachusetts/state.

Within a single find request, all the conditions are combined (and-ed). When you execute several find requests together, each request is treated separately (or-ed). Thus, a find with these two find requests would retrieve all records fulfilling find request one (Adams/last name), as well as all records fulfilling find request two (Massachusetts/state). In general, this type of find retrieves more records than a single find request. Note, too, that even if a record fulfills several find requests, it is only retrieved once.

When you enter Find mode, you have your first find request ready for you to type in the data to find. The keyboard appears, and you type the words or phrases into the appropriate fields. (Small magnifying glass symbols let you know that you are in Find mode and which fields can be used.) If you want to add another find request, you can use the find request icon to add on, as shown in Figure 7.46. You can also delete or duplicate a find request (although you cannot delete the last one—you just leave Find mode). Note that above the keyboard, the slider now lets you move through find requests rather than records.

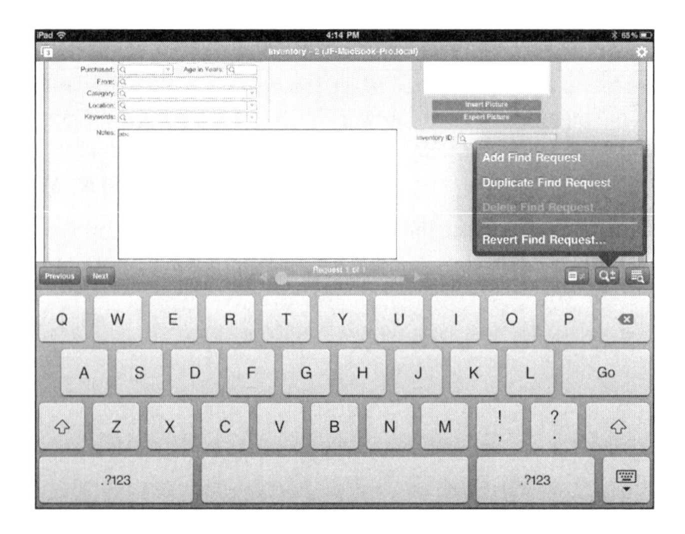

Figure 7.46
Manage Find requests.

When you are ready, tap Perform Find, as shown in Figure 7.47.

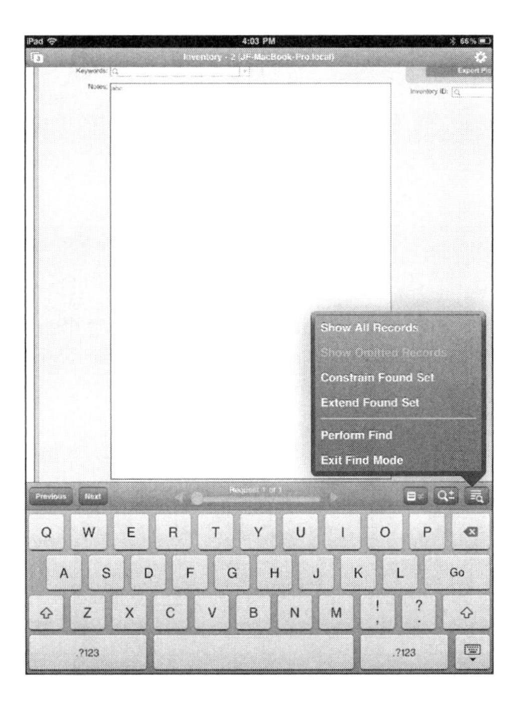

Figure 7.47
Perform the Find.

Other options in this menu let you work with an existing found set (the result of a previous find). You can create a new find request and apply it to extend or constrain the found set. Together with the options to omit records, you can soon learn how to home in on specific data you want to use.

There is even more to do when searching for data. The icon to the left of the Find icons, shown in Figure 7.48, enables you to include or omit matching records.

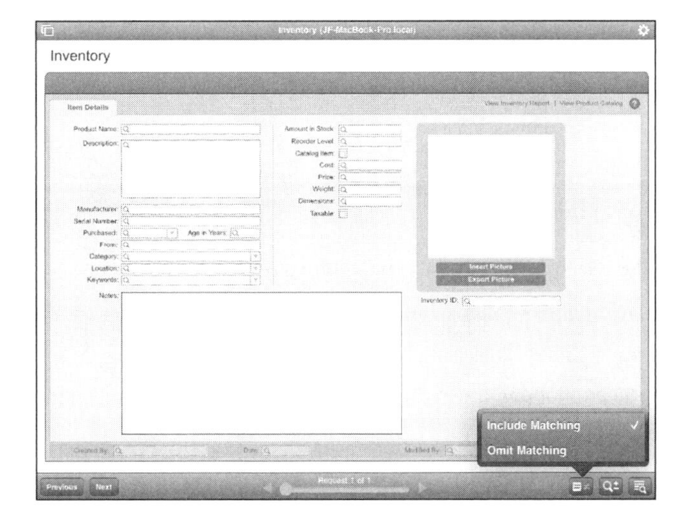

Figure 7.48
You can include or omit matching data.

The last point to note about find requests is that FileMaker Go keeps track of them at least in the short term (FileMaker Pro lets you save finds). When you enter Find mode, if you have find requests left from your previous Find, you are asked if you want to keep them or remove them. This enables you to go back to the previous find and expand or constrain it with the new one.

→ There is more on finding records in "Working with Custom Menus" in Chapter 8.

Sorting Data

One of the most common tasks people want to do with databases is to sort their data. You can do so easily from the icon in the bottom right (shown previously in Figure 7.42). This lets you choose the fields to sort on, as shown in Figure 7.49. As you tap each sort field, it appears at the top of the list so that you can develop a sophisticated multi-level sort.

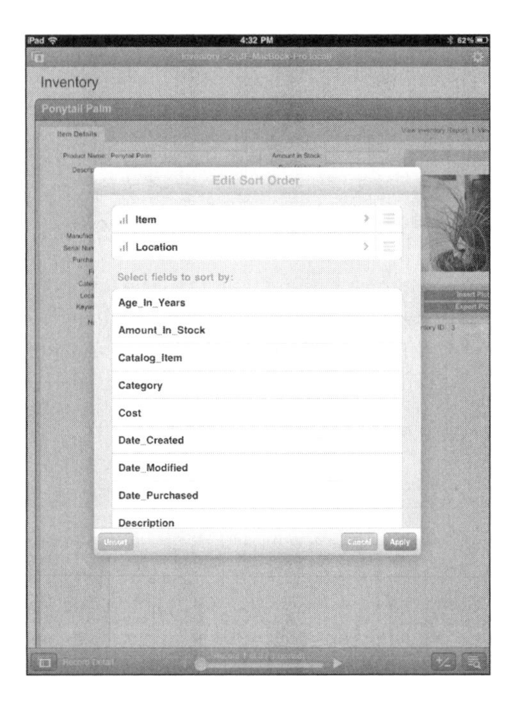

Figure 7.49
Sort data in FileMaker Go.

Tap a sort field at the top to specify its sort characteristics, as shown in Figure 7.50.

Note that you can use ascending or descending sorts, but you can also use value lists that specify an idiosyncratic sort order.

In addition, when you are done, you can drag the sort fields at the top to specify the order in which they are applied (see Figure 7.51).

Figure 7.50
Specify sort order for each field.

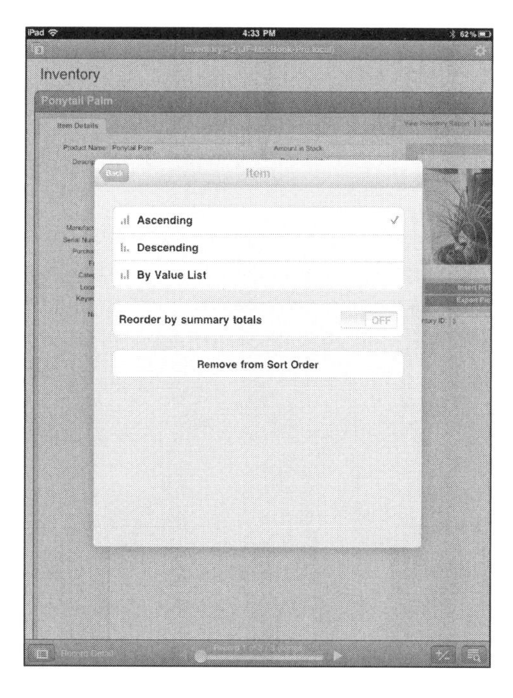

Figure 7.51
Drag the sort fields into the order you want them applied.

Finally, note that with complex sorts, as with complex finds, many FileMaker solutions encapsulate the complex sorts and finds into script that you can launch with buttons, so that the work of preparing the find or sort is already done for the end user.

Further Steps

If you have not used FileMaker or even a database, the next step is to experiment with a database. In many ways, FileMaker Go is a terrific way to learn about databases precisely because it does not let you create a database. You work with an existing database and do what you can. Download samples from the author's website, as described in the Introduction.

Also note a United States population database designed specifically for FileMaker Go. It is available on the author's website in the Chapter 7 folder. This is a summary of population data from a recent census. You can download it to your iPad or iPhone and experiment with searching and sorting. Because the database contains thousands of records, you get an appreciation of the power of FileMaker Go and the devices on which it runs.

8

IN THIS CHAPTER

- Controlling Database Access by Version
- Using Dynamic Reports Instead of Preview Mode
- Modifying FileMaker Go Menus
- Working with Functions and Scripts

Optimizing FileMaker Databases for FileMaker Go

In the previous chapter, you saw how to use FileMaker Go. Whether you are the end user or someone who is designing a database for end-users, that chapter gives you details of how people can use FileMaker Go.

This chapter is all about FileMaker Pro—what you can do in FileMaker Pro to make it easier (and, in some cases, even possible) to use FileMaker Go. Chapter 5, "Preparing FileMaker for Mobile Use," discusses overall issues you might encounter when preparing for mobile use, whether with FileMaker Go or with FileMaker Web Publishing on a mobile device. Now the focus is tighter: the FileMaker Go-specific adjustments you can make to FileMaker Pro.

→ Chapter 10, "Using Printing and Charting with FileMaker Go," explores functionality that you can put into FileMaker databases for display in FileMaker Go.

Understanding the Relationship Between FileMaker Pro and FileMaker Go

There is a common thread throughout this chapter: If you use current best practices and recommendations, you will already have included many of the features that make your FileMaker Go users happy and productive. Much of the advice in this chapter is not new because it reiterates best practices and recommendations.

Dealing with Old-Version Issues

These version issues may well be new to you and your databases. The reason is simple: FileMaker Go version 1 was released alongside FileMaker Pro 11. FileMaker Go is starting from scratch in many ways, whereas FileMaker 11 has more than two decades of history behind it. The developers at FileMaker work very hard to provide backward-compatibility. Rarely does a new release of FileMaker break existing solutions. Even when there is a significant change in FileMaker file formats (as happened with FileMaker 7), every attempt is made to provide compatibility and conversion utilities. Today, FileMaker Server supports files and features from FileMaker 7 through FileMaker 11—and that applies to FileMaker Server 7 through FileMaker Server 11. In cases where a new feature has been added (for example, the Web Viewer in FileMaker 8.5), older versions of FileMaker Pro manage to skip over that code so that although the feature might not be implemented, it does not cause a crash or unusual behavior.

This makes it possible for FileMaker databases to have long lives even as FileMaker versions advance. It also makes it possible for people to hang on to technically out-dated versions of FileMaker for years after their expiration date (in common with most software developers, FileMaker supports the current version and one prior version with technical support and discounted upgrade pricing).

Because FileMaker databases dating back to FileMaker 7 are current in the sense that they run on the latest FileMaker software, some of their design elements are supported but no longer encouraged. It is these that you might want to address. The closer your database is to state-of-the-art FileMaker techniques and conventions, the easier it is to make it run effectively on FileMaker Go.

If you are using some of the newer FileMaker features, people using older versions of FileMaker are not able to see them or use them, but the rest of the database works properly. As more and more modern features are added, you might want to prevent people from bypassing them with older versions of FileMaker Pro. Starting with FileMaker Pro 11, you can manage this situation by setting a minimal version for users of the file. Open the database file in FileMaker Pro, and choose File, Manage, Security. Click the File Access tab, as shown in Figure 8.1.

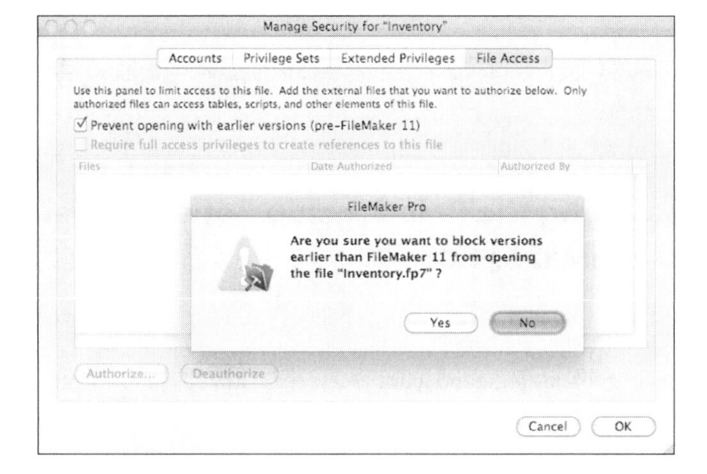

Figure 8.1
Limit access to the database file by FileMaker Pro version.

A FileMaker solution can have a number of files that work together. In that case, you can authorize these other files to allow these files to have access to the file you are working with. As soon as you click the FileMaker Pro 11 checkbox shown in Figure 8.1, the buttons enabling access for specific other files are enabled, as shown in Figure 8.2. (The reason that you have to click the FileMaker Pro 11 checkbox is because if people use previous versions, they are not able to recognize the restrictions you are setting up for the current file.)

Click Authorize and you are presented with a standard File Open dialog. Just navigate to the file that you want to authorize to have access to the current one and then click Open. Similarly, you can select a previously authorized file from the list and click Deauthorize. Remember, you are controlling access from these other files to the file you are working on.

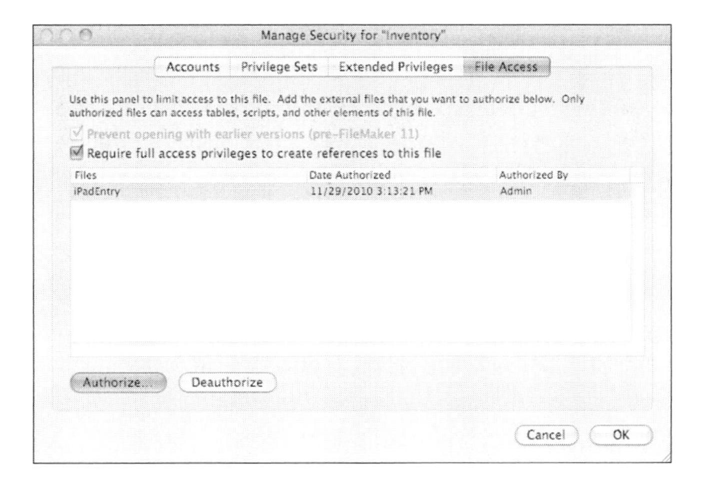

Figure 8.2
Authorize specific files to access your database.

What is happening behind the scenes is that the checkbox shown in Figure 8.1 makes a small change to the database file itself. After that change has been made, only authorized files can access the main file. If you uncheck that box, the database change is undone, and the authorizations mean nothing because they are not needed.

This process allows you to use new features in FileMaker Pro that can help to implement security and controls without having those features bypassed.

Use file access and authorization to see if you can safely use new features. Either use a test version of your FileMaker solution or set aside a time when users know there might be interruptions. Turn on the file access version settings and wait for reports from users. If you hear that people are unable to get in, you can find out who has old copies of FileMaker Pro; you also can find out what additional files need access to be authorized. Because access and authorization can easily be turned off, after you have an idea of the magnitude of the problem (if there is one), you can turn them off and then make a plan for how to proceed. Of course, the safest way is to work systematically with a test copy of the database, but you might not easily be able to identify which versions of FileMaker Pro may be used in your organization.

Starting from Scratch for FileMaker Go

If you are starting from scratch to build a FileMaker database that runs on FileMaker Go, it is a good idea to turn on file access immediately. That way, you can use all the features you want to use, and as soon as people start testing, they (and you) will discover if they have outdated copies of FileMaker Pro. You can always back up and turn off access version control.

If you are starting from scratch and you are an experienced FileMaker developer, take a look at the optimizations suggested in this section. They can help you to build better FileMaker Go solutions.

Refining Your Database Field Types

FileMaker has powerful built-in data type conversions. If you store data as text and you use it in a context that requires a date, FileMaker converts it. When you create a new field in a database table, it is automatically set to a text field by default, and with FileMaker automatic conversions, there is apparently little reason to change the field type.

In fact, one of the first things that you should do is to consider changing the field type. If you have specified a field type other than text, these are some of the features that are available to you:

- **Formatting.** You can format date and time fields using date formats, and number fields can be formatted with number formats including optional currency symbols, thousands separators, and decimal places.
- **FileMaker Go pickers.** On iOS, date and time pickers are automatically presented for input with date and time fields.
- **Automatic validation.** In the Fields tab of the Manage Database dialog, you can push validation onto the database, as shown in Figure 8.3. You can configure your options so that users (or perhaps only some users) can override the validation error. Also, note that this affects new data entry only. If you have legacy data in the database that could not pass the stricter validation rule, you can leave it there as long as your scripts can accommodate it (which, apparently, they can now).

None of these features are available if you simply leave the fields as text fields.

Although FileMaker can do your type conversions for you automatically, you might want to take a tip from experienced programmers and use explicit type conversions in your scripts. As your solutions get more and more complex (and as they are modified over the years by more and more people), keeping control of your data and data types can save you time—particularly in the debugging and testing stages of your project.

The text functions in the Specify Calculations dialog let you convert text to specific data types. Because any field type can be represented as text, this enables you to control conversion. Container fields may contain files or file references, but when displayed as text, the file name appears rather than text that might appear in the file inside the container.

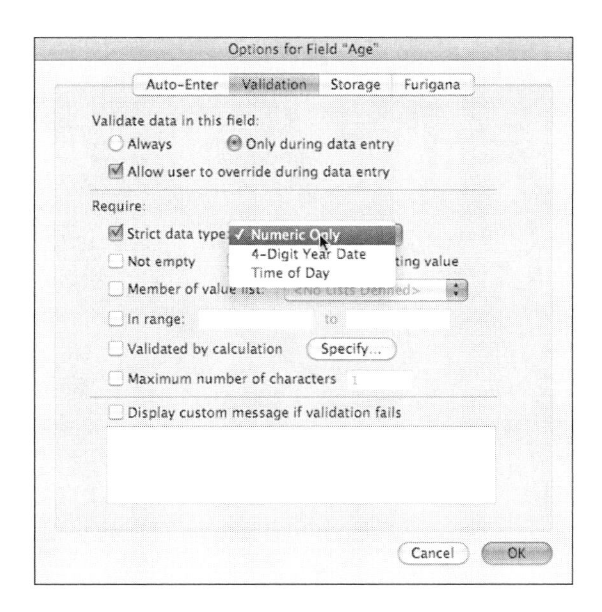

Figure 8.3
Specific field types allow stricter validation.

Use the Comment field in the Manage Database dialog to keep track of assumptions that you are making about the format of data in the field. Also, consider using naming conventions to identify the type of data in a field so that, for example, "dt" always is used to indicate a date type field and the spelled-out word "date" can be used to indicate a text field that happens to contain date information.

Reviewing Calculations

You can specify calculations in many places in FileMaker. They can be used to calculate the data in a calculation field; they are also used in the `If` script step so that you can calculate a true/false Boolean result for that script step. You find them when you are using privilege sets to set access for fields, layouts, and scripts in Manage, Security.

In all cases, you use the Specify Calculation dialog shown in Figure 8.4.

At the very top, you can set the context for the calculation. Because the relationships graph can contain multiple instances of the same table, you might need to indicate which one you want to use as the basis for the calculation. Having picked the context, FileMaker can then resolve the relationships for any related fields you reference in the calculation. This is often not an issue when you are providing a calculation within a table, but when you use a calculation to evaluate access permissions or a script `If` step, context might matter.

At the top left of the dialog, you can choose fields from the current or related table. As you double-click, the field is added to the calculation in the main part of the window. Double-clicking the operators in the top center adds them to the calculation, and, in the upper right, you can double-click functions to add them. You can also type your calculation directly into the calculation field. (Most people use a combination of these techniques.)

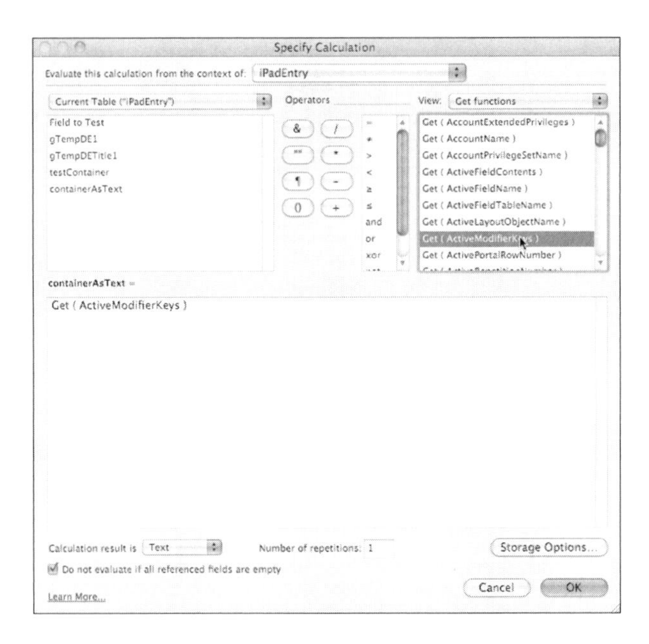

Figure 8.4
Specify calculations.

Some of the functions that are available in FileMaker do not evaluate on mobile devices. For example, the function that lets you get information about the function keys that are currently pressed cannot provide you with that information for a mobile device without a keyboard.

The issue is much the same as with some script steps that cannot be evaluated in certain environments. FileMaker handles all of these cases in the same way: It skips over functions and script steps that cannot be evaluated rather than returning an error.

Because calculations will not fail in most cases, you should take a look at them as you are planning to use an existing database with FileMaker Go.

Some environmental functions (those that return information about the current database environment) have new values for FileMaker Go. The functions themselves are unchanged, but the possible results are expanded.

Get (SystemPlatform)

There is a new return value for `Get ( SystemPlatform )`; it indicates mobile devices. As always with this type of function, when you review your calculations, look for tests that make assumptions that are no longer valid. For example, in the past you could check for the absolute value of `Get ( SystemPlatform )` to get a value of 1 if you were running on either Mac platform. That still works. However, if you went on to test for 1 to identify a Mac and then, with an `else` clause you assumed you were on Windows, your code now fails. (You could be on a mobile device.) It is always best to avoid `else` clauses in this type of return value.

Table 8.1 shows the current return values.

Table 8.1 `Get ( SystemPlatform )` RETURN VALUES

Platform	Result
Power-PC-based Mac	-1
Intel-based Mac	1
Windows CP, Windows Vista, or Windows 7	-2
Mobile device	3

Get (ApplicationVersion)

`Get ( ApplicationVersion )` lets you know two pieces of information:

- The FileMaker product that is running
- The version of that product

The result of this function is something such as the following:

```
Pro 11.0.2
```

```
Go_iPad 1.1.2
```

The product values are shown in Table 8.2. An easy way to test for them is to use one of the text string functions to test for the product value, as in this code:

```
PatternCount ( Get ( ApplicationVersion ) ; "Go" )
```

This identifies FileMaker Go as running on iPad or iPhone. If you attempt to match `iPad` you identify FileMaker Go on iPad. You can find FileMaker Go on iPhone by searching for the `Go` pattern and no occurrences of `iPad`; alternatively, you can look at the first value in the returned string and test to see if it is `Go` rather than `Go_iPad`.

Table 8.2 `Get ( ApplicationVersion )` PRODUCT VALUES

System Platform	Result
FileMaker Pro	Pro
FileMaker Pro Advanced	ProAdvanced
FileMaker Runtime (standalone)	Runtime
FileMaker Web Client	FileMaker Web Publishing
FileMaker Server Web Client	Web Publishing Engine
xDBC Client	xDBC
FileMaker Server	Server
FileMaker Go for iPhone/iPod Touch	Go
FileMaker Go iPad	Go_iPad

Working with Preview Mode

FileMaker Go only has two modes—Find and Browse. In fact, as pointed out previously, from a user's point of view, there is only one—Find and everything else (which turns out to be Browse). There is no Layout mode because you cannot modify the database structure with FileMaker Go, and there is no Preview mode. Preview mode prepares a FileMaker report for printing and generates summaries and subsummaries as needed.

Many people believe that without Preview mode, FileMaker Go is useless for their reporting purposes. Nothing could be further from the truth.

In fact, although Preview mode is still supported and used extensively in FileMaker Pro solutions, the introduction of dynamic subsummary reporting in FileMaker 10 has made it less critical. As is usual with FileMaker, this major change in features is accomplished without breaking existing solutions. Old scripts that perform sorts for layouts with subsummary parts will be around for many years, and when they are run on newer versions of FileMaker, they will still drop the user into Preview mode to calculate the summaries and subsummaries.

When you use the Layout assistant to create new layouts and scripts to manage them, you will see that beginning in FileMaker 10, there is a conditional test in the script so that it drops into Preview mode only when it is run on pre-FileMaker 10 versions. This works fine on FileMaker Pro, but when you are running on FileMaker Go, there is no Preview mode, and the error message shown in Figure 8.5 appears.

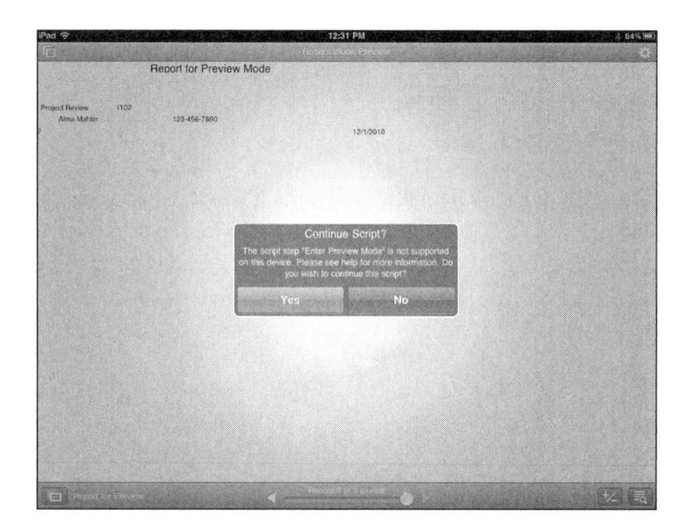

Figure 8.5
Preview mode does not exist on FileMaker Go.

Figure 8.6 shows a script generated by the Layout assistant with the code that handles the version test.

The key line of code here uses `Get ( ApplicationVersion )` to get the version string that consists of the product (shown in Table 8.2) and the version. By taking the result and using the built-in `GetAsNumber` function, all non-numeric characters (such as the product name) are dropped. This works well except that the version number is of the form 1.1.2, and the decimal points are included in the set of numeric characters. Thus, the `Substitute` function is used to take the decimal points out and replace them with the letter x—which is indisputably not a numeric character. With the substitution done, the numeric version number remains after the string is returned as a number.

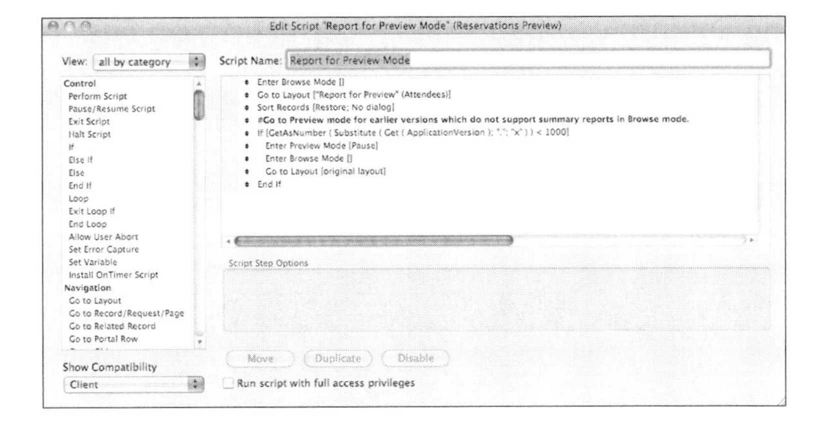

Figure 8.6

Scripts can test for FileMaker versions to decide whether or not to use Preview mode.

For example, for FileMaker Pro version 11.0v2, the `ApplicationVersion` string that is returned is the following:

```
Pro 11.0.2
```

After the substitution, the string becomes:

```
Pro 11x0x2
```

After conversion to a number, it becomes:

```
1102
```

The first version of FileMaker Pro 10 was Pro 10.0.0, and that is the first version that introduced dynamic subsummaries that can replace Preview mode in many cases. The steps outlined here produce a number of 1000 for the first version of FileMaker Pro 10. Thus, the test is for the result of `Get ( ApplicationVersion )` to be transformed (1102 in this example) and to be tested against 1000.

All is well until you consider FileMaker Go. Version numbers for FileMaker Go start at 1. Thus, the returned string on an iPad might be the following:

```
Go_iPad 1.1.2
```

After the substitution, it becomes:

```
Go_iPad 1x1x2
```

And after the numeric conversion, it becomes:

```
112
```

The test to see if the version number (112) is less than 1,000 passes so the script goes into Preview mode under the assumption that it is a version of FileMaker Pro earlier than 10.0.0 and that it therefore does not have dynamic subsummaries. As a result, the error message shown in Figure 8.5 is presented.

You might think that it is just a matter of changing this test, but that does not necessarily work. What you should do is change the test and make a minor adjustment to the script to use dynamic subsummary reporting, as described in the following section.

Using Dynamic Subsummary Reporting on FileMaker Go

Beginning with FileMaker Pro 10 and FileMaker Go 1, dynamic subsummary reporting makes your life easier in many ways. You still create reports as you always have done, with summary and subsummary parts, but now you can view them as lists or tables (not forms) and edit the data.

Figure 8.7 shows a report using Reservations data from the downloadable sample. The report is shown in a dynamic report.

Figure 8.7

Use a dynamic report.

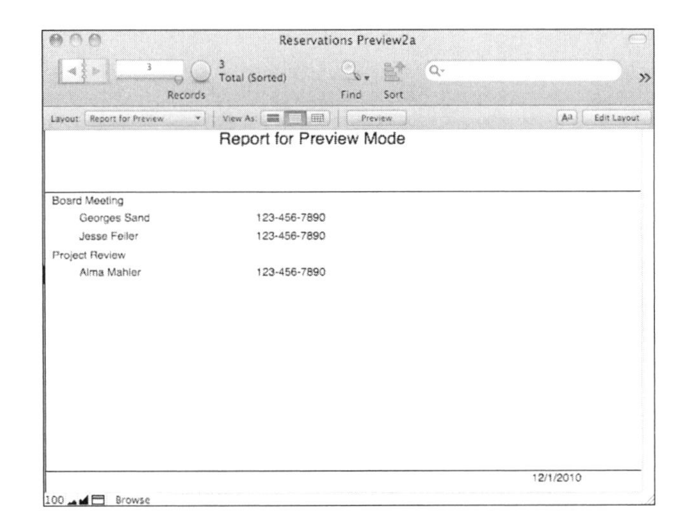

Dynamic reports must be viewed in Table or List view. The reason is simple: They group multiple records together, and in Form view, you look at only one record at a time.

Dynamic reports do not use a new construct; they are ordinary layouts that have subsummaries and summaries. They need to be sorted just as existing subsummary layouts require (so that all the records in a given grouping are in the right sort order).

When you create a layout to be used as a dynamic report, there are several considerations you should have in mind. Perhaps the most important one is that you should allow editing of fields that create groups of records. In Figure 8.8, you can see a field in each attendee record that contains the ID number of the event that person is attending. To make life easier, the ID number of each event is also displayed. Note that in Figure 8.8, the event ID for Test New Record 2, which is currently under event 1, is being changed to 3.

As soon as you complete editing that field (perhaps by tabbing out of it), the record immediately moves to its new subsummary part, as you can see in Figure 8.9.

This process is so much easier than entering data and then moving into Preview mode and then back into Browse mode that many people have stopped using Preview mode altogether.

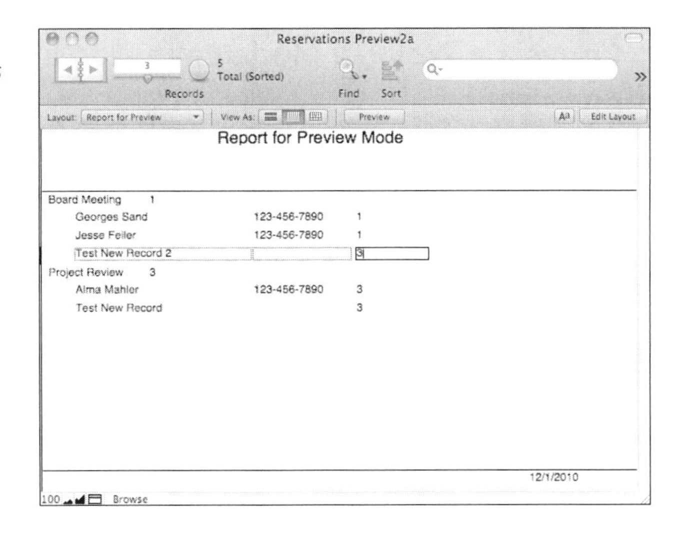

Figure 8.8
Display and edit event IDs in both event and attendee records.

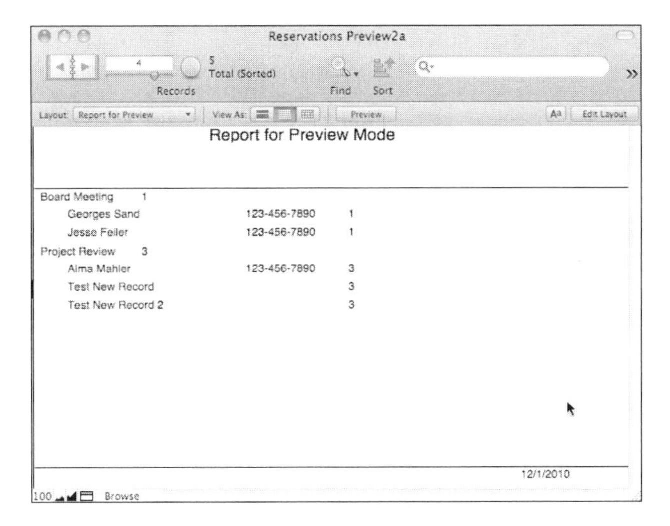

Figure 8.9
Records are automatically moved to the appropriate group.

4.11

Preview mode remains essential for pagination. There are no pages in the scrolling views of List or Table views, but sometimes you do need to produce paginated results, and in those cases, you still need Preview mode. However, remember that you can print to a PDF file, so "printing" does not necessarily mean paper.

Thus, the issue posed in the previous section in which a script attempts to drop a user into Preview mode on a device that does not support it requires several steps to be resolved.

How to Convert a Preview Mode Layout to a Dynamic Reporting Layout

Here is how you solve the problem identified in the previous section:

1. **In FileMaker Pro, open the database, go into Layout mode, and select the layout used in Preview mode.**
2. **To be safe, make a copy of the layout using Layouts, Duplicate Layout.** Use Layouts, Layout Setup to change the layout name by removing the word Copy and making it a meaningful name. Figure 8.10 shows the beginning of this process.

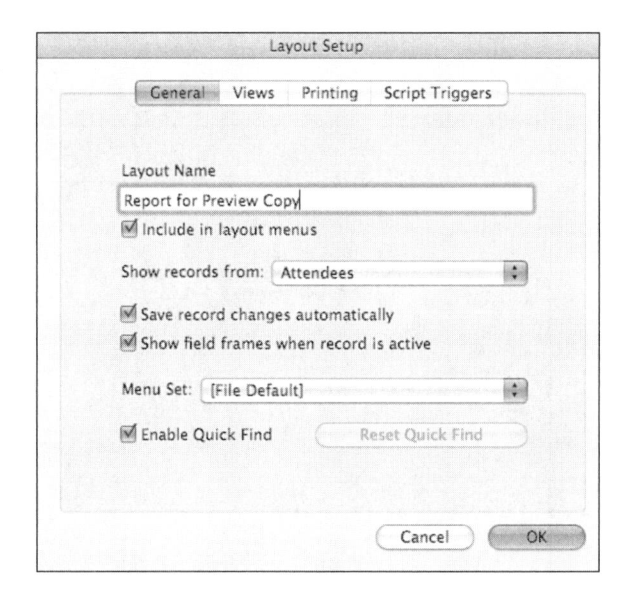

Figure 8.10
Make a copy of the layout and rename it.

3. **Remove the page number, as shown in Figure 8.11.** In this case, it is in the lower left, but it could be anywhere in the header or footer depending on the designer's wishes.
4. **Edit the script to use the new layout.** If the script was created by the layout assistant, the layout name appears only once—in the second line.
5. **Edit the script to change the test for Preview mode**, as shown in Figure 8.12. The entire section that tests for version number is now placed inside an `If` statement that tests for `Go` in the `ApplicationVersion`. If it is not found, the old code is executed. If it is found, you just use the original layout.

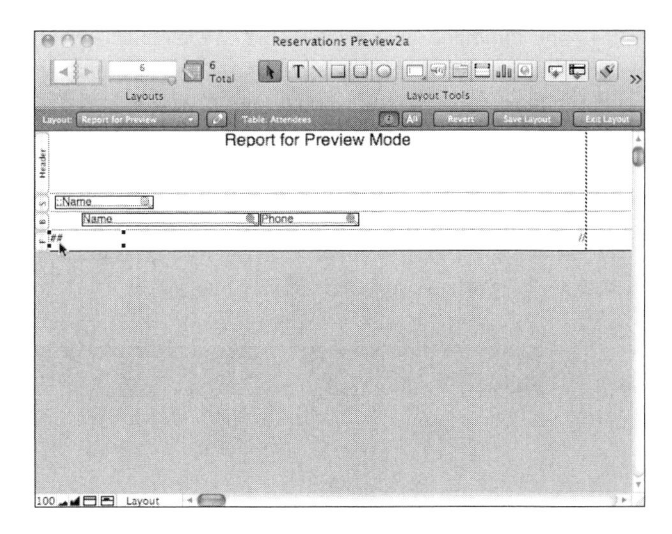

Figure 8.11
Remove the page number.

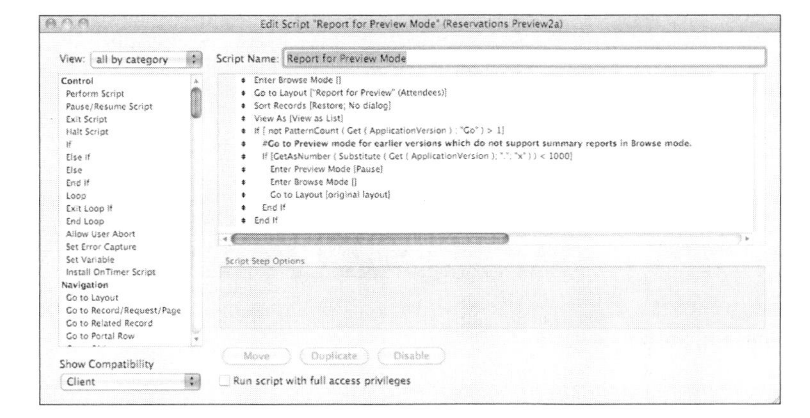

Figure 8.12
Modify the script.

Working with Custom Menus

You can create custom menus with FileMaker Pro Advanced; the resulting menus can be displayed on FileMaker Pro and FileMaker Go. Custom menus in FileMaker have two components:

- **You override existing menu commands.** You can change the name, and you can modify the behavior by adding a script.
- **Modified menus are placed in a *menu set*.** When you are editing a layout, you choose the menu set to use for that layout. Thus, it is the current layout that can determine the menus available on FileMaker Pro or FileMaker Go. You can also attach a menu set to a file so that it applies to all the layouts in the file unless you change it for specific layouts.

This section reviews the menu commands that are available on FileMaker Go and shows you how to customize them. There are no traditional menus at the top of the screen or window on iOS devices. Icons at the right of the toolbar and just above the keyboard when it rises up let you access a menu of commands. There actually are quite a few because the commands change depending on whether or not you are in Find mode.

Reviewing the Existing Menu Commands

At the right of the toolbar are the finding commands, as shown in Figure 8.13.

Figure 8.13
Finding commands are available in the toolbar.

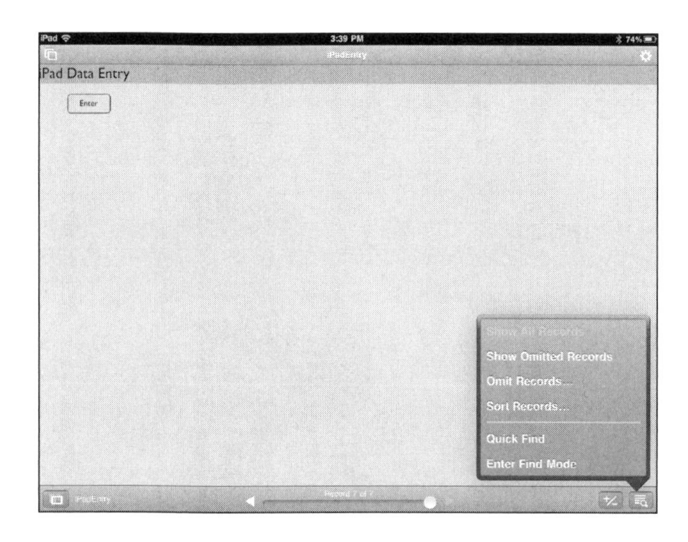

Showing and Sorting Records

The first four commands let you work with records:

- **Show All Records** is enabled if you have a found set. If all records are currently shown, this command is disabled.
- **Show Omitted Records** reverses the found set. If there is no found set, Show Omitted Records shows just that: no records.
- **Omit Records** lets you omit a certain number of records starting with the current one. After you tap Omit Records, you enter the number of records to omit, as shown in Figure 8.14. This command is very useful if you sort the records before executing a find so that similar records are together.
- **Sort Records**. For more details, see "Sorting Data" in Chapter 7, "Introducing FileMaker Go."

The second set of commands lets you use the two find mechanisms:

- **Quick Find**
- **Enter Find Mode**

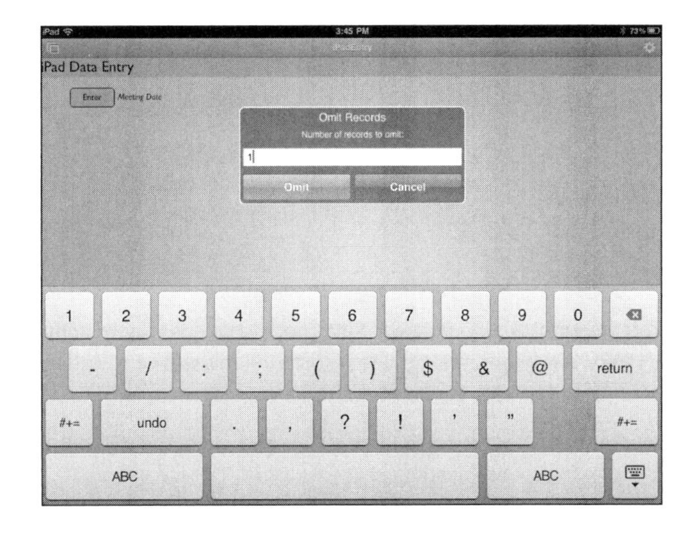

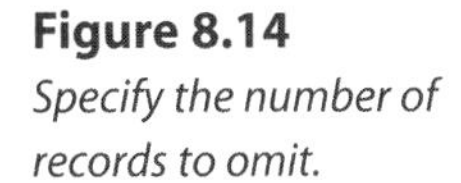
Figure 8.14
Specify the number of records to omit.

Working in Find Mode

When you are in Find mode, the same icon lets you fine-tune your search. The icon described in the previous section has new commands when you are in Find mode, as follows:

- **Show All Records** behaves as it does in the previous section.
- **Show Omitted Records** behaves as it does in the previous section.
- **Constrain Find Set** applies the current find request, but instead of executing it on its own (which is what would happen if you tap Perform Find further down in the menu), FileMaker Go applies it to the current found set so that you can find records within the found set. The result of constraining a found set is either the same size or smaller than the original found set.
- **Extend Found Set** performs the find specified by the newest find request and uses it to add those records to the found set. The result of extending the found set is either the same size or larger than the original found set.

Beneath the dividing line, you have two commands:

- **Perform Find** executes the find request without regard to the current found set. A new found set is created.
- **Exit Find Mode** returns you to Browse mode.

Working with Records

The second icon from the right with the +/- symbols lets you work with records. The commands are self-explanatory:

- **Add Record.**
- **Duplicate Record.**

- **Delete Record....** The ellipsis after the command indicates a further interaction. In this case, it is an alert asking you if you are sure.
- **Delete Found Records....** Again, you are asked to confirm the delete.

Beneath the dividing line, **Exit Record** lets you leave the record and release its update access to other users who might want to edit it.

Working with Find Requests

The second icon from the right when you are in Find mode changes to a magnifying glass with + and -. It lets you manipulate Find requests as you manipulate records in Browse mode. Find requests demonstrate the powerful search capabilities of FileMaker. You have three ways of searching:

- Quick Find lets you search for a word or phrase in any field in the database. This provides a search engine-like simple and speedy search.
- By default, when you enter Find mode, you see your layout with little magnifying glass icons in every field that is searchable. Type in the data for which you want to search and then tap Perform Find, Constrain Find, or Extend Find. Every record that matches all the data you have entered in all the fields is found. This is called a logical *and*.
- You can add additional find requests. Each one functions just as a single find request does in the preceding paragraph. However, FileMaker takes the records found from each found request and returns them all together, although no record is returned more than once. Multiple find requests are said to be *or* operations.

The commands in this menu in Find mode are as follows:

- **Add Find Request.**
- **Duplicate Find Request.**
- **Delete Find Request** is similar to Delete Record, but notice there is no indication of a need to confirm the deletion. The reason is that there is a difference between deleting data with a Delete Record command and deleting a find request, which is always a transient entity (find requests act on the database, but they are not stored in it).

Beneath the line, the **Revert Find Request** command acts on the current find request, if any. If there is no find request, the command is disabled.

Customizing Menus

The full range of menu customization that you can develop in FileMaker Advanced is designed for desktop use. (Remember that you design customized menus in FileMaker Advanced, but you can use them in any FileMaker product.) The dialogs that you access with File, Manage, Custom Menus enable you to do all of the following:

- **Redefine an existing menu command.** You can change its name, keyboard equivalent, and even its functionality (its *action*). To do this, you make a copy of the existing command and then work on that.

- **Create and modify new menus.** Using the same principle, you can create a copy of an existing FileMaker menu and add or delete its menu commands, which can be redefined, or just the standard commands.
- **Group your custom menus into a *menu set*.** The menu set can then be attached to the file.

It is important to remember this structure as you start to customize your menus because you often start by asking yourself how to do the first step—redefine an existing menu command. In order to get to it, you have to work up this list of tasks in order to get a redefinition of a single command. You start by creating or editing a menu set, then you go to a menu within it, and then you get to the commands. Along the way, there are features that do not apply to FileMaker Go with its constrained menu commands.

For example, you cannot create a new menu and place it in the menu bar because there is no menu bar. The menus pop up when users tap the icons at the right of the toolbar, and the icons (such as +/-) have specific meanings. Tapping +/- in order to sort records is the type of user interface known technically as *horrible*. Work within the menu command structure that you have in FileMaker Go and avoid changing the meaning of commands.

If you cannot add new commands and should not change the meaning of existing commands, what can you do to customize your FileMaker Go menus? One of the most common answers is extend the existing commands and make them smarter and more relevant to your user. A common customization is to change the New Record command so that it runs a script that not only creates a new record but does some database-specific initialization. You can do a lot of this using options in the Manage Database dialog, but at the moment the user tries to add a new record, a script can inspect the current record and the state of the database in order to make smarter initialization choices than is possible with options.

In a similar vein, you might want to replace the Duplicate Record command with a script that not only duplicates the record but also performs some database-specific initializations. The most common such customization involves duplicating a record that contains a field identifying a related record. There are cases where you want to duplicate the record but reset that foreign key to an empty field.

Customizing a menu command can solve these problems if you have them. Remember that the process is available only in FileMaker Advanced, and you have to skip over a number of steps that just do not apply to FileMaker Go.

How to Customize a Menu Command for FileMaker Go

Here are the steps involved in customizing one of the FileMaker Go commands. In this example, the New Record command is replaced with a script that adds a record and does any other initialization that you want to do. Note that because this process enables you to create new menus for FileMaker Pro, the first several steps contain commands that are irrelevant to FileMaker Go. It is only at step 7 that you actually start to customize a command, but do not worry—the initial steps go very quickly:

1. **Prepare the script you want to run.** You can start with a script that has a single script step: It creates a new record. When this script runs, it exactly duplicates the New Record command. You can then customize the script later. The advantage of this technique is that you attach the script to the command with FileMaker Advanced, and you can later on modify the script using FileMaker Pro.
2. **In FileMaker Pro, choose File, Manage, Custom Menus.** This opens the dialog shown in Figure 8.15.

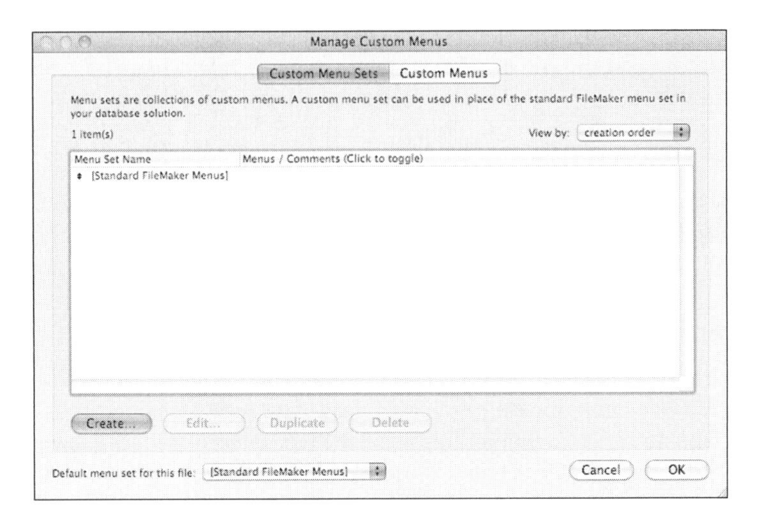

Figure 8.15
Open the Manage Custom Menus dialog.

3. **Click the Custom Menu Sets tab at the top and then click create in the bottom left to create a new menu set.** This opens the dialog shown in Figure 8.16.

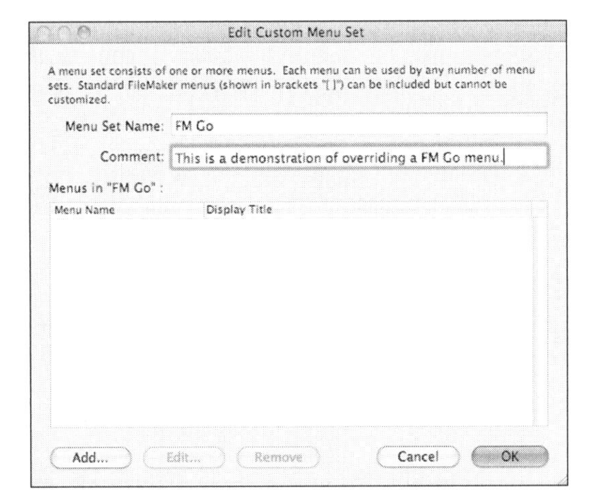

Figure 8.16
Create a new menu set.

4. **Name the new menu set and, before moving on, make certain to add a comment. Click Add to add a menu to the menu set.** The dialog shown in Figure 8.17 enables you to select the menus you want to add.

   You can use the standard keyboard shortcuts to make multiple selections. It can be a good idea to add the default copies of all the standard menus whether or not they appear in FileMaker Go. At least you have them in your menu set when you need them.

   The list of menus includes copies of menus that you can modify (they end in the word Copy). Menus in square brackets are default FileMaker menus that you cannot modify, but you can include the unmodified versions in your menu set.

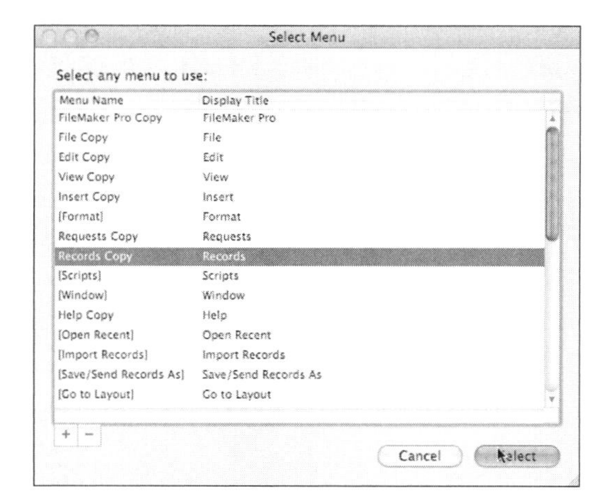

Figure 8.17
Add menus to your menu set.

5. **Click OK when you have added your menus to be returned to the Manage Custom Menus dialog, shown previously in Figure 8.15.**
6. **Select your menu set and click Edit to edit it.** The dialog shown in Figure 8.18 opens.

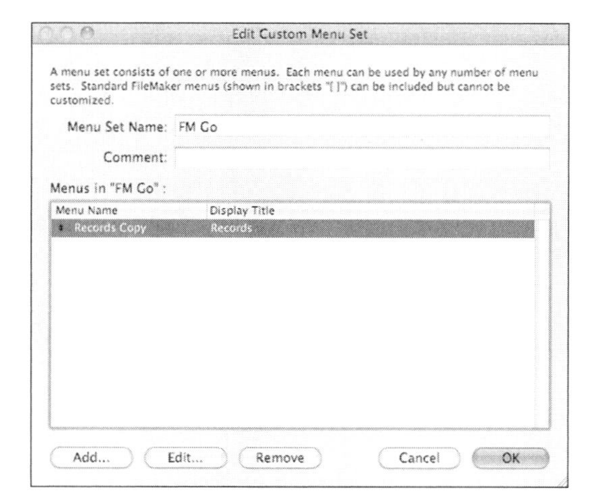

Figure 8.18
Edit your menu set.

7. **Select the menu in the menu set that you want to edit (that is, the menu containing the command you want to change), and click Edit.** The dialog shown in Figure 8.19 opens.
8. **Select the command to modify.** The menu name and menu title do not matter. They are not shown on FileMaker Go, so you can leave them as is. In the lower right of the dialog are the key components. In the pop-up menu, select Command. (The other two choices are Submenu and Separator; you can use them to construct your own menus, but on FileMaker Go, you can only modify the commands so just make certain the pop-up is set to Command.)

   The checkbox Based on Existing Command is set by default to the FileMaker command. You can uncheck the box, but it does not matter in this case.

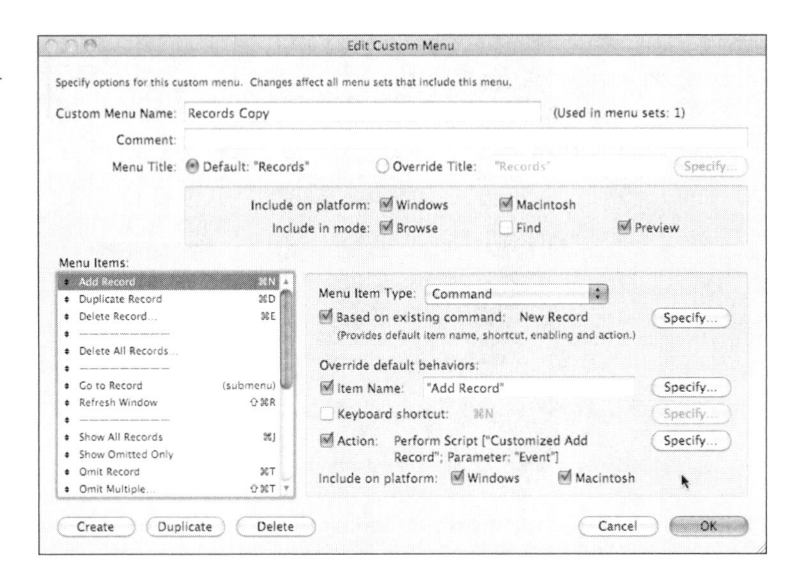

Figure 8.19

Choose the menu to modify.

You can change the item name if you want. If you are overriding the Add Record command, you might leave the name but add a script to modify the functionality. Leaving the default names keeps users comfortable, but if there is a compelling reason, modify the name.

Click the Action checkbox. Immediately the dialog shown in Figure 8.20 opens. Choose the action to perform. In this case, you want to select a script to perform. Note that you can pass in a parameter if you want.

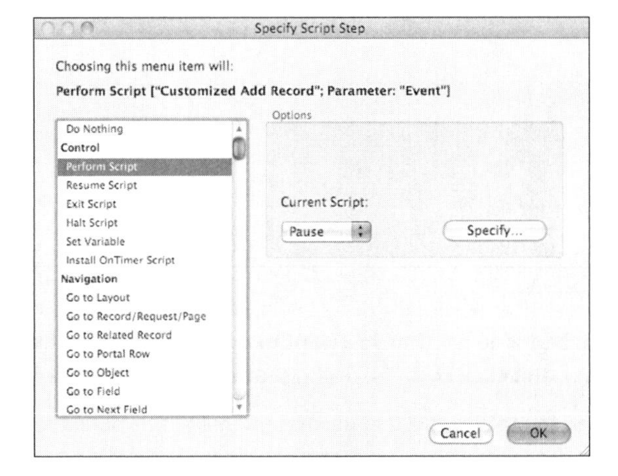

Figure 8.20

Choose an action for your command.

9. **Click OK to close each of the dialogs until you are back to Manage Custom Menus, as shown in Figure 8.21.** Finish up by setting the file's menu set to the new menu set you have created. You do this with the pop-up menu at the bottom of the dialog. Click OK and you are done.

After you have set the menu set for the file, you do not have access to other menus. However, you cannot lock yourself out. If you have access to FileMaker Advanced (and you have to have had access to FileMaker Advanced to manage custom menus), the Tools menu contains a Custom Menus submenu with all of your menu sets in it. You can use it to switch back to the default menus or any other menu set.

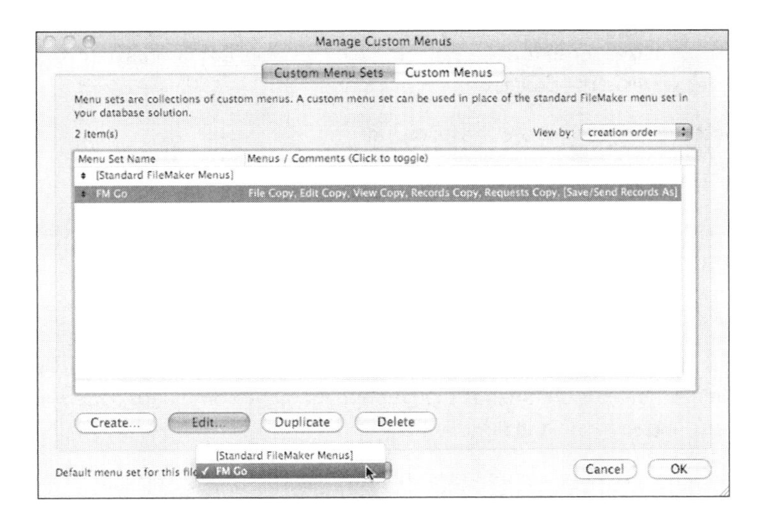

Figure 8.21
Set the file's menu set.

Managing Script Issues

When it comes to FileMaker Go, you can use scripts just as you would on FileMaker Pro. There are a few script steps that do not run on FileMaker Go, but you can avoid most problems by checking the script results, as described in this section.

Particularly if you are modifying a FileMaker Pro database to run on FileMaker Go, you might encounter scripts that were designed for the desktop environment and that run for a long time. (These often are end-of-day reports and analysis scripts that might be copying records or entire tables into denormalized analytical structures...or normalized ones as the case may be.) Long-running scripts can be problematic on FileMaker Go because it might hibernate at apparently random times. In fact, the times are not at all random, but you cannot assume that an unattended script will just chug away forever as it does on the desktop.

The simplest solution is to either run such scripts on FileMaker Pro or to use the FileMaker Server Scheduling features described in Chapter 6, "Introducing FileMaker Server." Scheduled scripts are processed on FileMaker Server itself, so FileMaker Go has nothing to do with them. You can argue that long-running batch-type scripts always should run on FileMaker Server using a schedule, and there is great merit to that position. Until FileMaker Go, it was mostly a theoretical concern. Now there is little question that scripts like that belong on FileMaker Server.

There are several script steps that should be used in scripts for FileMaker Go. They are not new by any means, and their use—such as the running of scripts on FileMaker Server—has always been a recommended best practice. Now, it is more important to pay attention to these script steps.

Get (LastError)

Errors often occur in scripts. Most of the time, FileMaker just does the right thing and you (and your users) might never know there was an error. However, it has always been good programming practice to check the results of any operation that is invoked from a script or program. `Get ( Last Error )` returns a number that describes the result of the last script step.

In some cases, it makes sense to set a variable to the result of `Get ( Last Error )` and then to test the variable a few steps later in the process.

The two most common values are the following:

- **0.** This means no error was encountered.
- **401.** This means that a find returned no records. Whether or not this is actually an error depends on the context.

With FileMaker Go, two other values become important:

- **3.** This means that the command is unavailable. On FileMaker Go, this applies to commands such as the spelling commands, commands that implement menu commands (Open File Options, Open Manage Layouts, and so forth), and Print Setup.
- **4.** This means that the command is unknown. On FileMaker Go, this includes commands such as New File (you cannot modify database structures on FileMaker Go), Enter Preview Mode (there is no preview mode), and Export Records.

The following section describes the practical difference between these two results.

Allow User Abort On/Off

This script step enables users to abort a script. They can do so in two circumstances:

- **User-initiated aborts.** On FileMaker Pro for Mac, pressing ⌘+. or on Windows pressing Esc aborts the current script but only if Allow User Abort is set to On. That is the most common setting. It is usually turned off in a script when it wants to start and finish a discrete process. It is usually turned off, for example, when you are pausing a script and posing a window as if it were a dialog. You do not want people to be able to abort the script or close the window except by clicking a Cancel button that you implement.
- **FileMaker-initiated aborts.** There are cases where FileMaker displays an alert and gives the user the option to abort the script. The most common case in which this happens is when error 4 (unknown command) occurs. If Allow User abort is off, the alert is presented, but the option to abort the script is not available to the user. For scripts running on FileMaker Server, if error 4 is encountered and Allow User Abort is on, the script stops. If Error 4 is encountered and Allow User Abort is off, FileMaker Server skips the script step.

Set Error Capture On/Off

This toggle lets the script suppress messages from FileMaker with regard to errors. The most important aspect of this on FileMaker Go is that when error 4 is encountered, the alert is not displayed if Set Error Capture is on. Turning set error capture on means that the script takes the responsibility for checking `Get ( Last Error )` and taking appropriate action. That might mean presenting an alert to the user, but it can also mean taking corrective action as well as possibly updating a log database.

Further Steps

Now that you have seen the basics of how FileMaker Go works, take a few minutes to experiment with databases of your own. If you have FileMaker databases, make a copy of one or two of them and then move them to FileMaker Go on iPad or iPhone. Before you start to make modifications, just take a look at what you have. You might get a sense for the extent of the work (if any) that is needed to move the database. In particular, you should look for things that need to be rethought for mobile devices. Also, see if you can get a feel for whether you need to modify existing layouts for dual use or if you need to duplicate and modify layouts so that you have two sets of layouts.

If you do not have existing FileMaker databases, try creating a starter solution and moving it to FileMaker Go. This gives you a sense of the general issues involved.

IN THIS CHAPTER

Designing a FileMaker Go Solution

This chapter helps you build your own FileMaker Go project for iPhone or iPad. Although the variety of FileMaker Go projects is vast, for many people they come down to two issues:

- Do you bring data *to* the user (an example of that is the catalog shown in this chapter) or do you get data *from* the users (the handheld estimator shown in this chapter is an example of that type of project)?
- Do you work from an existing FileMaker database or do you create something new?

You can draw a 2 × 2 grid with these two questions on the axes so that you have four cells. Most FileMaker Go projects can be situated in at least one of those four cells. Many projects wind up in several cells (a catalog that also lets you order is a good example of this).

These example applications provide a solid basis on which you can build. However, there is one important feature that is omitted: printing. See Chapter 10, "Using Printing and Charting with FileMaker Go," to find out more about how to add printing to these examples.

Introducing the Handheld Estimator and the Catalog

For many people, the first FileMaker Go project is a first in several ways, including your first handheld device project and perhaps your first project using FileMaker in any way. One of the lessons you learn very early on (and perhaps just by reading this chapter) is that you have to focus your project. In fact, this is not a bad strategy for even a large project, but when you are working with the limited screen size of a mobile device (there is never enough), the issues of conserving precious battery life (there is never enough), and the challenges of working with network signal that might not always be available mean that you should focus on the essentials and avoid wasting screen space, battery power, or even what may be a fluctuating and relatively expensive network signal.

One of the best ways to understand the handheld issues is to watch yourself as you use your devices and to watch others. Observing people who are productively going about their work and play is not particularly helpful; instead, watch the people who are confused or frustrated (and this might be yourself, as you observe your own adventures). Read the ads and descriptions of handheld apps on Apple's App Store, in magazines, and on the web. Read the reviews and watch for the recurring themes (one of the big ones is "it doesn't work"). Look at the screenshots in advertising and instructional materials to see how other people have addressed the issues of using a handheld device and limited screen size.

There are two projects in this chapter: a handheld estimator and a handheld catalog. The estimator is described in great detail. When it comes time to work on the catalog, you see that it is largely a matter of repeating the same steps you have already done. And when it comes to your third FileMaker Go project, things will probably go even faster.

The basic techniques do not differ from project to project. What does differ is the data that you use and the users for whom you are writing. The more you know about each, the better the end result will be.

The code for all the projects is downloadable from the author's website at northcountryconsulting.com, as well as the publisher's website at quepublishing.com.

Introducing the Estimator

The estimator represents a classic example of how a handheld device can transform an everyday transaction. The estimator is designed for people who need to estimate home maintenance projects, such as lawn mowing, snow plowing, basement or garage cleanouts, painting and decorating, and the like.

After the estimate is agreed to, the customer can sign the screen to approve the estimate, and you can use the printing and email features of FileMaker Go to give the customer a copy of the estimate. In many cases, that is sufficient for the records you need. You can also move the database to your computer using iTunes for Mac or Windows; after the database is on your computer, you can browse previous estimates.

Although the examples given here are based on home maintenance, the general design can be adapted for industrial or office maintenance and many other types of projects (such as dog walking).

This project assumes that the *user* is the person doing the estimating and, most likely, doing the work. The *customer* is the user's customer (the one who pays the bills and receives the service).

The heart of the estimator is several pieces of information that the user provides (this is an example of getting data from the user). Here are the items the user collects:

- **Task.** Painting, mowing, cleanouts, or even dog walking.
- **Location.** Where the work is to be done.
- **Picture.** The room to be painted, the lawn to be mowed, or even the dog to be walked.
- **Frequency.** One-time (as in painting a room, you hope), twice a day (dog walking), or when it is needed (mowing and plowing).
- **Price.** This is two data items—the amount and the units. Are you charging $35 per snowstorm or $100 per month for however many storms there are?
- **Customer contact.** Name and email (or phone or address).
- **Signature.** The customer's agreement to the estimate.

This is a great example of a project for a handheld device because so much of the data can be obtained directly from the mobile device. In the olden days, you might have written a full description of what the project task is. Today, you can take a picture of the garage or driveway and be done with it. Notice how many of the items in the list can be set up to require no typing. For example, you can use pop-up menus for the task and frequency.

In fact, the only items that definitely require direct input are the price and the customer signature. The customer contact can be a photo of a business card or street address, but it might also need to be typed in. This often depends on whether the customer will be billed. Many household maintenance services are cash-on-completion transactions.

As you think through a project, formulating the data to be collected in mobile-friendly ways (that is, ways that can rely on checkboxes and choices rather than on typing) can help make the result easier to use. For every data item you consider, ask yourself if you can rephrase the question so that it can be answered with a tap rather than by typing.

Also, consider how much data you need. For a project like the estimator, there might be some boilerplate text that only is needed on a printed version of the estimate. However, some types of projects might require more information and details. That degree of data entry might push you from iPhone to iPad for your FileMaker Go project.

A single-screen solution eliminates the need for navigation buttons, but even the smallest FileMaker Go project generally has more data than fits on a single screen (games can be an exception). The estimator project uses the simplest multiple-screen implementation: two layouts. As you can see in Figure 9.1, the basic screen has an info button (the *i* in the lower right). Tapping it switches to the alternate layout, which has a Done button that reverses the process. (This is the interface standard from Dashboard widgets on Mac OS X, as well as many other designs.)

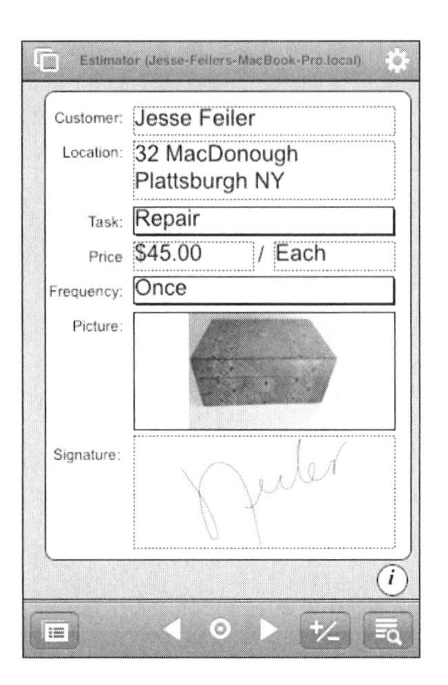

Figure 9.1
The estimator has two layouts.

Introducing the Catalog

The catalog is an example of the kind of project that brings data *to* users (as noted previously, it can be expanded to be bi-directional if you add ordering to it). One of the biggest challenges in this type of project is getting the data (see the sidebar, "Locate the Data"). However, if you are sitting on an existing hoard of data, this daunting first step is not a problem. In fact, it can be argued that if you are sitting on an existing hoard of data, you have a golden opportunity to make that data mobile.

LOCATE THE DATA

Data-driven projects, such as a catalog, need data to function. That scarcely seems worth mentioning, but it is very common for someone to have an idea for a wonderful project that consists of a database with information on a certain topic. Often the person who has the idea for the database project—and, often, who designs the database and the interface—assumes that other people will provide the data. Whether you are building an interactive website, a project for FileMaker, or anything else that requires data, do not assume that people will flock to your doorstep to enter their own data or other data that they might have.

Even if people are willing to enter data, that often becomes something to do when time permits, and time has a tendency not to permit. Is the consolidated database a way for local businesses to attract new customers? That may well be the case, but in that case, you are embarking on two projects: You need to get local businesses to provide their data and then you need to promote the database so that people actually use it. Until they use it, local businesses will not see the value of their participation and the whole endeavor can fall apart.

Freelance consultants who work with databases have seen these projects over and over. The idea might be great, but the need to sell it twice (to data providers and then to data consumers) doubles the task. Fortunately, FileMaker Go and handheld devices are on your side. If your idea is to promote local businesses to visitors to the region, you have to overcome the natural resistance of many people to explore an area they do not know. With FileMaker Go, you can give people the assurance to explore because the geolocation tools prevent all but the most determined from getting lost.

The same basic principles of developing for handheld devices apply to a catalog: minimize data entry, for example, in favor of data choice such as checkboxes, radio buttons, and pop-up menus.

With a data-driven project such as a catalog, a new set of factors enters into the picture. You are putting your data on a mobile device, and that device will be moved around. In addition to being moved around, chances are probably greater that it will be lost or stolen than if your project is deployed on a desktop computer. (Mobile phone and laptop computers are among the most frequently lost or stolen items of value.)

If you are working from an existing database, you might want to consider restructuring it for a mobile device. If you are updating the data in the database, this might be impossible. However, if you are building something like a catalog that only presents data, you might want to consider not putting some data on the mobile version of the database. Your desktop database might have data such as profit calculations and even customer lists that you do not want to expose to the vicissitudes of a mobile device that can be lost or stolen. Accordingly, compartmentalizing the database into a mobile version versus a desktop version might be the safest route. Almost as good can be using FileMaker security mechanisms to block access to certain fields when they are accessed from a mobile device. You see how to do this as you implement the catalog.

The catalog project has two layouts. If you are familiar with FileMaker (or any other database, for that matter), you will recognize them by their functionality. In Figure 9.2, you see a layout shown in a standard FileMaker list view. At the top, the header section of the layout contains some controls. The body of the layout is the list of records. If the list is long enough, a scrollbar appears when you open this database in FileMaker Pro. On FileMaker Go, you can swipe up and down to quickly scroll the list.

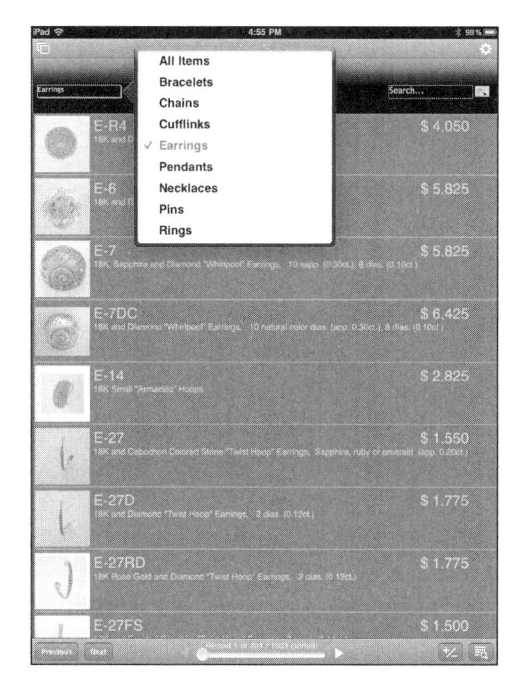

Figure 9.2

The catalog has a list view.

Tapping an image in the list switches to a form view, as shown in Figure 9.3. This view presents a full-screen layout for a single record. Notice in the upper right that a button lets you return to the list. This enables you to hide the status bar at the bottom of the standard FileMaker Go window. Whether you want to do so or not depends on your users and what they expect. If your users are likely to be familiar with databases in general and FileMaker in particular, the status bar is a big help because it enables simple navigation as well as access to basic commands, such as find and record adding or deletion. Many of the standard commands are not relevant in a catalog type of project because you might not want users to add or delete records from the catalog. Thus, in this project, the status bar is hidden, and the necessary commands are presented as user interface elements on the layout. This type of choice is entirely up to you (with advice from your test users, of course).

Figure 9.3

You can view individual records in a form view.

Building a Handheld Estimator for iPhone

Start by thinking through what it is you want to do for your estimator. This is a simple example, and in a traditional development environment (not to mention a traditional book), you might build on the bare bones of the simple example to make it more complex and customized. With mobile devices, those additions to the simple case might very well unfocus your project. Change anything that does not work, but always resist the temptation to complicate the project.

→ The steps in building the estimator build on basic FileMaker techniques that have been described throughout this book. Some additional basic techniques are used; you can find them described in more detail in *FileMaker Pro 10 in Depth* and in FileMaker's own documentation on filemaker.com.

411

Although this example shows the iPhone, if you follow the autosizing steps, you will discover that it works equally as well on the iPad.

These are the steps involved in building the estimator:

- **Create the database with FileMaker Pro.** Add the database fields, the options to automatically set initial values, and build the value lists for data entry.
- **Build the front layout.** There is only one main layout because there is only one screen. You use the automatic positioning features to accommodate the rotation of the user's iPhone.
- **Build the back layout.** Implement navigation controls to switch back and forth between the main layout and the credits.
- **Set up security and default behavior.** Create an On Open script to start a new estimate.
- **Implement printing.**
- **Test and revise.**

Creating the Database

The brief description of the project presented earlier in this chapter is the type of analysis you should do before you begin. This need not be a lengthy process, but even the most experienced developers and designers start with a rough sketch (this can be a pencil sketch on a scrap of paper). Think about the layout and the data that you need. As you sketch, you might realize that you need some extra fields (and that you can eliminate others). For many simple projects like this, ten minutes can be enough for the initial design.

Many people find it most productive to set aside two five-minute periods for this project. Sketch out your idea, and then take a break. Come back to it with fresh eyes, and, perhaps, with someone else to take a look at it.

How to Set Up the Database

Like all of your FileMaker Go design work, you use FileMaker Pro or FileMaker Pro Advanced to set up your database:

1. **Create a new FileMaker Database.** Use File, New Database to create the database; when requested, name the file. In this example, it is called Estimator. Figure 9.4 shows the new database in Table view.

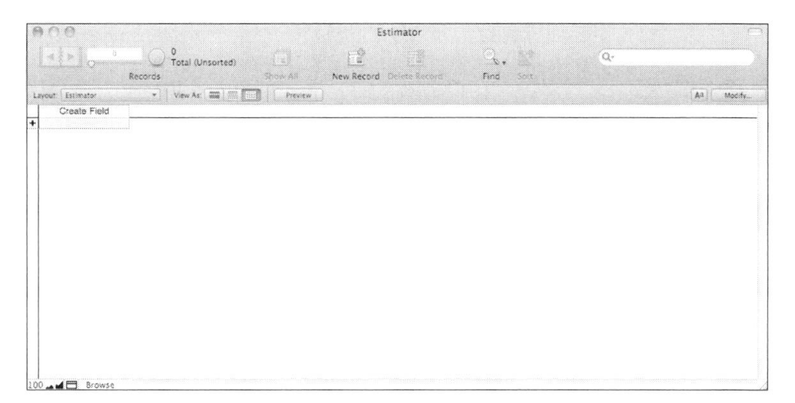

Figure 9.4
Create a new database.

2. **Manage the database.** Although you can create new fields in a table view, it is easier for most people to use Manage Database, as shown in Figure 9.5. Open this dialog with File, Manage, Database or ⌘+Shift+D on Mac or Ctrl+Shift+D on Windows.

   If you click the Tables tab, you see that you have a single table in the database. By default, it has the same name as the file. If you want to change the name, you can click the table, type in a new name at the bottom, and click Change.

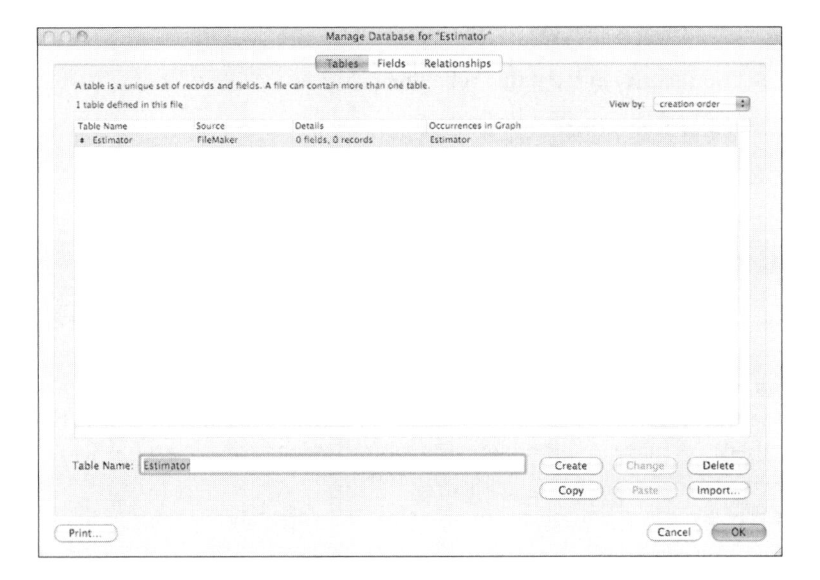

Figure 9.5
Open Manage Database.

3. **Add the fields.** Select the Fields tab at the top of Manage Database. Add the fields shown in Table 9.1 and set their field types accordingly. You can use the Options button in the lower right of each field to set a default initial value. Whether or not you do so depends on the types of jobs you estimate. Remember that you can change this value by going back to Manage Database so your default value in the winter might be plowing and the default value in the summer might be mowing.

Table 9.1 FIELDS FOR ESTIMATOR

Format	Type
Task	Text
Location	Text
Frequency	Text
Picture	Container
Price	Number
Price Units	Text
Customer	Text
Signature	Container

Figure 9.6 shows the created fields. Note that comments are provided for each field. That is always a good habit to get into.

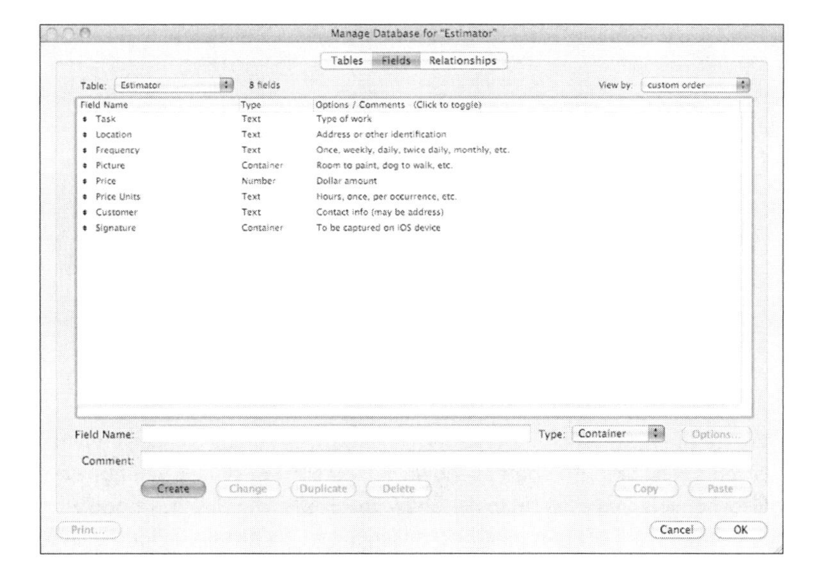

Figure 9.6
Create the fields.

Note that you do not have to do anything to the standard container fields to allow people to sign in them or to take photos—FileMaker Go and the iOS device take care of everything.

4. **Create value lists.** To minimize typing, create value lists for any input that lends itself to that type of interface. Choose File, Manage, Value Lists to open the dialog shown in Figure 9.7.

   In "Create a Basic iPhone Layout," you see that you can allow users to update the value list as they enter data or to enter one-time-only values for the data field without changing the value list. This means that you do not have to think of every possible value now.

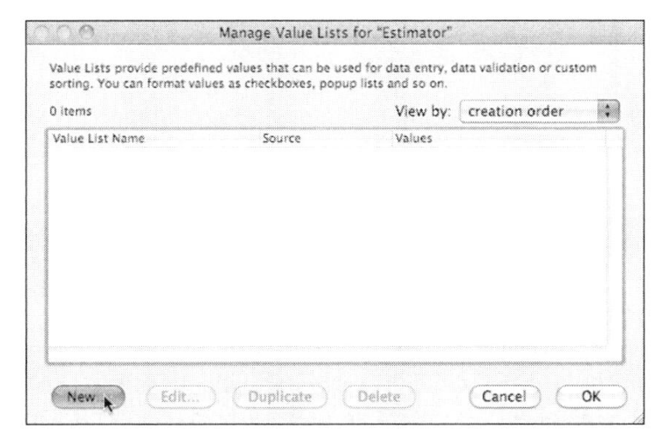

Figure 9.7
Manage value lists.

5. **Create the Tasks value list, as shown in Figure 9.8.**

6. **Create Frequency and Price Units value lists in the same way.** Depending on your work, values for frequency might be Once, Weekly, As Needed, or similar frequencies. Price units might be per hour, one-time, monthly, and so forth.

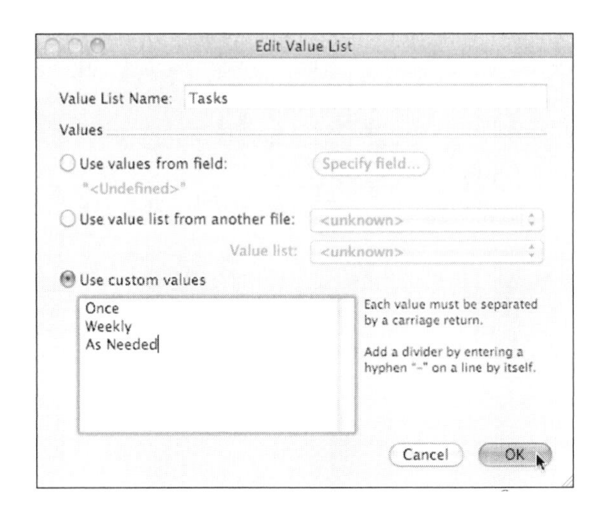

Figure 9.8
Create the Tasks value list.

7. **Create the front and back layouts.** Go into Layout mode by choosing View, Layout Mode. Then choose Layouts, New Layout/Report as shown in Figure 9.9. By default, your one default table in the database file when it is created is set to the Show Records From value. It is a good idea to override the default layout name (Front is a better name than Layout #2, for example). For the layout type, choose Blank Layout and then click Next. Repeat the process to create the Back layout.

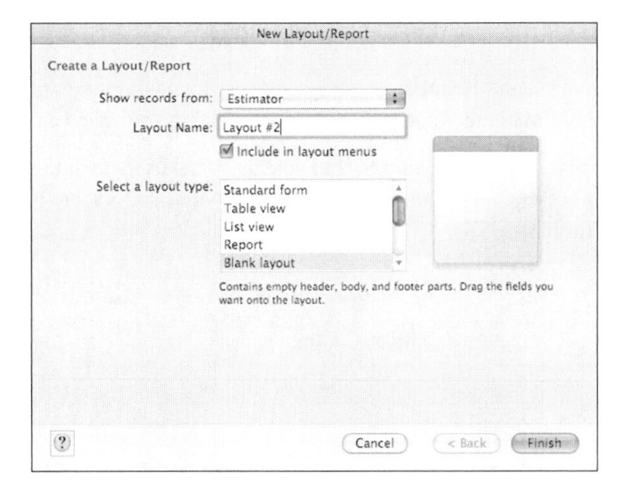

Figure 9.9
Create the layouts.

Building the Front Layout

The front layout is where most of the action takes place. Begin by going into Layout mode. Use the recommended fonts and sizes (17 to 22 pixels), as described previously in Chapter 8, "Optimizing FileMaker Databases for FileMaker Go."

How to Create a Basic iPhone Layout

The basic tools are provided for you with the Layout assistant, but you need to adjust a few settings you might not have used before.

1. **In Layout mode, set the graphic ruler units to pixels.** Choose View, Graphic Rulers and click the units indicator in the top left. It cycles through inches (in), centimeters (cm), and pixels (px). You can actually use any units that you like, but because the specifications for the iPhone and iPad interfaces are given in pixels, using those units gives you the best control.

   The resolution for the retina display of the iPhone 4 is different than that of the earlier iPhones. iPhone 4's screen dimensions are 640 × 960; other iPhones are 320 × 480. For iPhone 4, multiply the pixel dimensions given in these steps by two: for example, 45 pixels becomes 90 pixels on iPhone 4.

   Figure 9.10 shows the graphic rulers and the control in the upper left to set the units.

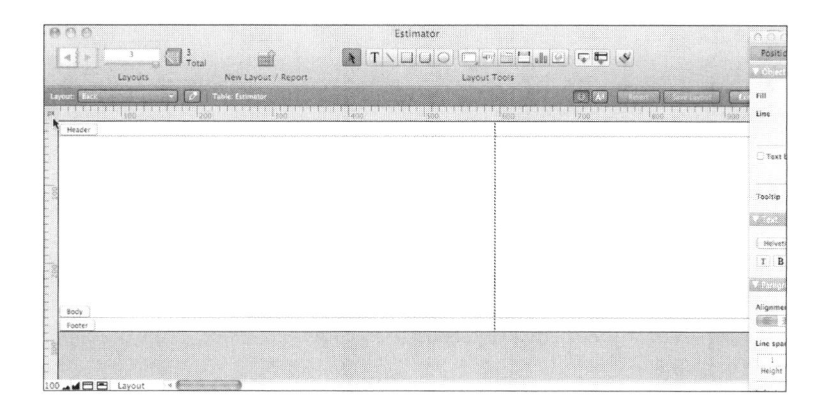

Figure 9.10
Set the graphical rulers units to pixels.

2. **Create the background for the top of the layout.** On iPhone, the bar at the top of most screens is 45 pixels high; the width is 320 pixels in portrait mode. (This is typically a *navigation bar*, but it can also be a *toolbar*—those are the terms used in iOS.) You can use a solid color, but if you have access to a drawing program such as Photoshop, creating a gradient with the lighter color at the top and the darker color at the bottom can provide a better effect. If you are using Photoshop or a similar program, save the image and import it into your layout with Insert, Picture. FileMaker can import Photoshop files (.psd), as well as JPEG or PNG files.

   The width of the image does not matter because the autosizing feature in FileMaker expands it as necessary.

3. **Set autosizing for the top of the layout.** If you have not done so already, open the Inspector using View, Inspector or ⌘+I (Mac) or Ctrl+I (Windows). With the top image selected, set autosizing, as shown in Figure 9.11. This anchors the image to the top and both sides of the screen so that it moves and expands or contracts as you rotate the iPhone.

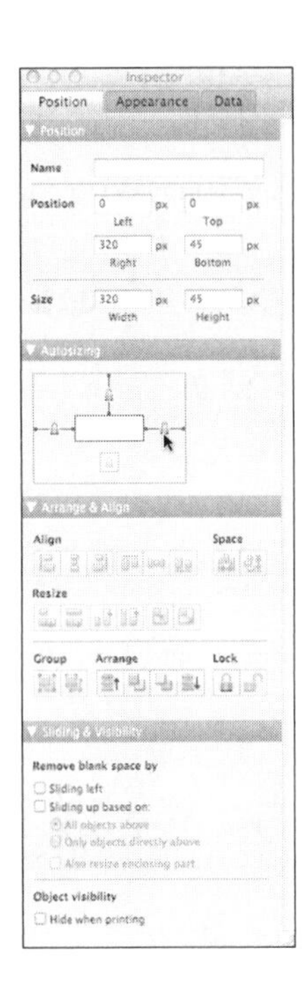

Figure 9.11

Set autosizing options.

4. **Create the background for the main part of the layout.** The total screen size is 320 × 480, so subtract the 45 pixels of the top image, leaving 320 × 435. It can be a solid image or a subtle background (check out the various Apple apps for background ideas). Remember that the more complex your background is, the harder it is to see the data. Add the image to the layout. Set the same autosizing options as you did in step 3.

5. **Add the Info button.** Before you forget, add the Info button (it is in the lower right of Figure 9.12). Typically, the lower-left or -right corner is best for this button. This can be an image from Photoshop or another drawing program, or you can construct it with FileMaker by drawing a white circle, placing a bold and italic lowercase ***i*** in it, aligning those objects on their centers, and then grouping them to create a single object.

   The button will be implemented when you create the Back layout later in this chapter.

6. **Add the Task field.** Using Insert, Field or the toolbar icon shown in Figure 9.13, add the Task field. It is a convention in many iPhone interfaces to place data entry fields on a rounded rectangle, as you see in Figure 9.13. Multiple fields and their labels can be grouped in a single rectangle as is shown here.

   To draw a rounded rectangle, use the rounded rectangle tool in the Status toolbar or choose Insert, Graphic Object, Rounded Rectangle. Resize and reshape it as you want. You can add fields on top of it and group the fields and the rounded rectangle together. Rounded rectangles used in this way are often white.

Figure 9.12
Add the Info button.

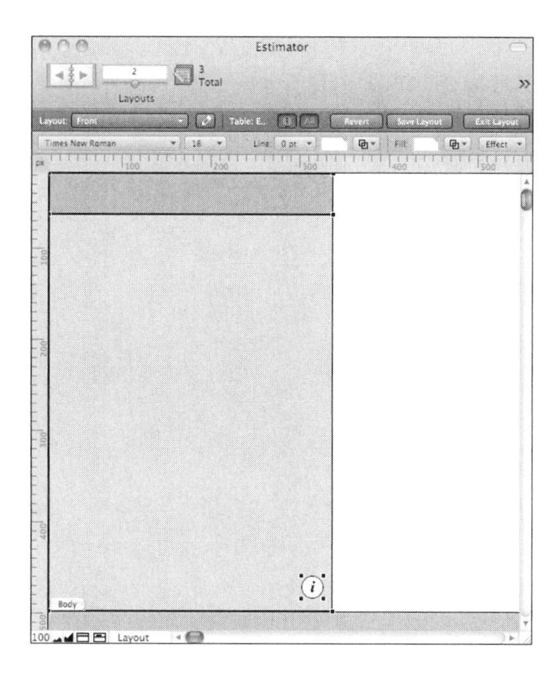

Figure 9.13
Add the Task field.

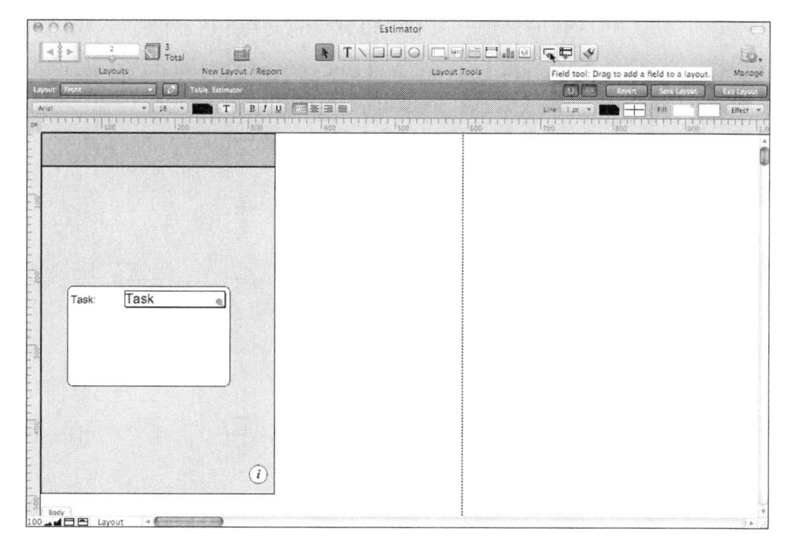

7. **In the Data tab of the Inspector, set the field type to pop-up menu, as shown in Figure 9.14.** When you have selected the pop-up menu, you are then prompted to choose the value list, as you see in Figure 9.14.

   Note that this is the moment when you can specify that users can enter new values to the value list or, on a one-time basis, they can enter values that are not in the value list. These options are set for a specific use of a value list in a pop-up menu or other interface element; the options are not part of the value list itself.

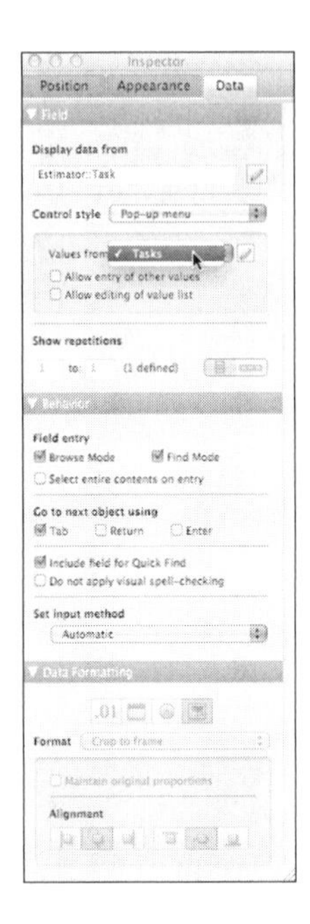

Figure 9.14
Use a pop-up menu for the Tasks field.

8. **Implement the Frequency and Price Units fields in the same way.**

9. **Implement the Price field.** Use the Edit box control style pop-up menu on the Inspector's Data tab, as shown in Figure 9.15. At the bottom, set the Data Formatting Format to Currency. Choose your symbol, the decimal places, and the optional thousands separator.

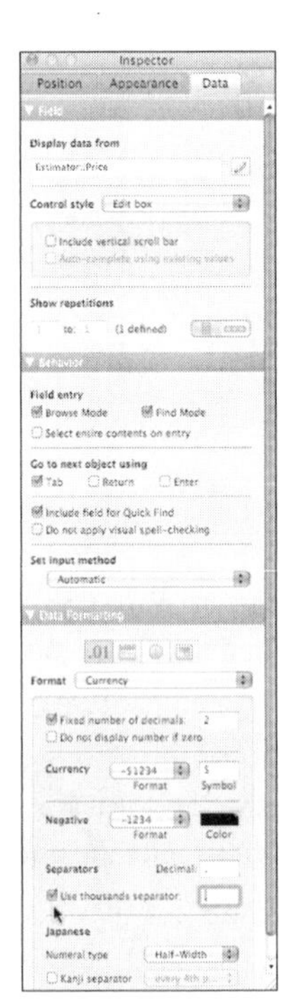

Figure 9.15
Adjust currency settings for the Price field.

10. **Add the picture field.** This is a container field, and it needs no customization here.

11. **Add the signature field.** This, too, is a container field, and you do not need to do anything else now.

12. **Rearrange and resize the fields as you want.** Remember that to reposition a field without changing its size, anchor it to the top or bottom as well as the left or right. To resize a field, anchor it to the top or bottom as well as both the left or right in most cases. You can also anchor it to all four sides, but if you do that with more than one field in the layout, you might wind up with problems.

    Figure 9.16 shows how you can position a field label that is not resized (left) and how you can position and resize the actual data entry field (right).

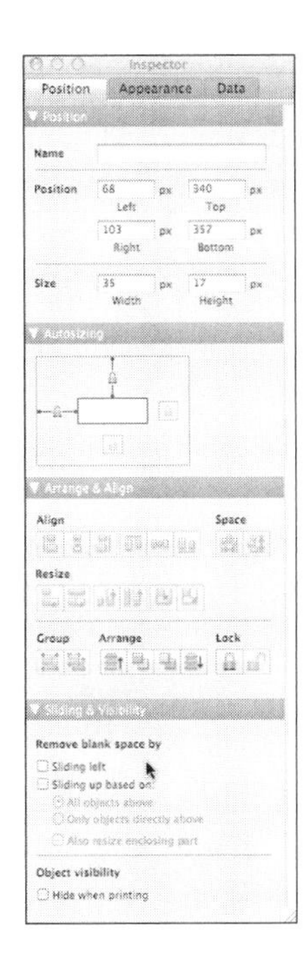

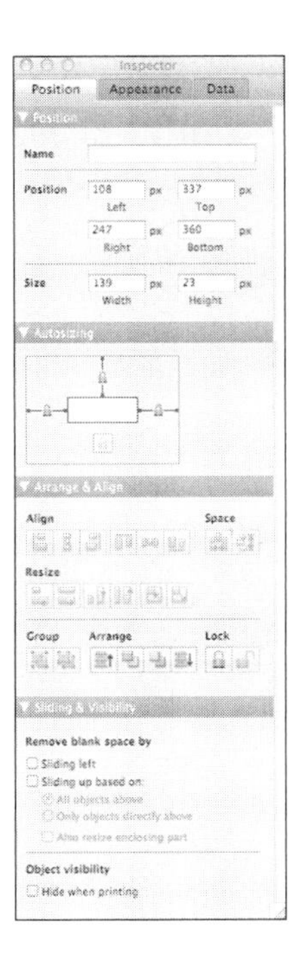

Figure 9.16
Set autosizing for fields and labels.

Building the Back Layout

The Back layout can be much simpler than the front layout. The one essential is a Done button. You can draw a standard FileMaker button using the toolbar and type Done on it, as you see in Figure 9.17. Placing it in the same relative position as the Info button on the front might help your users find it. With the Done button selected, implement it in the same way as you did the Info button. However, in this case, the layout you want to go to is Front layout.

The simplest way to implement the rest of the back might be to add graphical elements such as your logo and perhaps some styled text.

Figure 9.17
Implement the Done button.

Implementing the Info Button

Select the Info button on the Front layout and choose Format, Button Setup. Choose Go to Layout, as shown in Figure 9.18. The layout you want to go to is the Back layout you just created.

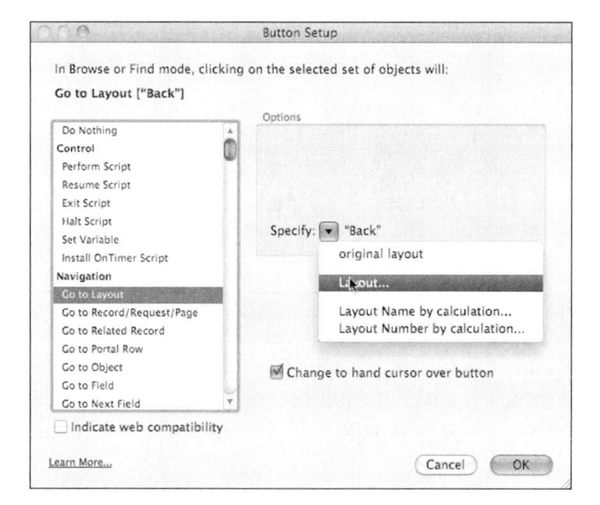

Figure 9.18
Implement the Info button.

Setting Up Security and Default Behavior

If you will be the only person using this database, you can leave things as they are. However, it is a good idea to implement a new user and privilege set.

How to Implement Customized Security

These steps help you implement customized record-level security:

1. **Choose File, Manage Security to open the dialog shown in Figure 9.19.** Click the Privilege Sets tab to edit privilege sets.

Figure 9.19
Manage security for the new database.

A privilege set enables you to set field-level security. You can also set security using custom menus as well as the menu sets (All, Minimum, and Editing Only) in the lower right of the window. Because FileMaker Go implements only a limited number of menus, setting field-level security explicitly might be the best way to manage your database if you need to worry about security.

2. **Select Custom Privileges in Data Access and Design for records, as shown in Figure 9.20.**

Figure 9.20
Create a new privilege set.

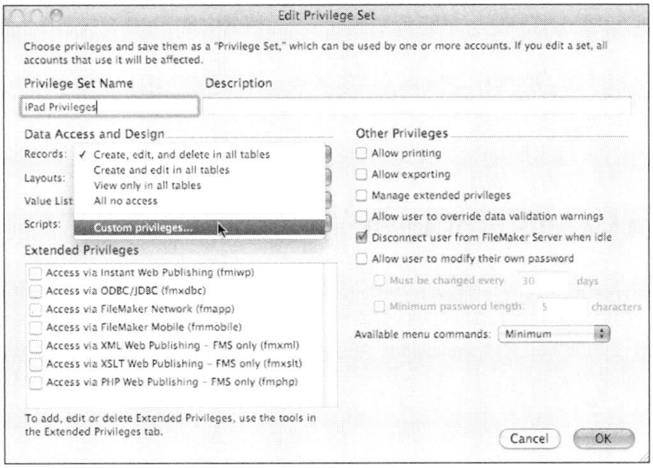

This opens the Custom Record Privileges dialog shown in Figure 9.21.

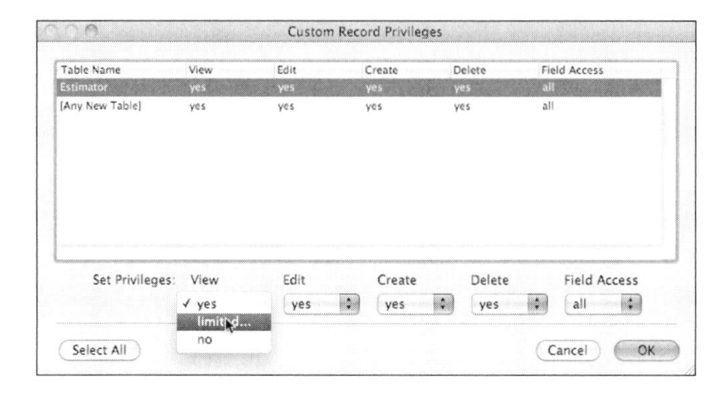

Figure 9.21
Begin to set custom privileges.

3. **Choose Limited for View or Edit, as you see in Figure 9.21.**
4. **This opens a standard calculation dialog.** Define a calculation that returns a True value if you want the user to be able to view or edit the record. You can use the values shown previously in Chapter 8 to allow (or disallow) access using a mobile device.
5. **If you have created a new user, use File, File Options to automatically log in that user.** You can also use a startup script in the same dialog to log in one user on a mobile device and another on a desktop device.

Testing and Revising the Project

Install your database on your iPhone for testing. Either email it to yourself and open the email on your iPhone or transfer it using iTunes. Test that everything works as you expect it to. Then, move on to the next step: Try to break it. Try entering an invalid number in the Price field, for example. Make certain that you can take pictures for the Picture field, as well as select images from your photo library on the iPhone.

Try everything with the phone in various orientations. Make certain that changing orientations does not distort data or images (if this happens, recheck your autosizing options).

Building a Catalog for iPad

If you have an existing database, why not move it to iPad? Whether you decide to download the database to users' iPads or let them connect to it over the web, your data can be viewed subject to any constraints that you provide with FileMaker. All that people need is an iPad and the FileMaker Go app.

THE ALEX SEPKUS CATALOG

This section is based on an actual database project implemented for Alex Sepkus, a jewelry firm located in New York City. That project followed the trajectory described in this section. A large FileMaker database with years of data runs on an office server to support the company's local area network. The basic goal was to implement the catalog on iPads to be used in jewelry stores.

The initial implementation was done in Bento. The relevant data from the FileMaker database was exported into Bento. This enabled users to experiment with various interface ideas themselves. It also addressed what can be a common issue, particularly with database images from old databases. The resolution and screen size of iPad might reveal imperfections in the images, such as specks of dust that were never seen in printed catalogs even if they were printed on high-quality paper using the best inks.

The Bento implementation served as a prototype and enabled implementation of the FileMaker Go solution to proceed separately. Although this was necessary for this project (Bento existed on iPad, but FileMaker Go was not yet available), the idea of prototyping an iPad project quickly with Bento and—most important—reviewing database data and images on iPad's high-resolution screen remains a good solution in many cases.

Note that the downloadable code for this example is a simplified version of the production catalog shown in Figures 9.2 and 9.3.

The screenshots in this section are used with permission of Alex Sepkus.

These are the steps involved in building the catalog:

- **Create or modify the database with FileMaker Pro.** Whereas the estimator project provides an empty database into which users can add data, a catalog starts from an existing database. This step describes some changes you might make to an existing database.
- **Build the layout header part.** In order to provide a standardized look, you might want to create a consistent header part in your layouts.
- **Build the list layout.** This is a standard FileMaker list layout designed for iPad.
- **Build the form layout.** This is a standard FileMaker form layout designed for iPad.
- **Set up security and default behavior.** Create an On Open script to start a new estimate.
- **Test and revise.**

Working with an Existing Database

When you are working with an existing database, it is not easy to provide a step-by-step guide to this first part of the process. The steps depend on what the current database looks like and what you want to do with the data in your new catalog project.

What follows in this section is a list of items to consider. Pick and choose which ones apply to your project, and feel free to free-associate to think of new issues to consider:

- **Public data.** If the data in your database (and ultimately in your catalog) is public data that is subject to no security constraints, you do not have to worry about security.
- **Live data.** If your data is subject to change both from your mobile device and from other users, and if those changes need to be reflected immediately to all users of the database, it is hard to escape sharing the database using FileMaker Pro (for up to ten simultaneous users) or FileMaker Server (for more than ten simultaneous users). If your users are mobile and may access the database at any time, this generally requires hosting the database either on a hosting service (search filemaker.com for "FileMaker hosting") or on your own network.
- **Securing data.** If some of your data is confidential (profit calculations, for example), you can restructure your database so that the confidential data is in a separate file (not just a separate table) that is related to a table in the primary database. You can then include the less-secure data in a FileMaker database that is accessed by FileMaker Go either locally on the iPad or over the network. Alternatively, consider using the security shown previously in this chapter in the "Setting Up Security and Default Behavior" section of the estimator project.
- **Preventing copying and printing.** The Save/Send command from Actions (the gear in the top right of the FileMaker Go interface) enables users to send the database via email to anyone. Turn this off in File, Manage, Security by editing or creating a new privilege set, as shown in Figure 9.19. In Other Privileges at the right, turn off Allow Printing and/or Allow Exporting if necessary.
- **Decide on the type of interface.** If you are providing mobile access to a database that people are used to working with, the FileMaker Go interface is perfect; it implements many familiar FileMaker Pro concepts. If you are using a different type of interface, you might need to allow people to enter data into a global field that are used in scripts (this technique is illustrated in the following sections). Thus, although you might think that you do not want people to edit data in the database file, in fact you might be forced to handle this on a field-by-field basis.

Building the Layout Header Part

Think about the controls you want to have in the header part for your list and form layouts. To provide a consistent look, they should be fairly similar. In this example, they are identical with one important difference. From the list view, you can click any record in order to go to its form layout. From the form layout, you need a button to return to the list layout. The button to go from the list layout to the form layout is part of each record in the list.

If you want to use a consistent header part, create it as part of a layout and then duplicate the layout.

The layout shown in Figures 9.2 and 9.3 has a title in the middle of the header part and two interface controls.

On the left, a pop-up menu enables you to browse specific sections of the database. You select the section that you want to see with the pop-up menu and a script trigger automatically finds the relevant records and sorts them. Desktop interfaces often let you choose a category or set of records and then click a Go button to carry out the action. In this environment, choosing a pop-up item can be sufficient to start the process.

On the right, a search field lets you search for a word or phrase. This enables you to hide the Status bar at the bottom of the FileMaker Go window if you want to. It also lets you customize the behavior of searching and controlling the database. You can use these interface elements as jumping-off points for your own. (And, of course, you might prefer to use the out-of-the-box FileMaker Go tools, which is definitely faster and simpler for you.)

How to Create a Common Header Part

The common header part appears at the top of both the list and form layouts. You do this work in FileMaker Pro (or FileMaker Pro Advanced):

1. **In Layout mode, set the graphic ruler units to pixels.** See step 1 in "Create a Basic iPhone Layout" earlier in this chapter.
2. **Create a new layout using Layouts, New Layout/Report.** Name the layout something meaningful, and for the layout type, choose Blank Layout. You build the header inside this layout and then duplicate the entire layout, so you might want to call this one List (or Form). After it is duplicated, you can change the name of the duplicate to Form (or List).

   By default, the new layout has header, body, and footer parts.
3. **Create a graphic for the header background.** This can be a rectangle you create in FileMaker, or you can use Photoshop or a similar tool to create an image with a vertical gradient as described in Step 2 of "Create a Basic iPhone Layout" earlier in this chapter.

   This example uses an iPad. The screen is 768 × 1024 pixels.
4. **Set autosizing as in Step 3 of the previous steps.**
5. **Create the body background as in Step 4 of the previous steps.**
6. **Create a sort script.** The pop-up menu and the search field both automatically sort their results. Create a sort script to do whatever sort you want to use. (Yes, this means that you are determining the sort order rather than your users. In some cases, that is entirely appropriate because some users only use a single search order—alphabetically, by part number, or the like.)
7. **Create a new database field to support the pop-up menu.** This is a global text field, as shown in Figure 9.22. Create the field as a text field, and then use Options in the lower right to set the global storage. The value selected in the pop-up menu is stored in that field, and a script trigger then carries out the canned search and the sort.
8. **Create a value list for the pop-up menu items as you did in steps 4 and 5 of "Set Up the Database" earlier in this chapter.** The value list consists of the catalog categories. For jewelry, these might be necklaces, bracelets, and rings; for a catalog of books, the categories might be fiction, non-fiction, and biography.
9. **Create a script that is triggered when a value is set in the pop-up menu.** The script is shown in Figure 9.23. It conducts a search based on the value in the pop-up menu, and it runs the sort script you created in Step 7 of this section.

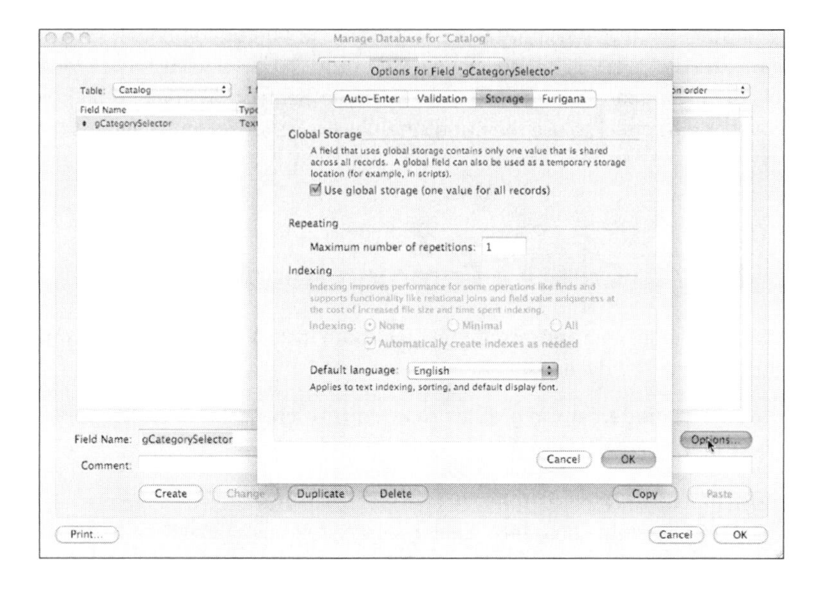

Figure 9.22
Create a global text field for the pop-up menu.

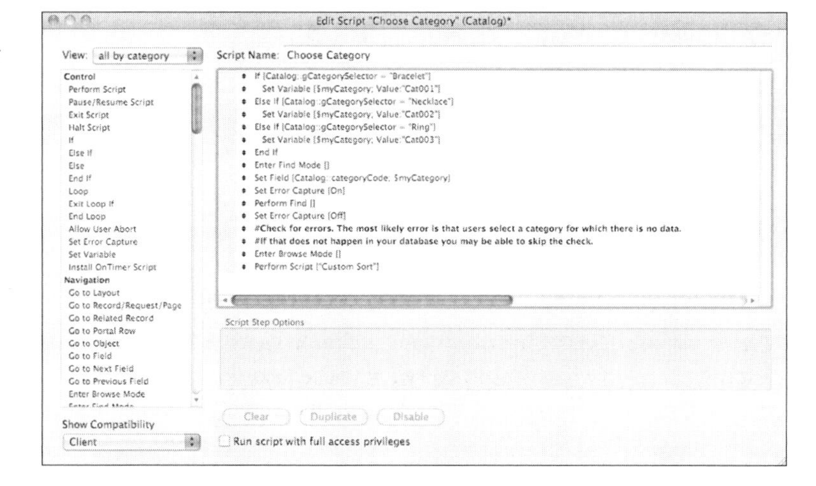

Figure 9.23
Create a script to support the pop-up menu.

10. **With the pop-up menu selected, choose Format, Set Script Triggers to open the dialog shown in Figure 9.24.** Choose On Object Modify, and, on the next screen, select the script from the previous step.

11. **For the search field, create a global text field. Display it as an edit box and place a search button next to it.** Implement a script to be attached to the search button. This script can perform a QuickFind and automatically sort the data. Alternatively (and potentially faster), if you know what field should be searched on, you can set that field and do a standard find.

    The built-in FileMaker Go interface lets users choose QuickFind or specific fields to find. Depending on your data and your users, the field to search on might be obvious or you might need the flexibility that allows users to customize their search. In fact, you might need both.

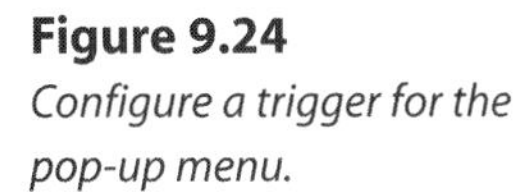

Figure 9.24
Configure a trigger for the pop-up menu.

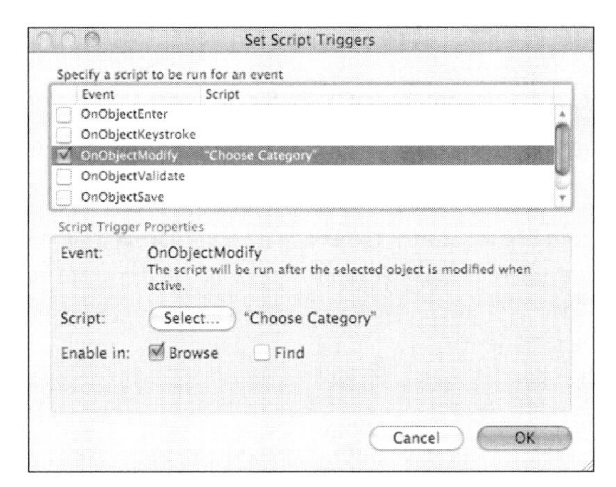

Building the List Layout

Copy the layout with the header part in it and focus on the body part. Add whatever fields are appropriate for the list. Remember that the form layout lets you show much more data. The more data you put into the list layout, the fewer items can be displayed on a single screen. Most people prefer more records on a screen with less information on each one because a tap can take you to the detail form layout.

The rules are the same as in all autosizing rules:

- To move a field (including a container field), anchor it to a side and top (or bottom) of the layout.
- To resize a field, anchor it to three or four sides of the layout.

Select some (or all) of the elements in the list body part and make them hot so that a tap takes users to the form view. You do this by selecting the elements and choosing Format, Button Setup and choosing Go to Layout. This sends you to that layout for the given record in the list. If you want, you can choose to perform a script, but frequently you just want to go to the form layout.

Building the Form Layout

If you have not already done so, duplicate the header layout and change the duplicate name to Form layout or some other meaningful name. In Layout mode, use Layouts, Layout Setup to change the name.

Construct the body part of the form layout as you normally would. The only additional item to add is a button in the header to go back to the List view.

Testing and Revising

Test is the same as for the iPhone estimator project. Test that everything works properly, and that when you deliberately make errors that they are handled correctly.

Further Steps

The examples in this chapter provide a jumping-off point for many other projects. Here are some suggestions:

- **Implement an Add button to create new records.** You can also use the Status toolbar at the bottom of the FileMaker Go screen, but you might want to perform special scripts as you add new records.
- **Experiment with placing the estimator database on a server so that you gain experience with using FileMaker Go and server databases.** You can serve the database from your copy of FileMaker Pro (on a Mac or Windows computer). This way, when it comes time to work with server-based databases, you have had the experience.

10

IN THIS CHAPTER

- Expanding the Reservations Example
- Implementing Printing Features
- Adding Charting to the Reservations Example

Using Printing and Charting with FileMaker Go

For many years, the heart of computer interfaces has been What You See Is What You Get (WYSIWYG, pronounced wizzy-wig). The concept became very important as graphical user interfaces and high-quality graphics became more and more prevalent (this was in the 1980s). With WYSIWYG, what you saw on the screen was what would be printed out on your printer. (You really don't want to know what printing and displaying was like on personal computers before WYSIWYG.)

WYSIWYG is a great concept, but particularly when you are dealing with large amounts of data, it is not a perfect solution. Printing out something that tries to exactly match what is shown on a computer display really can never be WYSIWYG. When it comes to colors, printed colors are reflected light, and on a computer display, color is transmitted. They are two different physical processes, as you see when you combine all the colors in the spectrum. For transmitted light, the result is white, but for reflected light (ink and paint, for example), the combination of all colors is black.

More to the point, how do you print interactive elements that appear on a screen? Scrollbars work on a screen, but there is no printing technology in existence that makes scrollbars printed on a piece of paper do anything other than just lie flat on the printed page.

In part because databases, including FileMaker, tend to deal with large amounts of data, it is the database world that has often pushed new technologies for printing and displaying data. For databases, the challenge of printing is far greater than the challenge of printing a one-page flyer for a yard sale.

FileMaker Pro has built-in features for handling the presentation of data for printing. Among these features are the following:

- Page numbering
- The ability to dynamically change the size and location of data elements depending on the size they need to display properly
- Options to omit some items from printing (this is used, for example, to omit buttons on the screen that say Click Here for More Details)
- Creation of charts from data in a database

In all of these cases, you use FileMaker Pro to set up your database; FileMaker Go then obeys those instructions and prints the data properly. This chapter focuses on what you need to do on the FileMaker Pro side to get the most out of your FileMaker Go printing.

Adding More Features to Reservations

Databases have powerful reporting tools because they so often need to present complex and voluminous amounts of data. The Reservations example is a good start for you to use to explore FileMaker Go and its features, but in order to use it to drive reports, it needs a bit more complexity. (Don't worry; this is exactly the type of complexity that you often use in real-life solutions.)

As presented in the first part of this book, Reservations lets you keep track of events and attendees. This is a one-to-many relationship between events and attendees, which means that for a specific event, you can have many attendees. For each attendee, you provide the name, email address, and phone number. You also provide the number of reservations and the number of people who actually showed up.

If you play around with the example for a little while, you see that there is a problem: You need to enter the name and phone number for each attendee of each event. If someone attends two events, you have to re-enter all the information. As a result, it is very easy to have inconsistent data from the two data entries.

This is a very common database situation, and there is a simple solution. What you want is a many-to-many relationship that lets many events relate to many attendees. You enter each event once, and you enter each attendee once. Because there is only one entry for an event or for an attendee, you do not have to worry about inconsistencies. A single database record cannot be inconsistent with itself.

You implement a many-to-many relationship with a third table—a *join table*. A join table provides a link to the event table and another link to the attendee table. Together, those links define a single reservation: a reservation for an attendee at an event. In addition, information about the reservation itself is stored in the join table. The date of the event and the name of the attendee are stored in the event and attendee tables along with other information, but the number of reservations (and the number of people actually showing up) is stored in the join table. That is because that information is part of the join and not part of either the event or the attendee table.

How to Add a Joins Table to Reservations

You need to not only add a Joins table but also modify the interface:

1. **Using FileMaker Pro on your Mac or PC, open Manage, Database and select the Tables tab.** Create a new table called Joins and add the fields shown in Figure 10.1.

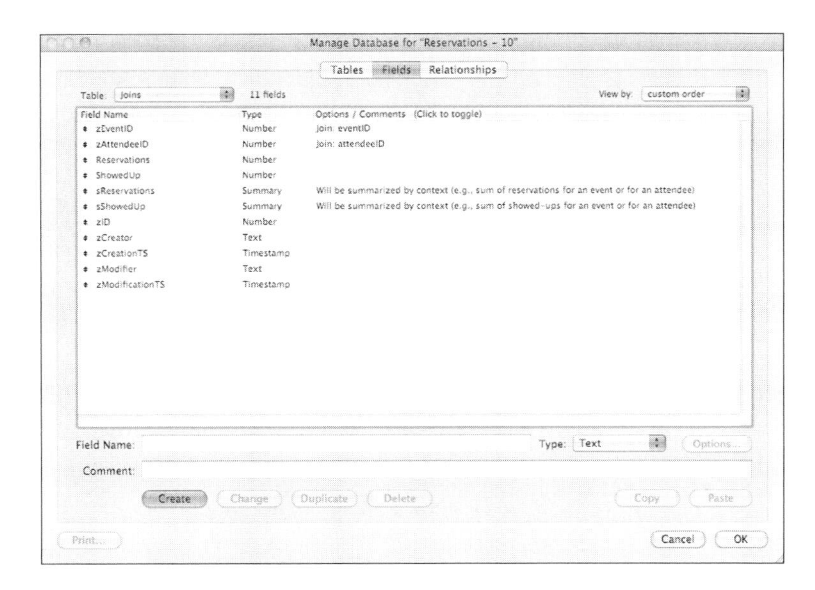

Figure 10.1
Build a join table.

Naming the tables is not difficult: Events and Attendees basically name themselves. If you approach the example from the database side, the many-to-many table is Joins. If you are approaching it from the side of the actual solution, you might call the many-to-many table Reservations, because that is the real-world object the table stores. Some database designers use a naming convention in which join tables have some indication in the database structure. If there were dozens of tables in this solution, having a table called join_Event_Attendee would make the matter clear.

2. **Delete unneeded fields from Events and Attendees.** As you can see from Figure 10.1, the Joins table now contains the fields for the reservations and attendees along with the corresponding summary fields that provide context-sensitive totals (that is, totals for the event or for the attendee on which the fields from the Joins table are placed). Because these four fields are now provided in Joins, you can remove them from Events and Attendees.

3. **In the Relationships Graph, relate attendees::zID to joins::attendeeID and events::zID to joins::eventID.** The result is shown in Figure 10.2.

4. **Change the portal in the Events layout to use the Joins table.**

   The Events Layout from the first version of the Reservations example contains information about the event at the top; a portal displays all the joined records from Attendees. Now that you have a Joins table, that portal needs to be changed to display records from Joins rather than Attendees. Just go into Layout mode, double-click the portal, and change its base table, as shown in Figure 10.3.

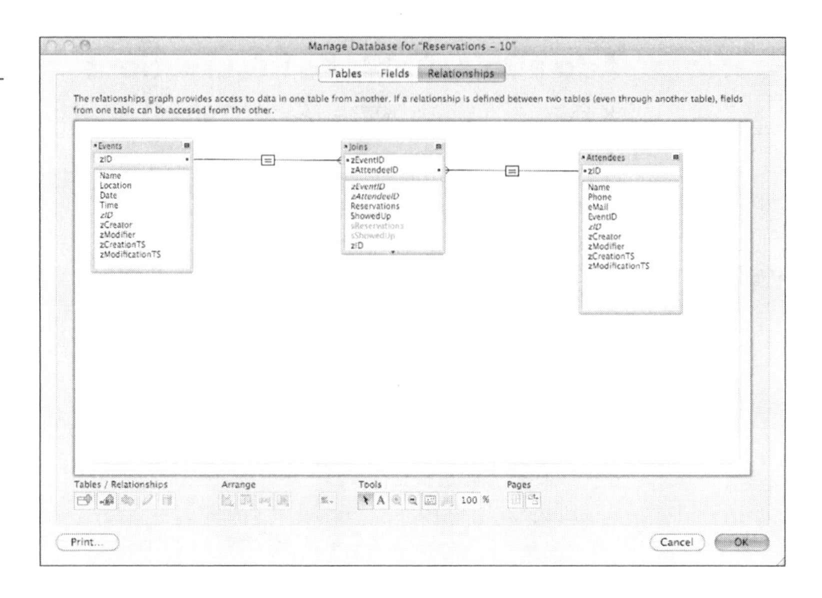

Figure 10.2
Create the Joins table relationships.

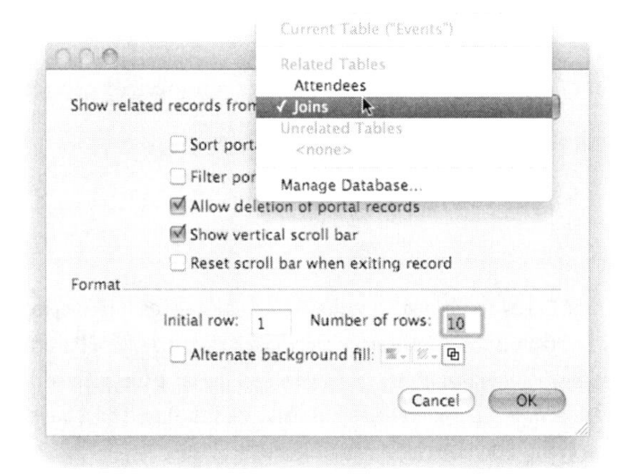

Figure 10.3
Use the Joins table to power the portal on Events.

5. **Create a value list for Attendees.** The Attendees table is related to the Joins table, so the other fields in the portal are still valid with one exception. Because the Joins table relies on the ID numbers for attendees and events, instead of the name of each attendee, you need that ID. That is the Attendees::attendeeID field. You need a way to let people enter that number.

   Using Manage, Value Lists, create a new value list called Attendees, as shown in Figure 10.4. Choose the first radio button (Use Values from Field) and click Specify Field.

6. **Set the value list to display values from the attendees::name field but to set values for attendees::zID, as shown in Figure 10.5.**

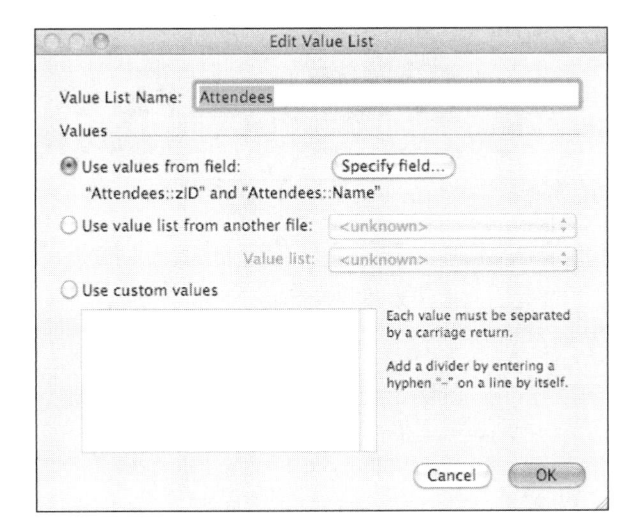

Figure 10.4
A value list for attendees.

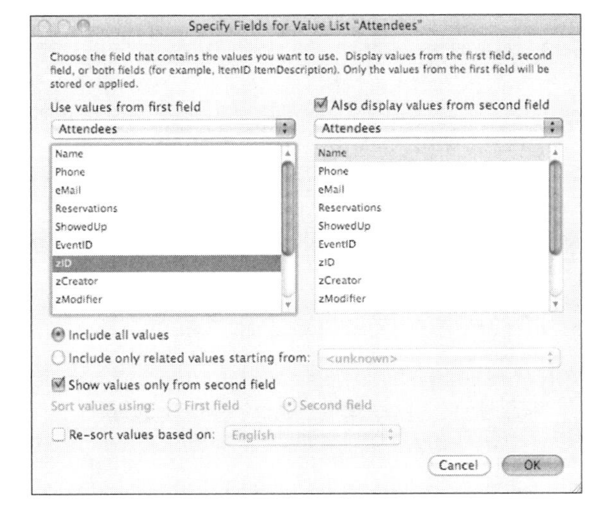

Figure 10.5
Configure the value list to display attendee names and set attendee zIDs.

7. **In Layout mode, attach the value list to the attendees::zID field, as shown in Figure 10.6.**
8. **Test the interface in Browse mode, as shown in Figure 10.7.**
9. **As a final step, go to Layout mode and use Layouts, Layout Setup to name this layout FMGo Events.**

Follow these steps to create a new layout that shows an attendee and a portal with its events, as shown in Figure 10.8. Name that layout FMGo Attendees.

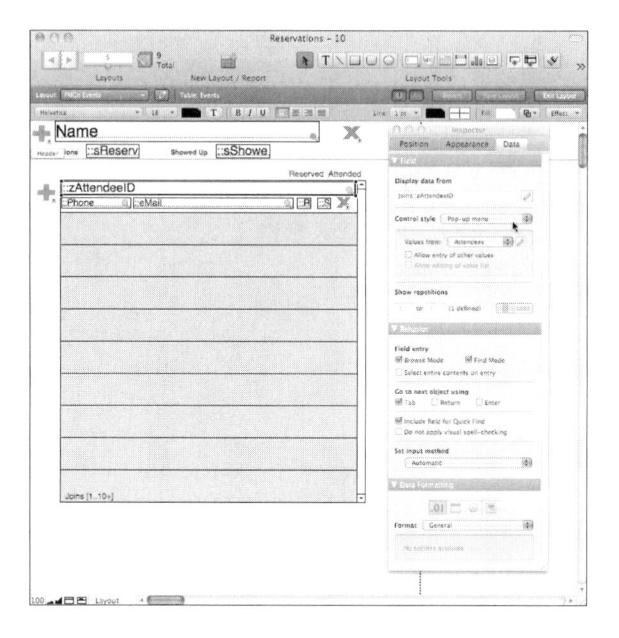

Figure 10.6
Attach the value list in Layout mode.

Figure 10.7
Test the interface.

In the two layouts you have already built, you have a layout that displays an event along with a portal containing its attendees (FMGo Events) and a layout that displays an attendee along with a portal containing the events that attendee is signed up for (FMGo Attendees). The Reservations example is becoming very slightly more complex, but it now is much closer to being a real-world solution.

What will make it even more useful (and easier to use for testing and experimentation) is a pair of layouts that present lists of attendees and events. This enables you to switch from a list of all events or attendees to the details of a specific event or attendee. Building those lists is not difficult, and it sets you up for the main event of this chapter—using FileMaker's printing features.

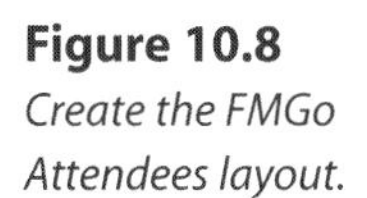

Figure 10.8
Create the FMGo Attendees layout.

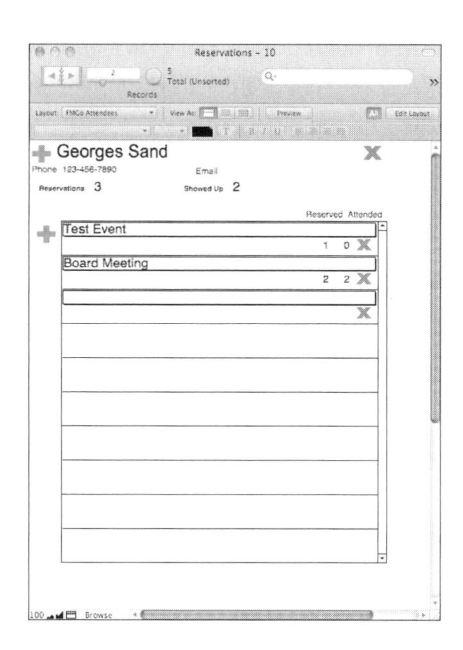

How to Build an Event (or Attendee) List

The structure of an event or attendee list is simple because you can use FileMaker's build-in List view. Here is how to build the event list; the same technique can be used for the attendee list:

1. **Using FileMaker Pro on your Mac or PC, enter Layout mode and choose Layouts, New Layout/Report to open the dialog shown in Figure 10.9.** Name the new layout, and select the Joins table as the basis for the layout. Choose the List layout.

   You can modify all these settings later on by going into Layout mode, so do not agonize over them at this point. Also, note that in the assistant, you are able to add fields from other tables to the layout.

   Click Next.

Figure 10.9
Use the assistant to create a new layout.

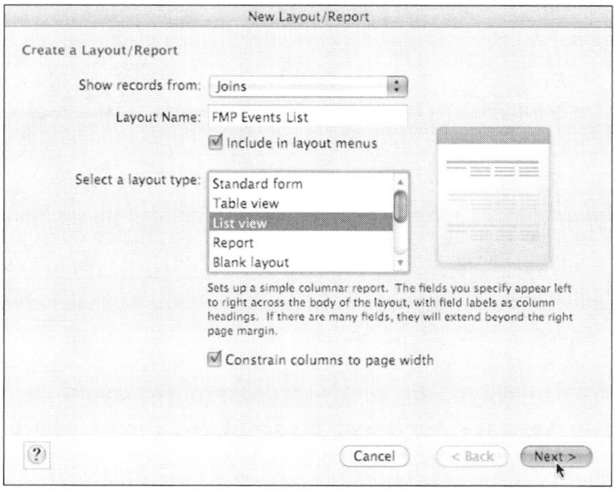

2. **Select the fields you want to add from the report, as shown in Figure 10.10.** The Joins table is the base table, and the only fields from that table you want are the two summary fields.

   When you build an attendee list, you still use the Joins table as the base, and you add these same fields to the attendee list.

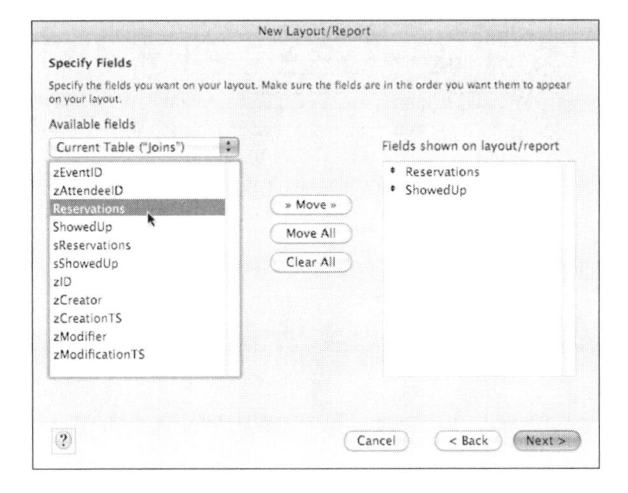

Figure 10.10
Add fields from the base table.

3. **Use the pop-up menu to move to the Events table, and move fields such as the name, location, date, and time to the layout, as shown in Figure 10.11.**

   When you create an attendee list, you use fields from the Attendees table instead of the Events table in this step.

   Click Next.

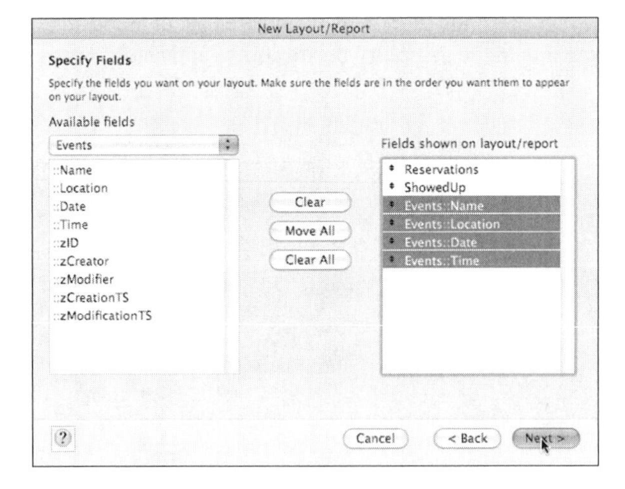

Figure 10.11
Add related fields.

4. **On the subsequent screens, you can set a sort order for the layout, choose a theme, specify a header or footer, encapsulate everything in a script, and choose whether to see the results in Browse or Layout mode.** Remember that you can always come back to these settings, so for now you might want to just click Next and, on the last screen, Finish.

5. **The layout is created with a header and, beneath it, a body for a single record in the list.** Rearrange the fields as you see in Figure 10.12. Remember that the body is repeated for each element in the list so you normally want to reduce its size significantly, as you see in Figure 10.12. There is not much point to a list if each record in the list takes up the full screen.

   Although you are reducing the size of the body part, remember that if this list is going to be displayed on a mobile device, you probably want to enlarge the font size. (You can decide at this point whether to make separate layouts for mobile devices and for FileMaker Pro.)

Figure 10.12

Lay out the list body.

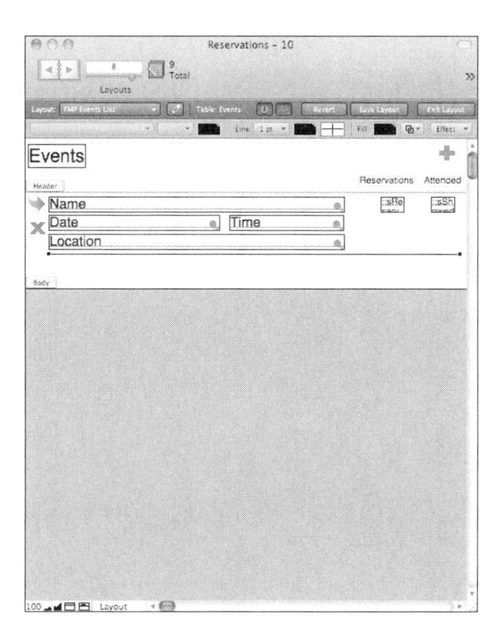

6. **Add any other buttons and elements you want, as shown in Figure 10.12.** Normally, buttons that affect the entire list (that is, buttons to add new events and summary fields) belong in the header. Buttons that apply to a single record (such as a button to go to the FMGo Events layout or a button to delete that record) belong in the body. They are repeated for each record in the list automatically.

   You might want to add a dividing line at the bottom of each record. In Figure 10.12, that line is moved down a bit and the body is deeper than needed in order to show those elements in the figure. In practice, you probably want the dividing line to be one pixel below the bottom of the last field in the body, and you do not want extra space beneath it. Remember that the extra space appears on every single record in the list.

Implementing Printing Features

Although you can use basic printing to print the display, using the FileMaker printing features makes your printing much more efficient and makes it look better. After you have a solution such as the Reservations database described here, you can make a few adjustments so that printing works optimally.

To adjust printing settings, start from the layout that you want to print and enter Layout mode in FileMaker Pro on your Mac or PC.

How to Hide Interactive Elements

Those buttons that let you navigate to other layouts are terrific when you are sitting at a computer (or holding a mobile device in your hand). But when you are printing a report, there is no reason to show those interactive elements. Here is how to hide them.

1. **Open the Inspector with View, Inspector or ⌘+I (Mac) or Ctrl+I (Windows), as shown in Figure 10.13.**

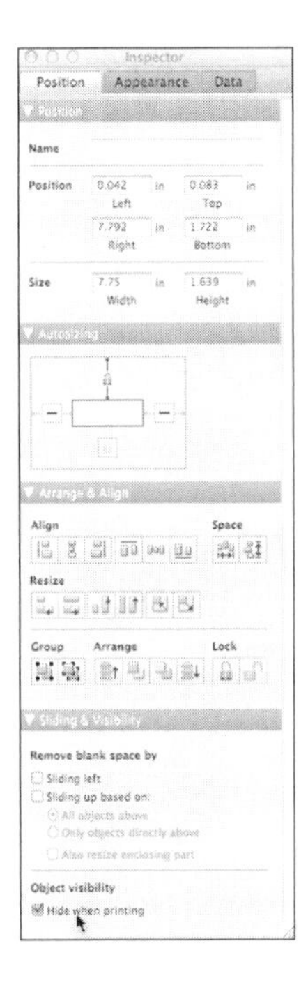

Figure 10.13
Open the Inspector.

2. **Select the layout elements you do not want to print.**
3. **Select the Position tab.**
4. **At the bottom of the Inspector window, select Hide When Printing.**

If you go into Browse mode and enter some data, you can move to your list, as shown in Figure 10.14. Deliberately omit some of the data (such as the location), and you see that there is a lot of blank space in the list. You can remove that blank space.

You can allow data entry into a list layout, but be aware that adding field labels can pose some issues when you start to optimize for printing. Specifically, it is likely that the labels will appear next to blank fields. That might be what you want, but many people leave data entry to Form view for a single record so that the data itself without labels shows up in the list view. There are many ways to deal with these issues, but just remember to think about them in the context of your own solution.

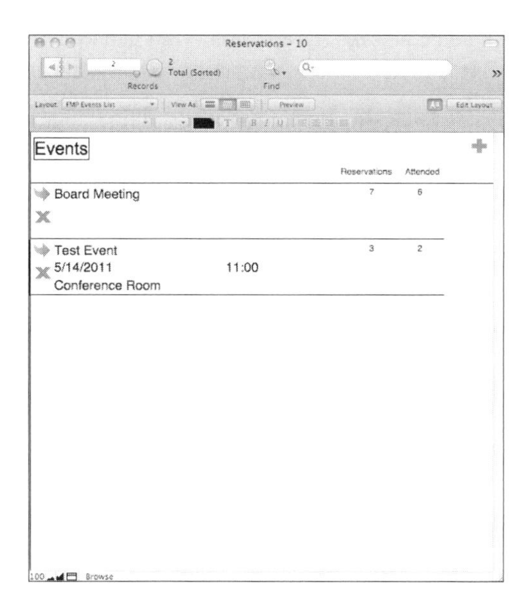

Figure 10.14
Layouts can contain unnecessary blank space.

How to Set Moving and Sliding

You get rid of the extra white space in a layout by using the moving and sliding settings. Start by opening the inspector, choosing the Position tab, and selecting the items you want to close up:

1. **Use the Remove Blank Space settings at the bottom of Sliding & Visibility on the Position tab of the Inspector, as shown in Figure 10.15.**

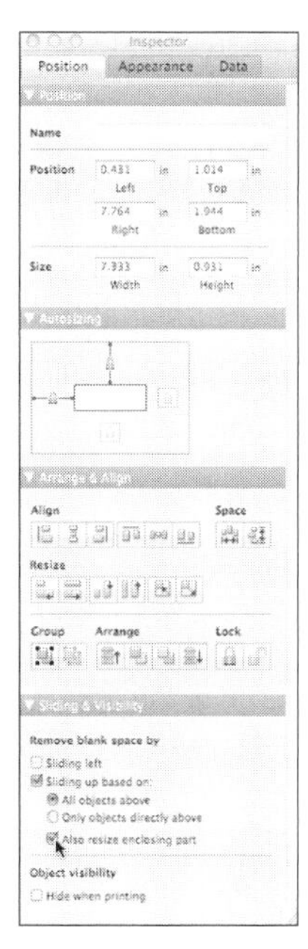

Figure 10.15
Remove blank space.

2. **Adjust the settings for your selected objects.** Be careful with sliding left. That can close up gaps, but if you have columns (such as the summaries for reservations and attendance), they are no longer aligned.
3. **Be certain to check Also Resize Enclosing Part.** This is the setting that actually removes blank space.
4. **Test your settings by going into Preview mode, as shown in Figure 10.16.** After you are satisfied, further test your settings by printing from an iPad or iPhone. That is the ultimate test.

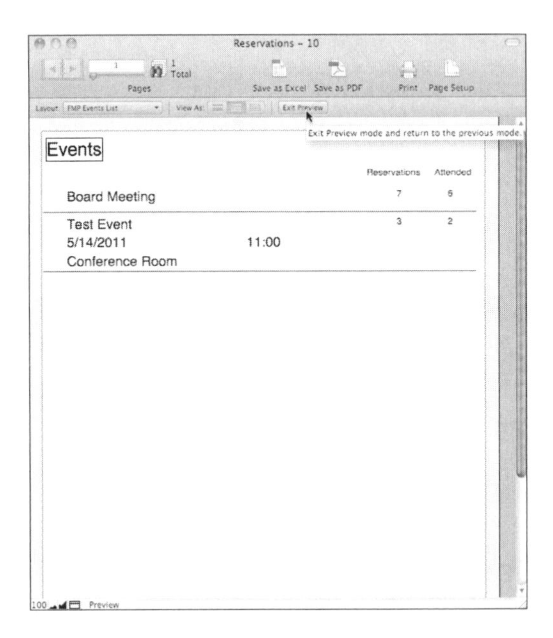

Figure 10.16
Test the results in Preview mode on FileMaker Pro.

Adding Charting to Your Database

You use FileMaker Pro to set up the printing features that you use on a mobile device. In a similar way, you can use FileMaker Pro to set up charts on your layouts. FileMaker draws very attractive charts, and they are even more attractive on an iPad than they are on a desktop display.

There is nothing that you need to do specifically to create a chart for your iPhone or iPad. FileMaker Go presents the chart based on the settings you have provided for your layout in FileMaker Pro.

FileMaker and the iPad are an unparalleled set of tools for displaying data in an understandable way.

→ Charting was introduced in FileMaker Pro 11. The definitive reference is available as a downloadable PDF document on the FileMaker site at http://help.filemaker.com/ci/fattach/get/21300/1268417531/redirect/1/session/L2F2LzEvc2lkL0VZTjdraG1r.

→ To make the most of your charting tools and the FileMaker data that can support them, you might want to read any of Edward Tufte's books on charting and data displays. These include *Visual Display of Quantitative Information*, *Envisioning Information*, and *Visual Explanations*. They can help you understand the role that charting can play in communicating information. It is much more than decoration.

Further Steps

The message for further steps in this chapter is much the same as it is in many other chapters in the book: Try it. Take your own databases, the sample databases for this book, or create your own databases and then explore the printing options. Similarly, experiment with charting. Here are some of the specific issues you might explore:

- **Sliding/printing with grouped objects.** If you use the Align menu to group objects together, they slide as the group. The most common group is the combination of a data entry field and its label, so you might start there.
- **Compare autosizing with sliding.** If you are using autosizing to move or resize objects to handle device orientation, think about whether those objects should also slide when they are printing. Sliding and autosizing are similar but different techniques.
- **Charting with test data.** If you are not familiar with charting, experiment with some test data to get a feel for how the different chart types work. You might find it easiest to do these experiments in FileMaker's Table view, which looks like a spreadsheet. You might also want to use Numbers (from iWork) or another spreadsheet program to explore the various possibilities.

11

IN THIS CHAPTER

- Introducing the Personal Database
- Exploring Bento
- Using Bento Libraries

Using Bento and Bento Libraries

FileMaker launched Bento in 2007. The first release (a public beta) has been updated since then, and in 2011, the current version is Bento 4. In some ways, Bento is like iPad: Both are innovative products, and people tend to describe them in terms of previous-generation products. This is scarcely a new phenomenon; radio was first described in its predecessor's context as "wireless," and in the early days, people understood that television was radio with pictures.

FileMaker's array of database products is centered around the traditional FileMaker Pro desktop application. For more demanding tasks, you can use FileMaker Pro Advanced or FileMaker Server, and for more focused tasks, you can use Bento. All the products bring a focus on databases and ease of use. In addition, all the products integrate well with mobile devices.

With Bento 4, there is a new aspect to mobility for Bento. Although it has always been possible to store location information in a database field, there is now a specific location field. This is a field that not only can store location, but it can be automatically updated by a mobile device with the device's location. After the field contains data, Bento can then automatically map it.

→ For more on location fields, see Chapter 12, "Using Bento Records, Fields, Forms, and Tables," **p. 269**.

This chapter introduces you to the basics of Bento. The underlying functionality is the same on Mac OS X, iPad, and iPhone. On Mac OS X, you have access to a few additional features than you have on the iOS devices, but do not worry: No matter where you have created or edited your Bento libraries, the libraries and the data are the same in all three environments.

As you saw in Chapter 2, "Introducing the FileMaker Architecture," FileMaker is a set of applications that work with the FileMaker architecture. The applications FileMaker Server, FileMaker Go, FileMaker Pro Advanced, and FileMaker Pro all read and write FileMaker files using the FileMaker architecture. In this book, *FileMaker* is used to refer to the file format and architecture. When features are identified as belonging to FileMaker, they can be accessed using several—or all—of the applications. Features that are available only through a single application (most often, this is FileMaker Pro) are identified in that way. As an example, if you want to write scripts, you use FileMaker Pro or FileMaker Pro Advanced. The scripts are stored in FileMaker databases and can be accessed by FileMaker Server, FileMaker Go, FileMaker Pro, or FileMaker Pro Advanced.

Working with a Personal Database

FileMaker describes Bento as a *personal database*. Although you can share and synchronize data with other users of Bento, it is primarily for one person, and your sharing might be with your various devices. Personal databases are not new; chances are you use several of them every day. Personal databases are centered on one person's data, and they typically focus on one specific task. With Bento, you have the ability to create and modify personal databases for whatever task you have at hand.

What are these databases that you use every day? They might be your contact manager, your date book, or even the page cache in your web browser. The database is there, hidden inside the software that you are using. Most people pay little attention to how the software implements its functionality—if you are worrying about the implementation, chances are either that you are a developer trying to learn new techniques or that you are a frustrated user trying to figure out why it is not doing what you want it to do.

WHAT IS IT ABOUT DATABASES?

Just as some people are not comfortable with numbers, others are uncomfortable with the very concept of databases. For them, there is something perfectly normal and understandable about dealing with actual data, but the abstraction involved in manipulating a database can be daunting.

Bento comes with templates you can use as-is, and you might never need to think about databases at all; you just enter your own data. If you tend to be database-phobic, have no fear: You do not need to think about the database concepts if you do not want to. It might make you more comfortable to know that you use databases every day without knowing it (and without getting your hands dirty), and perhaps that might encourage you to dip your toe into the database waters using Bento. But you don't have to do so; just pick a template and enter your data if you want.

With Bento, there is another aspect of the concept of a personal database that is important. Bento users often create their own databases. Do not panic; creating a database with Bento often consists of choosing File, New Library From Template and then clicking on the template you want: Recipes, Inventory, Expenses, Event Planning, Student List, or any of the others that are built into Bento.

If you want to, you can use File, New Library From Template Exchange. This connects you to the Bento Template Exchange on FileMaker's website, where you can choose a template that has been contributed by other Bento users (you can find the Template Exchange from your browser at http://www.filemaker.com/templates).

Whether you are using a built-in template, a template from the Template Exchange, or a library of your own making, you are probably the person who both chooses the specific library and who enters the data. In the world of databases outside these personal databases, it might be someone else who tells you which database to use and who makes the choices about its design and functionality.

With Bento, it is up to you, and that means you can do what you want. If you change a database layout, you will not have people yelling and screaming at you (or applauding you) because something is different. It is true that you can share your Bento libraries with a few people, but most often they are members of your family or work group.

Because you can do what you want, you can do something that is considered bad practice on other databases. You can make changes to the database layout and structure as often as you want to. You can refine your approach to storing and using your data as you learn more and as your needs change.

For example, if you are using the built-in Projects Bento template, as you go along you might find that the Project Name field is not big enough to enter a new project. Centennial Parade and Community-Wide Yard Sale does not fit in the Projects template. There is no real problem with this. Bento lets you type in the text, as you can see in Figure 11.1; as soon as you click out of the field, Bento truncates the text with three dots, as shown in Figure 11.2.

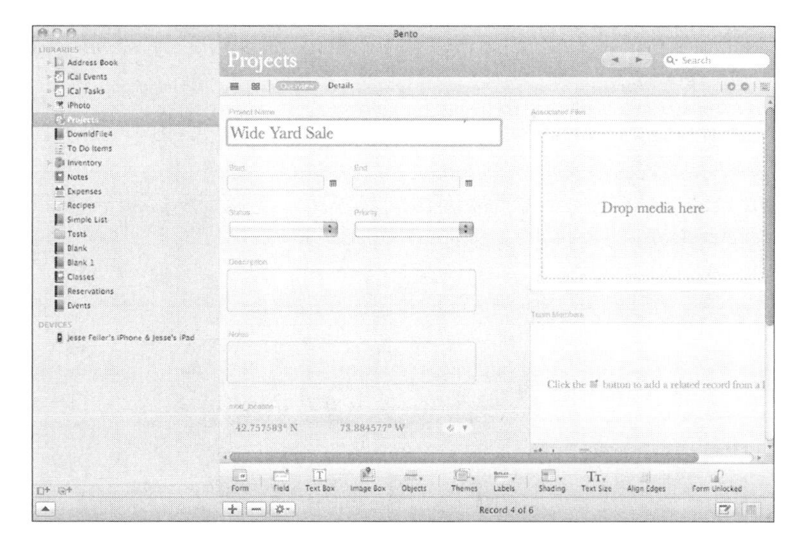

Figure 11.1
You can type in text that is too big for a field.

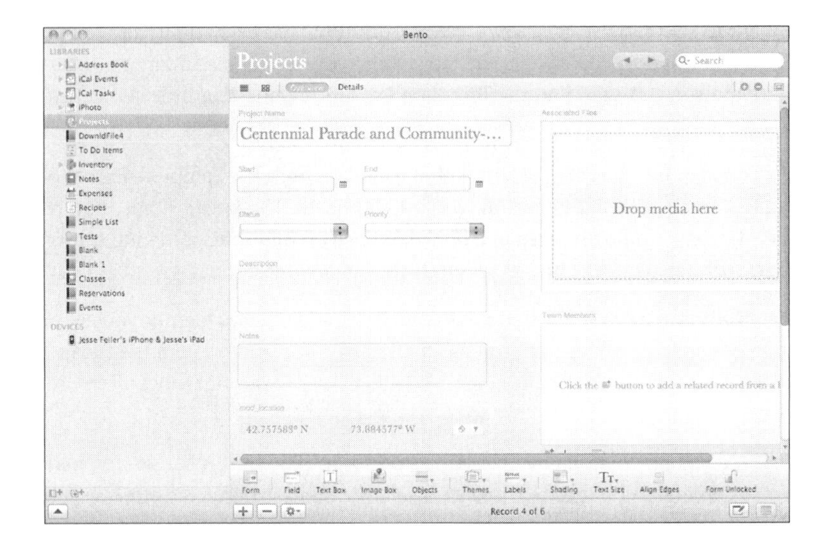

Figure 11.2
Bento automatically formats too-long text.

Because most of the time you are the main user of the database, you can now choose what to do:

- You can leave things as they are. It is your data, and you probably know or have a hunch that project names this long will be a rare occurrence. Doing nothing is a great option in many cases.
- You can enlarge the text field. Click on its border, and you see resizing handles, as shown in Figure 11.3. Resize the field, and Bento automatically moves the other fields out of the way. Doing nothing is a great option in many cases, but having Bento do the work for you by rearranging the fields on the form is a close second.

Note that with Bento, you can modify field layouts as you continue to view the data in the field.

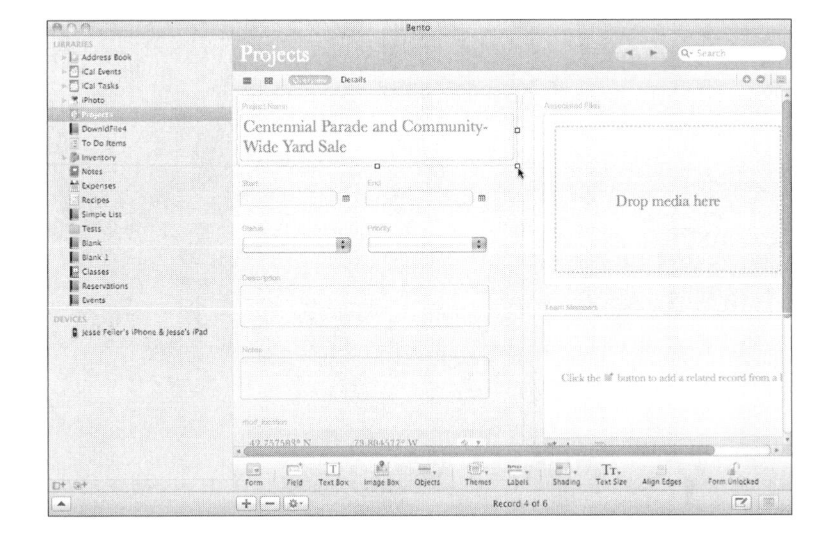

Figure 11.3
Rearrange and resize Bento fields.

Because Bento is designed for easy and repeated modifications, you can use it not only to help you organize your life, but also as a prototyping tool for databases that you might implement later on in FileMaker itself.

This section has illustrated just a few of the things you can do with Bento as you use it and modify your libraries. The rest of the chapter provides more details to help you use Bento.

Talking Bento

Bento has its own terminology (do not worry; it is not extensive or complicated). In fact, wherever possible, Bento uses common language for common concepts. Some of these common terms and concepts have very specific meanings in Bento, as described next.

Database

Bento has its own database—one for each user account on a Mac. This means that if you have an account, your mother has an account, and your neighbor who comes over to use your Mac a lot also has an account, there are three accounts on your Mac, and no one's files are visible to anyone else unless they explicitly permit it.

Each of those users can use his own copy of Bento. The first time a user runs Bento, it creates a library in the user's own directory (called the *Home* directory). In the Finder, your Home directory is shown in the Finder window in the sidebar, as you see in Figure 11.4.

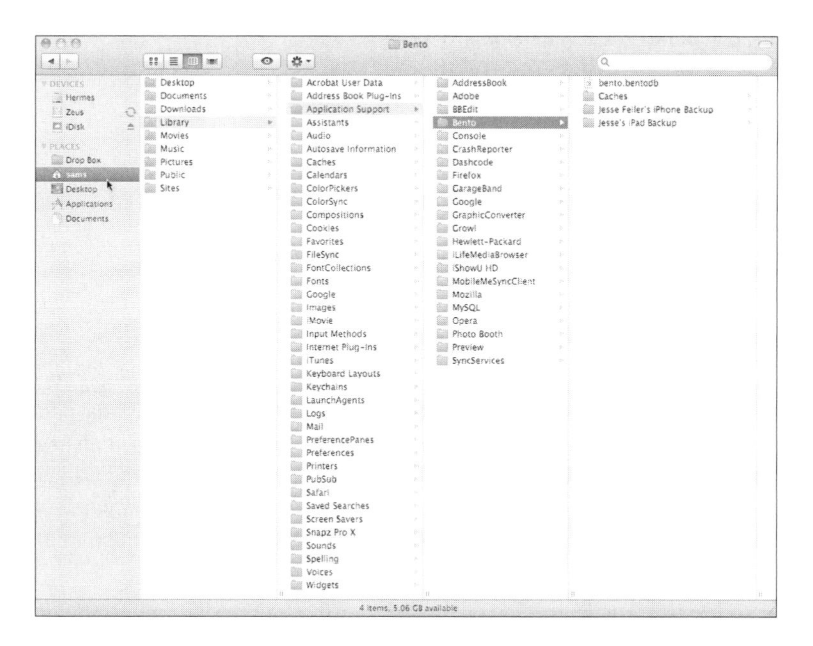

Figure 11.4
Bento stores its database in your Home directory.

You can also get to your Home directory in the Finder by choosing Go, Home.

The Bento files are stored in a subdirectory: Home, Library, Application Support, Bento. Do not move them or rename them; you will break Bento.

When you back up your Bento data (as you should do regularly), you choose File, Back Up Bento Data, and you can specify the filename and location for the backup. If you need to restore from a backup, you choose File, Restore from Bento Backup and specify the file and location to use. You never go into the Bento folder itself, but it is important to know that it is there and that there is a separate folder in each user's Home directory.

In general, you never go into any Application Support folder unless specifically asked to do so by a technical support person. This also means that if you get the urge to clean up "unused" files that take up disk space, you bypass the files in Application Support. Cleaning up your hard disk makes sense only if you know what you are cleaning up.

Libraries

Inside your Bento database, you have *libraries*. A Bento library is pretty much the same as a *table* in a relational database such as FileMaker or MySQL. Furthermore, Bento libraries as well as relational database tables look like spreadsheets—they consist of rows and columns. You create new libraries in Bento using *templates*. These are libraries that are already put together and that might or might not contain data. Bento comes with several templates built into it, and the template exchange (described earlier in this chapter) lets you share libraries with other users.

If you want to start from scratch, you still start from a template: the Blank template. There is no New Library command in Bento. You always work from a template.

Records

Inside a Bento library (or a relational database table) are rows of data. They are frequently displayed like a spreadsheet. Each row represents one set of data values—another inventory item, another student in the class, or another temperature and humidity reading, for example.

Fields

The columns in a table represent specific data *fields*; in the world of relational databases, they are often referred to as *columns*. Either term is fine. A field can be something such as an inventory item's name or the temperature value for a specific reading. Whereas it is possible to rotate a spreadsheet, it is not easy to rotate a table or library. This usually is not a problem because you generally know what data you are storing and how you want to view it.

Collections

Within a library, you can create *collections* of records. A collection is just that: a set of records that you have selected into a group. One simple way of doing this is to create a collection in a Bento library and then to drag records into it. When you drag records into a collection, they remain in the basic table, but they also appear in the collection.

This can be a useful way of working. For example, if you have a library containing information about members of a club, you can create a collection called Hospitality Committee. Just drag the names of the relevant members to that collection, and you have an easy way of working with them.

Smart Collections

Throughout Mac OS X, you find examples of *smart* groupings of data, and Bento is no exception: You can create *smart collections*. A smart collection starts when you use Bento's search tool. You can search for records based on various criteria. When you have completed the search, you can save it as a smart collection.

What you are actually saving is the search itself and not the found records. If you search for all inventory items that are desks, you might find 18 different desks. Tomorrow, after some deliveries and some sales, you might find that the smart collection has 23 desks. It is the same smart collection (that is, the search criteria have not changed), but the actual contents of the smart collection have changed. Contrast this with ordinary collections that consist simply of the records you have placed in the collection; the contents of a collection change only if you add or remove records.

Working with the Bento Window on Mac

Bento has a single window in which you work. Most people have gotten used to having several windows open at a time even within the same application—you can edit several Pages documents in Pages, for example. A quarter of a century ago when the Mac was first introduced, this was a great feature. However, over time, things have changed. On the Mac itself, tools such as Spaces, Expose, and Dashboard help you manage your various open windows. In addition, more and more apps such as iPhoto work in a single window, and that window can take up the entire screen. (Starting in Mac OS X 10.7 "Lion," Mission Control combines these tools.)

When you open Bento, you see the Bento window displaying the most recent library you have used, as shown in Figure 11.5. When the Bento window is open, it always displays a library and its data. You cannot have a Bento window open that does not show a library (and chances are, you would not want to do so). To select another library, just click it in the list at the left.

The Bento window displays a single library, but it can display two views onto that library (a *split view*) as you see in Figure 11.5. At the top, each record is shown in a *grid view*, and at the bottom each record is shown in a *form view*. At the left, you can see the other libraries in your Bento database and the field inside the current library.

You can close the window; to re-open it, use Windows, Bento, (⌘+ 0 (zero)). The only other available command is File, New Library from Template Exchange, which opens the Template Exchange in your browser. From there, you can download a template and install it in Bento.

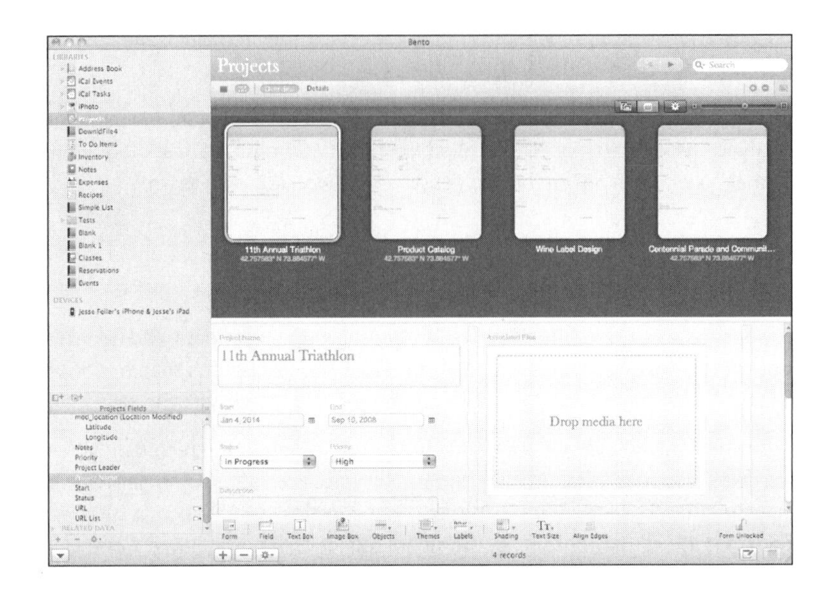

Figure 11.5
Bento uses a single window.

Working with Bento on iPhone

The Bento window on Mac OS X provides you with a single interface to manage your Bento data and libraries. On iPhone, there is not enough room for a comparable interface element. Accordingly, on iPhone, you have a Home screen and additional screens for managing your libraries and data. Figure 11.6 shows the Bento for iPhone Home screen.

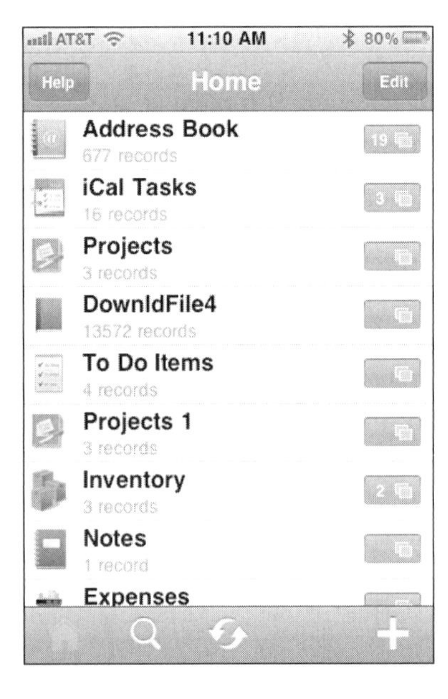

Figure 11.6
Use the Bento for iPhone Home screen.

To open a library from the list, tap its name. At the right of each library is a small image that indicates the number of collections in that library. Tap the collections image to open a list of collections.

At the bottom of most screens is a toolbar with four items on it, as follows:

- **Home** takes you to the Home screen, as shown in Figure 11.6.
- **Search** (the magnifying glass) takes you to the Search screen, as shown in Figure 11.7.
- **Sync & Setup** (the two rounded arrows) is described in Chapter 3, "Managing Data on the Move."
- **Add** (the +) at the bottom right lets you add a new library when you are on the Home screen. When you are looking at records within a library, + adds a new record to that library.

Figure 11.7

You can search in all your libraries at one time.

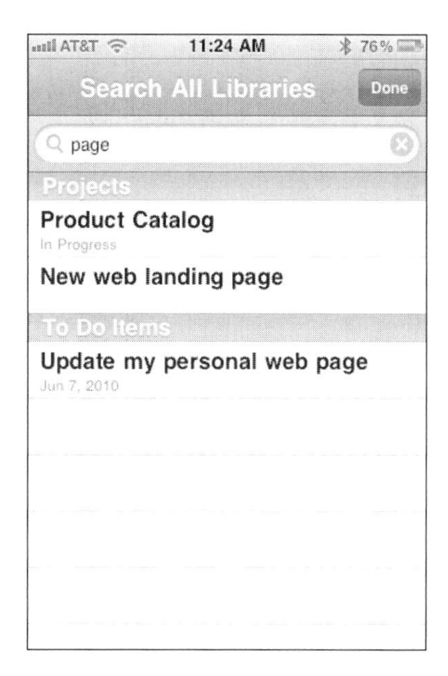

Notice that this search command is different from searching with Bento on Mac OS X. It searches through all your Bento libraries on the iPhone.

Working with Bento on iPad

On iPad, there is much more space than on iPhone, but there is still less space than on Mac OS X. Furthermore, there are differences in the way people interact using a mouse and using touch. The Bento for iPad interface uses the basic structure of Bento for iPhone but with a new implementation.

The list of libraries that is at the left of the Bento for Mac window and that is on the Home screen for iPhone appears in a popover window on iPad. Figure 11.8 shows the Libraries popover when the iPad is horizontal, and Figure 11.9 shows the Libraries popover when the iPad is vertical.

To open a library, just tap its name.

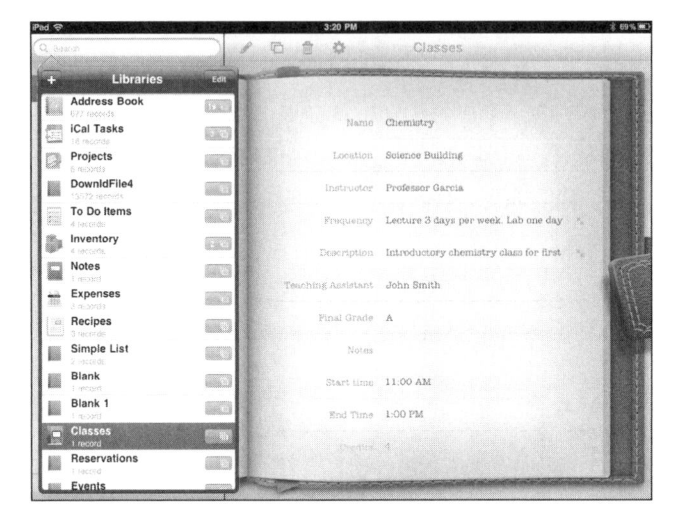

Figure 11.8
Libraries are shown in a popover on a horizontal iPad.

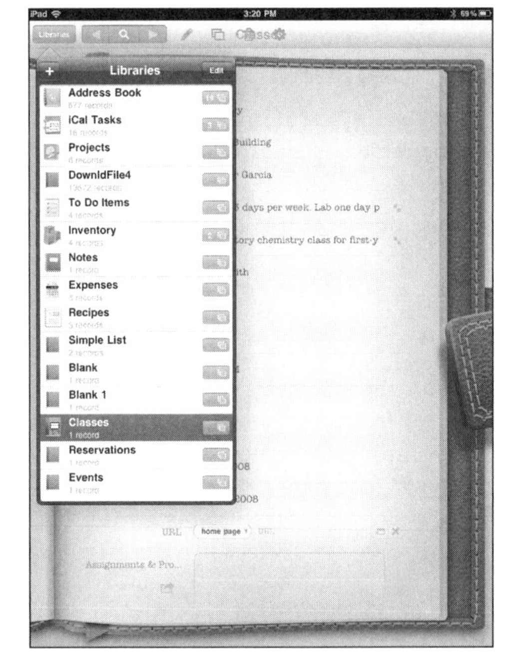

Figure 11.9
Libraries are shown in a popover on a vertical iPad.

Working with Libraries

The basics of working with libraries are the same in all of the versions of Bento:

- You can reorder libraries in the list of libraries.
- You can create new libraries.
- You can delete existing libraries.

In addition, on Mac OS X, you can organize libraries into folders.

Working with Libraries on Mac OS X

At the left of the Bento window, the Libraries & Fields Pane enables you to manage your Bento libraries and work with the fields in the current library. Generally, when you open Bento, the Libraries & Fields Pane is visible, as shown previously in Figure 11.6. If it is not visible, choose View, Show Libraries & Fields Pane to open it.

When the Libraries & Fields Pane is shown, you can show or hide the Fields pane at the bottom; by default, it is shown. To hide or show it, use the arrow in the lower left of the Bento window frame. When the Fields pane is shown, the arrow points down, as you saw previously in Figure 11.5. When the Fields pane is not shown, the arrow points up, and you can use it to show the Fields pane. Figure 11.10 shows the arrow about to be used to show the Fields pane.

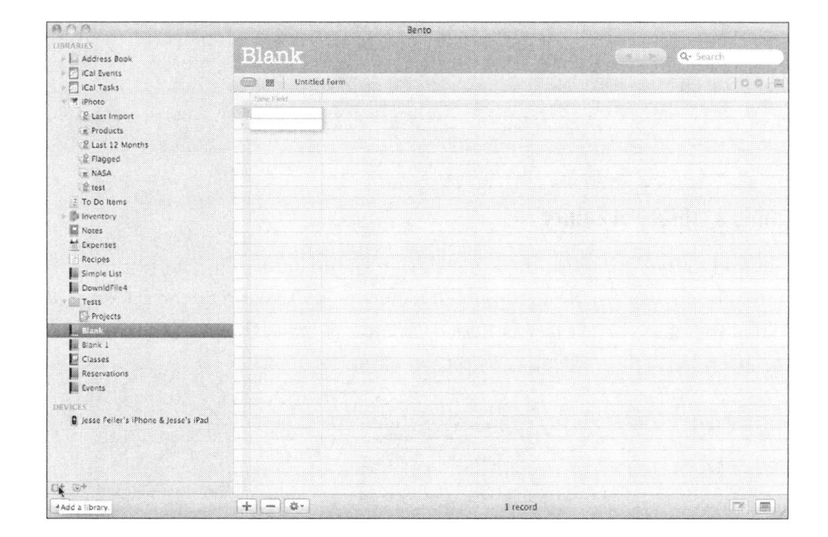

Figure 11.10
Show or hide the Fields pane.

Organizing Libraries with Folders

At the top of the Libraries & Fields Pane is the list of your libraries. You can organize them into folders, as you see in Figure 11.5. When you have libraries in a folder, you can close or open it with the triangle to the left of the folder name so that your list of libraries is easier to navigate. (In Figure 11.5, the Address Book, iCal Events, iCal Tasks, iPhoto, and Tests folders are closed.)

You can drag folders and libraries up and down in the list. If you want to drag a library into a new folder, drag it on top of the folder so that the folder name is highlighted, as shown in Figure 11.11. Projects is being dragged into the Tests folder; in addition the iPhoto folder is open so you can see its contents.

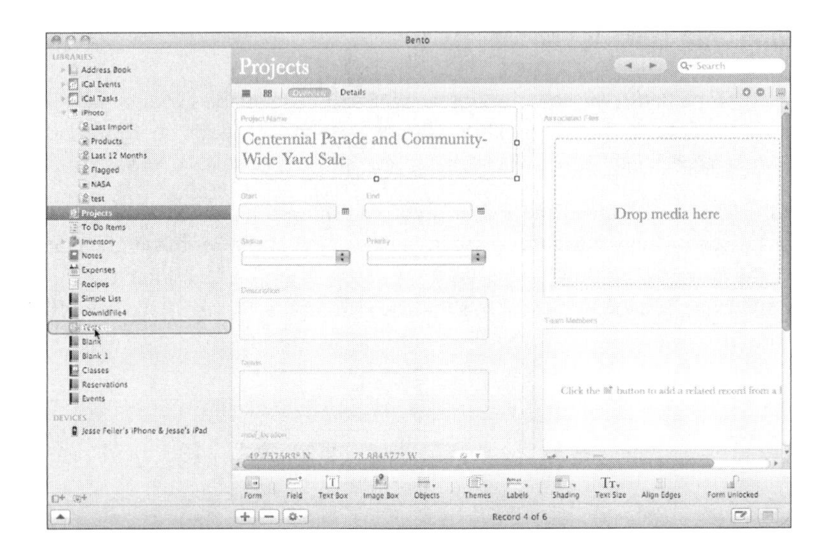

Figure 11.11
Drag a library into a folder.

Renaming a Library or Folder

Libraries and folders appear in the list at the left of the Bento window, but their data appears in the main section of the Bento window. No matter what view you are looking at, the folder or library name appears at the top. To edit the name of a library or folder, click the name at the top of the data window. The field is selected as you see in Figure 11.12, and you can edit the name.

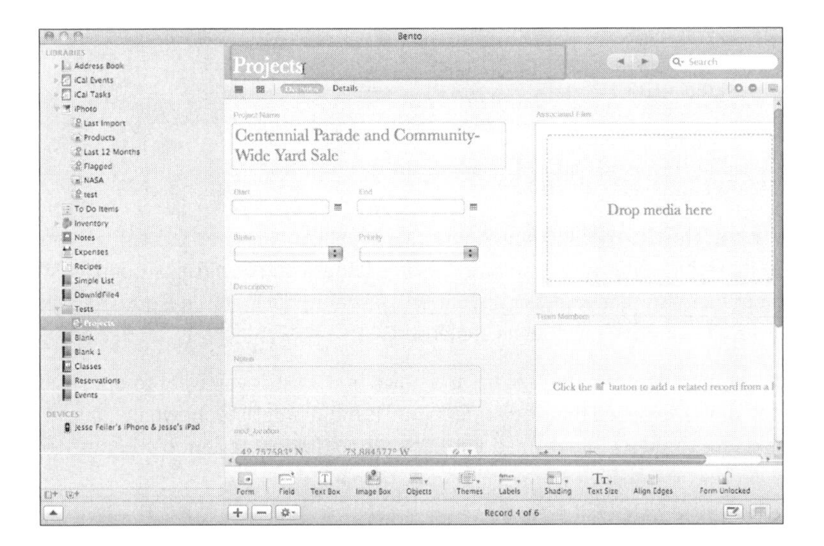

Figure 11.12
Rename a library.

Rearranging Libraries in the List

If you just want to rearrange your libraries, drag the library to the place in the list where you want it. Figure 11.13 shows a library being moved to a new location in the list. Compare it to Figure 11.11, which shows moving a library into a folder rather than just to a new location in the list.

Figure 11.13
Rearrange libraries.

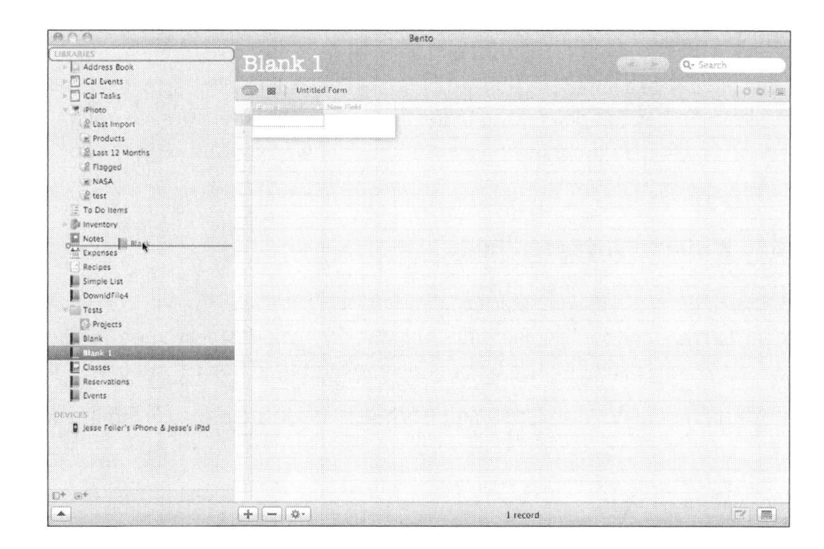

Creating a New Library

Adding new libraries is simple. You can choose any of these commands in the File menu:

- File, New Blank Library
- File, New Library from Template (⌘+L)
- File, New Library from Template Exchange (⌘+[option]+L)

You can also use the button at the lower left of the libraries list, as shown in Figure 11.14. It provides a one-click interface to File, New Library from Template.

Figure 11.14
Add a new library.

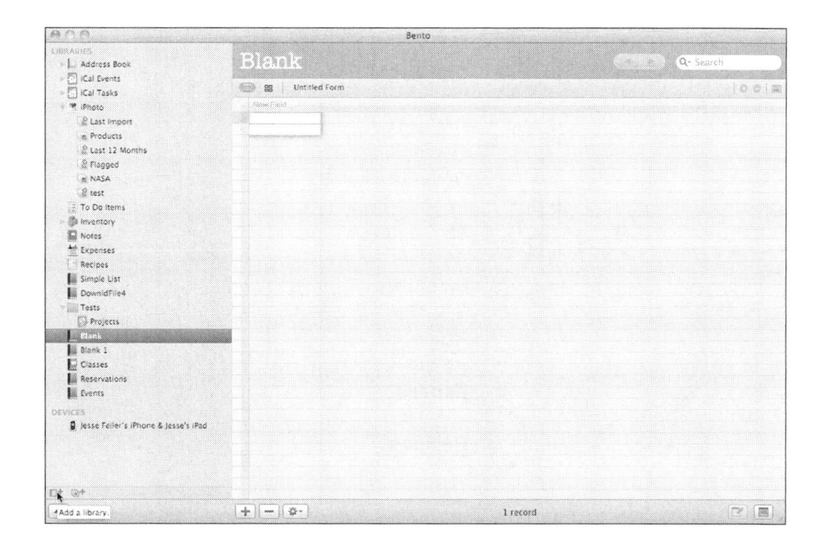

Deleting a Library

If you want to delete a library, select it and choose File, Delete Library; to delete a library folder, choose File Delete Library Folder. This deletes the folder as well as all libraries within it. You might want to drag the libraries out of a folder before deleting the folder; that way, you can keep the libraries and delete the folder.

Working with Libraries on iPhone

One of the differences between Mac OS X and the iOS devices is that on iOS devices you explicitly enter an edit mode to manage your list of libraries. On Mac OS X, you just rearrange libraries by dragging them with the mouse, the trackpad or mouse ball, but on a touch-based device, you have only your finger's touch.

In practice, this is a minor distinction, and it is one that you are probably familiar with by now.

Editing Existing Libraries

To begin editing libraries, you tap the Edit button in the upper right of the Home screen as shown previously in Figure 11.6.

When you go into Edit mode, you can delete libraries, add new ones, or rearrange their order as you see in Figure 11.15. Delete fields with the icon to the left of the name and rearrange them with the icon at the right of the name. Add new libraries with + at the bottom right of the screen.

Note that Address Book and iCal libraries cannot be deleted from the list of libraries. You manage them with Bento for Mac using File, Address Book, iCal, and Photo Setup.

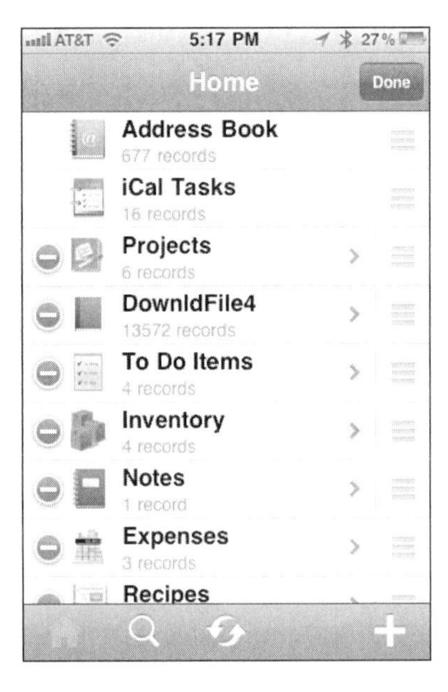

Figure 11.15
Edit libraries.

Tap a library name to edit its name and fields as you see in Figure 11.16. You can delete a field using the red icon on the left and reorder the fields list using the icon on the right.

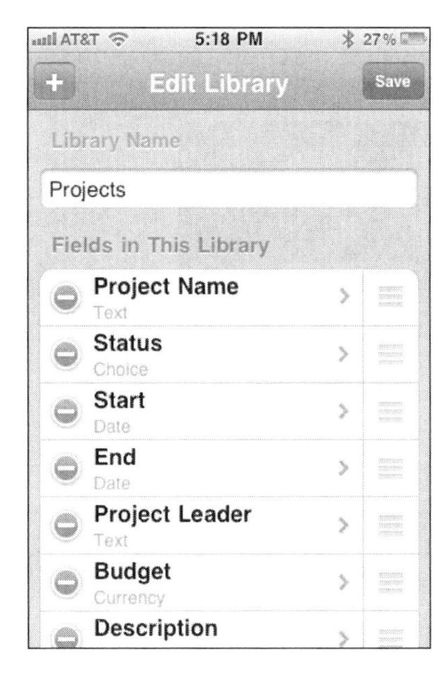

Figure 11.16
Edit field names and options.

Tap the field name to change its name and set options as you see in Figure 11.17.

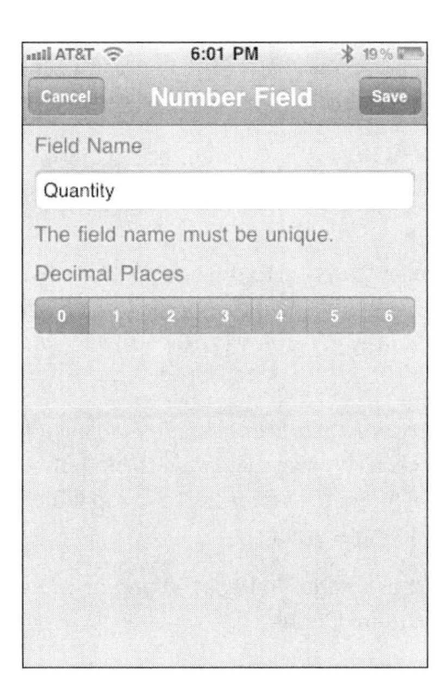

Figure 11.17
Edit field names and options.

When you are editing your libraries list, a Done button appears at the upper right of the screen. Use it to return to your list view, as shown in Figure 11.15.

Creating a New Library

Use + at the bottom right of the Home screen to create a new library. As you see in Figure 11.18, the various templates are shown in a Cover Flow display. Swipe through them and choose the one you want to use. Notice that there is a brief description of each library above the Create Library button.

Figure 11.18
Create a library from a template.

The Template Exchange is not available from Bento for iPhone. The templates available in Bento for iPhone are similar to those available in Bento for Mac, but they are sometimes customized for iPhone.

Working with Libraries on iPad

When you use Bento for iPad, you have a similar structure to the one you have with Bento for iPhone; you edit the list of libraries to perform most library operations. The difference is that on Bento for iPad, the list of libraries appears in a popover, as shown previously in Figures 11.10 and 11.11, whereas on iPhone you get to the list as a separate screen.

Because there's more space on iPad, the popover and its list of libraries can include functionality that is spread over several screens on iPhone. The list of libraries is always visible at the left of the screen when the iPad is horizontal; when it is vertical, you might have to tap the Libraries at the top left of the screen. In all of the following sections, you need to have the list of libraries visible.

Use the Edit button at the top right of the list of libraries to begin editing, as you see in Figure 11.19. The interface is larger, but its functionality is the same as on iPhone.

Figure 11.19
Edit libraries on iPad.

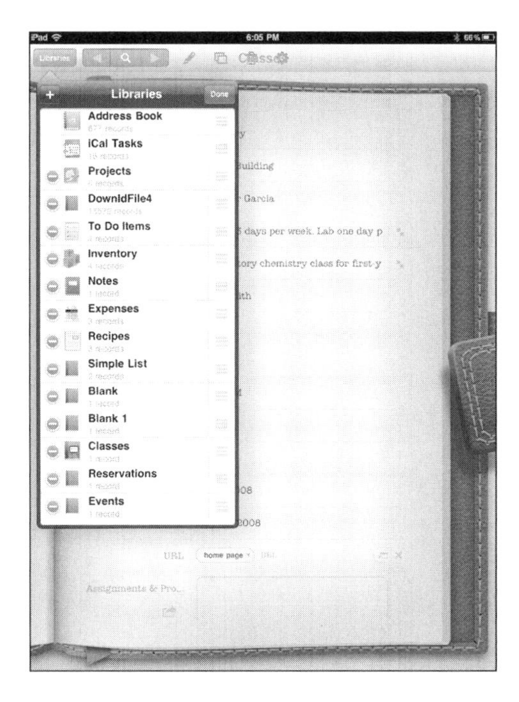

Further Steps

The next chapter delves into the Bento records and fields: the places where your library's data is stored. You can move on right away, but you might want to browse the library templates that are already available to you either built into Bento or in the Template Exchange. The commands are as follows:

- File, New Library from Template (⌘+L)
- File, New Library from Template Exchange (⌘+[option]+L)

You do not even have to create a new library. Just browse the options using either of those commands to get a feeling for what the developers of Bento and other Bento users have done.

Look at the templates not only as tools to help you accomplish your goals but also as learning experiences. If you want a Bento library to help you manage your collection of (anything) or a Bento library to help you keep track of anything from passwords to blog postings, you can find one that has already been created.

Browsing the existing templates might broaden your horizons in two specific ways:

- As you see the kinds of things that other people have done with Bento, you might get even more ideas of your own.
- Even if you have an idea, looking at how other people have approached similar projects can also give you more ideas.

12

IN THIS CHAPTER

Using Bento Records, Fields, Forms, and Tables

In Chapter 11, "Using Bento and Bento Libraries," you saw the top-level view of Bento. In this chapter, you look beneath the surface at the data you enter and use that is stored in the Bento database and its libraries. You need to know the basics, but it is the data that matters to you.

In this chapter, you see how to use Bento's features to work with your data on your Mac, your iPhone, or your iPad. What changes do you have to make to accommodate these different devices with their different screen sizes and shapes? How do you customize your interfaces to accommodate mouse-driven as well as touch-driven interfaces?

You just use Bento, and let it do the work.

Taking Advantage of Bento's Built-In Tools for Mobility

FileMaker is one of the most impressive and full-featured databases available today. Its easy-to-use and easy-to-customize interface is unmatched. You can do almost anything you can imagine (and more) with databases when you are using FileMaker. And, yes, that is not an exaggeration. Because you can use heavy-duty databases such as Oracle, DB2, and Microsoft SQL Server as external data sources for FileMaker, FileMaker can be a front-end using xDBC connectivity. (Unfortunately, external data sources are beyond the scope of this book, but you can read about them in *FileMaker Pro 10 in Depth*.)

With FileMaker Pro and its mobile companions FileMaker Go and FileMaker Web Publishing, you can customize your database and its interface for mobile devices. If you use database scripting, you can modify scripts so that they run differently on mobile devices than they do on traditional computers.

When you start thinking about using mobile data, you see another side of Bento, and it is one that might amaze and impress you. As a personal database, Bento is designed to make you productive as soon as you download or create a library from a template. Yes, you can customize the libraries, but you do not have some of the options for fine-tuning that you have with FileMaker Pro. This means that within your Bento design choices (and there are many), the particular details are handled by Bento, and they are handled appropriately on each version of Bento: Bento for Mac, Bento for iPhone, and Bento for iPad.

Looking at Fields

The heart of your Bento solution is your data. Data is stored in *records*, each of which contains *fields*. You can structure fields without entering any data. When you set up a Bento library to track your garden, you can create fields such as these:

- Plant name
- Date planted
- Overall success (this is a great use for Bento's *rating* field, so that you can decide if the flowers were attractive and plentiful enough, the fruit delicious, or the berries attractive enough to the birds you like)
- ...and more

This structure requires absolutely no data, and many times you set up your library just by identifying the fields. As you start to enter data, you might come to realize that you need different fields, and so you modify the structure as you enter data.

Forms can display the data that is stored in the library fields. Because of this tight linkage, forms need to be referenced in the fields section of this chapter, although those references are kept to a bare minimum.

Using Bento Field Types

Every field in a Bento library has its own *field type*. By default, all fields are *text* type fields. They can contain anything that you type into them. If you use a field in a calculation, Bento automatically converts the text to the needed value.

Whether you are using Bento or another database, it is generally a good idea not to rely on these automatic type conversions. If you use a more specific type of field, you can fine-tune its storage and display. For example, if you use a currency field, you select the currency symbol to use. Then, when you type in the number, the field provides the appropriate currency symbol automatically. Furthermore, using specific field types can enable automatic editing and validation of data entry. In FileMaker Pro, this can mean not allowing you to type text into a currency field. On Bento for iPad and Bento for iPhone, matters are taken further: The on-screen keyboard for a currency field is a numeric keyboard, so there are no letter keys. You have only the digits zero through nine, a decimal point, and a ± button that switches the value from positive to negative and back again.

These are the Bento field types. As you can see, a number of them are designed specifically for integration with Address Book, Mail, and iCal. In those cases, Bento directly accesses the data from Address Book and iCal. You can view it and modify it, but the data itself is stored in Address Book and iCal:

- **Location** fields display a location that can be mapped. They are so important in the world of FileMaker on the Move that they have their own chapter: Chapter 13, "Working with Location and Media Fields."
- **Text** fields store any characters you can type. These fields sort in alphabetical order. Among other things, that means that if they start with numbers, they sort in an alphabetical rather than a numerical order: 1,10,11,100,150,2,3,4,5,6, and so forth.
- **Number** fields are for numbers. One of their features is that when they are sorted, they sort in true numerical order: 1,2,3,4,5,6,10,11,100,150, and so forth.
- **Choice** fields provide a multiple-choice option. When you create the field, you define the possible choices; the user chooses one when entering or editing data. On Mac OS X, choice fields are implemented with checkboxes. On iOS, they are implemented with a picker, as shown in Figure 12.1.

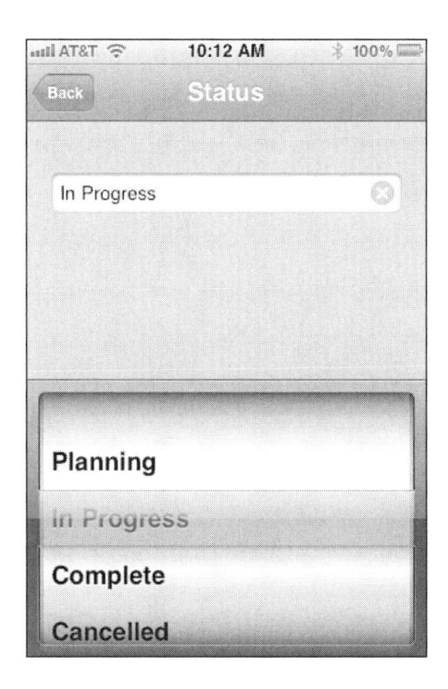

Figure 12.1
On iPad and iPhone, choice fields are implemented with a picker.

- **Checkbox** fields in Bento are either checked or not; you can use them for On/Off settings. Note that on Mac OS X and some other operating systems, checkboxes can be grouped together so that any number of checkboxes in a single group can be checked. On Bento, it is just one checkbox that is either on or off.
- **Media** fields can contain a pdf, audio, graphic, or movie file. When you sync your Bento library, contents of media fields are synced.
- **Time** fields are specially formatted to show a time of day.
- **Date** fields contain a month, day, and year, as well as an optional time of day.
- **Duration** fields store a period of time using a combination of numbers and letters. For example, 4d10h represents four days and ten hours. You can use duration fields with calculations to compute durations based on time or day fields.
- **Currency** fields store a number and prefix it with a specified currency symbol. The symbol is set for the field, so you can have multiple types of currency fields within one Bento library.
- **Rating** fields display a set of stars. You click one of the stars to store the rating. Figure 12.2 shows a rating field in action. The third star has been clicked, providing a rating of seven (out of ten).

  The number of stars used in the set can be set in Bento for Mac (you can choose one to ten stars); the default is five. Bento for iPad and Bento for iPhone use that setting.

Figure 12.2
Use a rating field.

- **Address** is a composite field that contains street, city, state, and postal code. Bento is able to map address fields.
- **Phone number** fields store the components of a phone number. On Bento for iPhone, you can automatically dial the number in a phone number field.

- **Email** fields store an email address and can be used to generate an email message. Bento checks that the address is formatted correctly (it can still be wrong).
- **URL** fields store a URL. Bento can open the URL if you choose.
- **IM Account** fields are used to store instant messaging account names.
- **File list** fields contain a list of files. The files do not sync across the iOS devices.
- **Message list** fields contain messages, RSS articles, and notes from Mail.
- **Automatic counter** fields contain a unique number for each record in the library; the number is automatically incremented with each new record. If you delete records, you might have gaps in the numbers. Automatic counter fields are particularly useful if you will be exporting data from Bento to another environment where record numbers are needed.
- **Encrypted** fields are text fields that are protected from unauthorized display. All encrypted fields in a library are locked or unlocked together with the same password. The data is also encrypted when it is stored on disk.
- **Related data** fields store data from another library. They are described later in this chapter in the section, "Working with Related Data."
- **Simple list** fields are like small spreadsheets with rows and columns that can be accessed separately. They can be displayed and edited with Bento for Mac as well as Bento for iPad and Bento for iPhone. Figure 12.3 shows a simple list field in Bento for Mac. Note that you can change each column's data type; you can also show or hide a summary row at the bottom of the list.

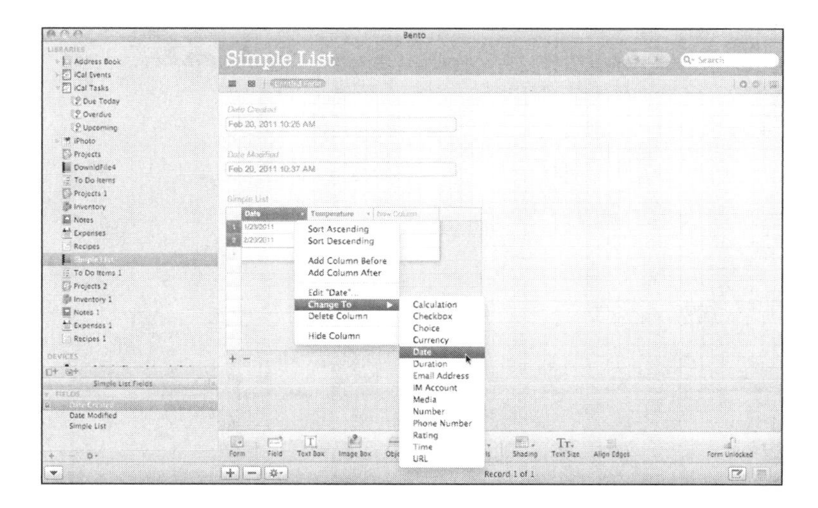

Figure 12.3
Use a simple list field on Bento for Mac.

Using Bento for iPhone, the same simple list is displayed on several screens as you see in Figure 12.4.

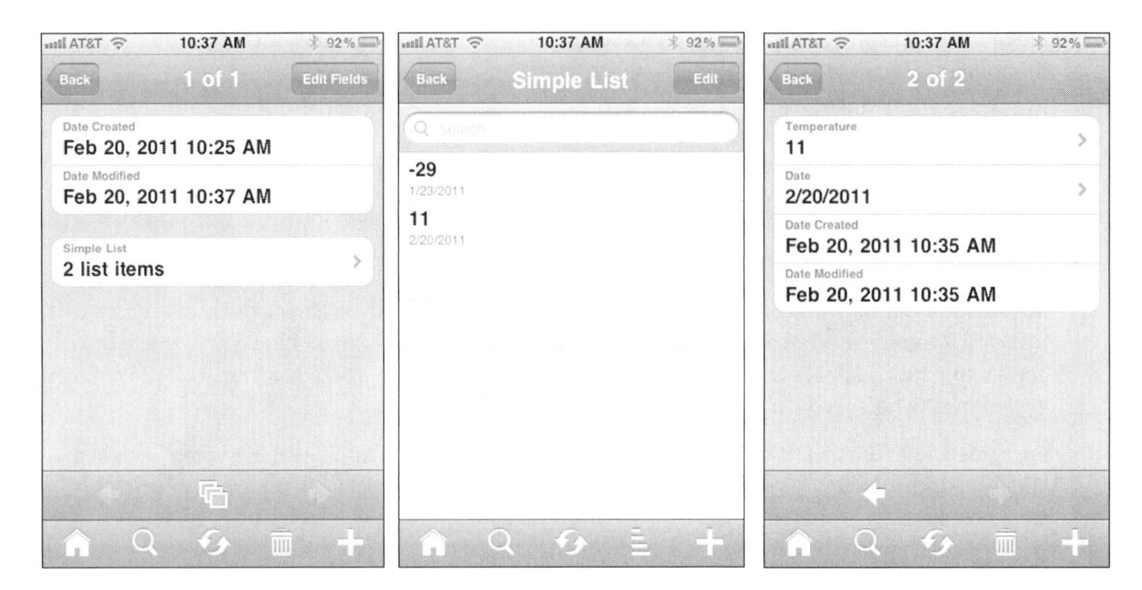

Figure 12.4
Use a simple list field on Bento for iPhone.

With Bento for iPad, you have more space and the ability to control how columns are sorted as you see in Figure 12.5.

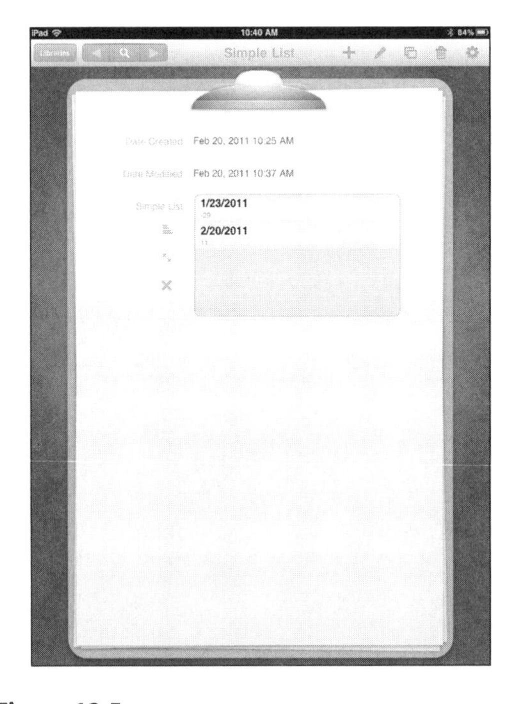

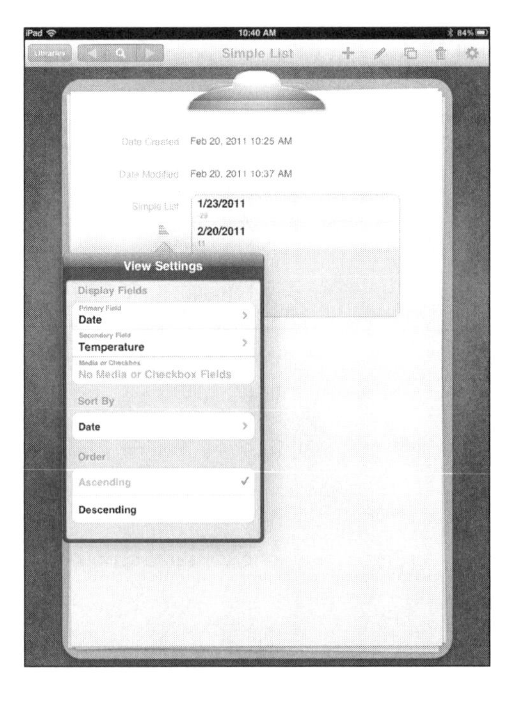

Figure 12.5
Use a simple list field on Bento for iPad.

Editing Fields on Mac OS X

On Mac OS X, you work with the structure of fields in the Libraries & Fields pane at the left of the Bento window. Select or create the library you want to work with, and then make sure the Libraries & Fields pane is shown by choosing View, Libraries & Fields Pane. If necessary, use the triangle at the bottom left of the Bento window frame to show the Fields pane at the bottom of the Libraries & Fields pane, as you see in Figure 12.3.

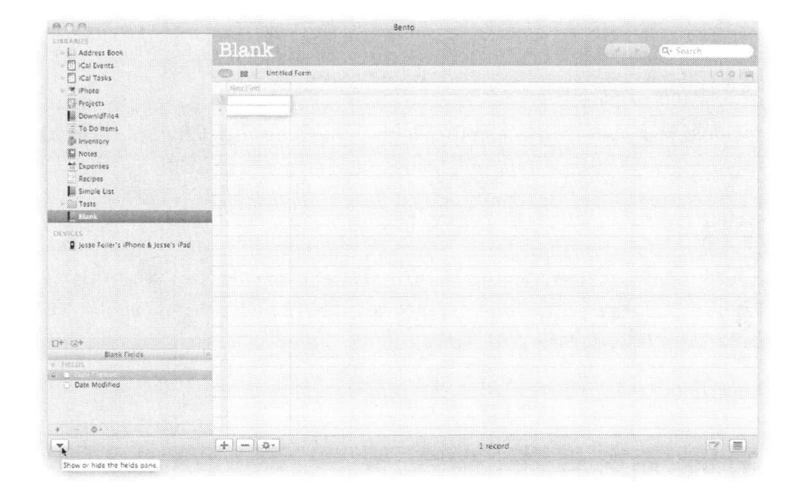

Figure 12.6
Work with fields in the Fields pane.

Figure 12.3 shows a newly created library using the Blank template. As you can see, there are two fields in the Fields pane. Every library has these two fields: Date Created and Date Modified. Bento takes care of filling them in appropriately, so you do not have to do anything.

Adding a New Field

Click the + at the bottom left of the Fields pane. This enables you to select the field type and provide a name for it. If you prefer the menu bar, you can choose Insert, New Field, or you can use ⌘-N. Each field type has additional settings that you can configure. For example, Figure 12.4 shows the settings for a new date field.

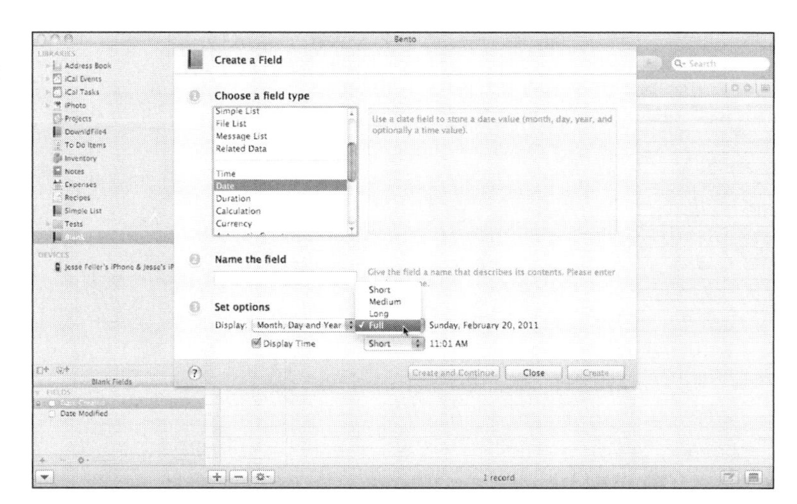

Figure 12.7
Provide a name and select a field type as well as field-specific settings.

Modifying an Existing Field

You select the field you want to use, and then you can make your changes using the same interface, as shown previously in Figure 12.7. Use Edit, Edit Field or open the Action pop-up menu by clicking the gear below the Fields pane. The Action menu not only lets you edit a field, but you can also change its type or duplicate it.

If you select one of the two standard fields shown in Figure 12.3, you see a small padlock to the left. These are among the fields that are locked; you cannot rename, redefine, or delete them.

Deleting a Field

Select the field you want to delete in the Fields pane. Then click – below the Fields pane or choose Edit, Delete Field.

The Delete command changes depending on what is selected. If you do not see a Delete Field command, make certain you have selected a field. If a library is selected at the top of the Libraries & Fields pane, the command is Delete Library; if a field on a form is selected, the command is simply Delete.

Editing Fields on iPad

With Bento for iPad, tap the pencil at the right of the toolbar to edit the open library, as shown in Figure 12.8. Note that a new toolbar appears below the standard toolbar. If it covers the first line of data, just drag the list of fields down below it. The toolbar contains the fields in the current library that are not used in the form you are browsing.

Adding a New Field

Tap New Field at the left to create a new field in the library. Select the field type from the picker that appears next, as shown in Figure 12.9.

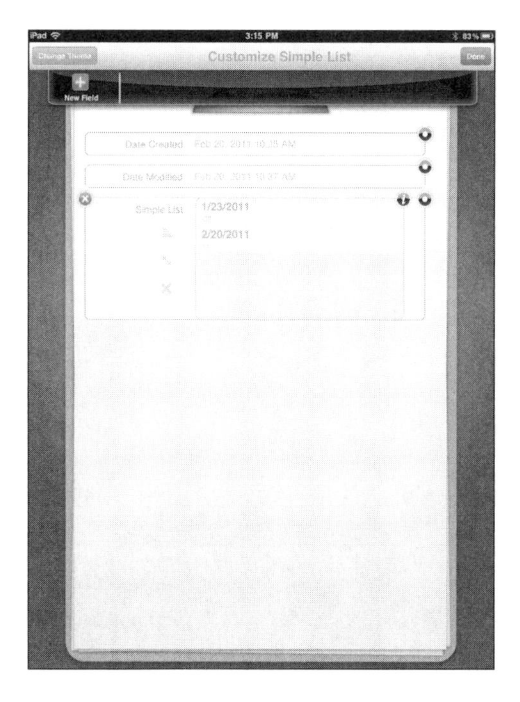

Figure 12.8
Work with fields with Bento for iPad.

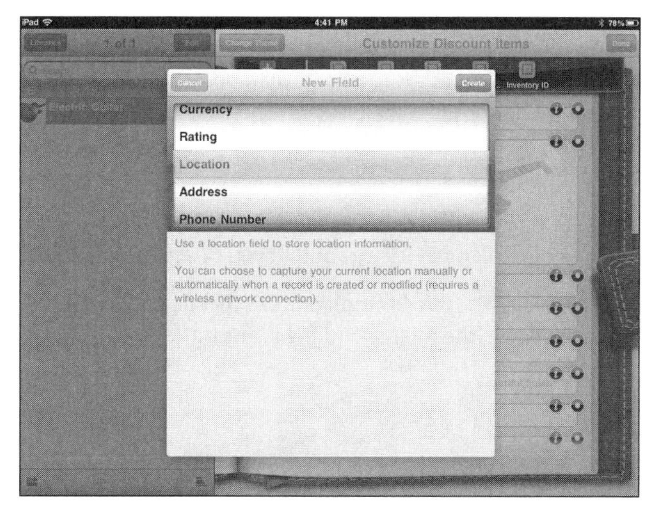

Figure 12.9
Choose the field type on iPad.

Tap Create, and then enter the field name, as shown in Figure 12.10. Some field types ask for additional information, as you also see in the figure.

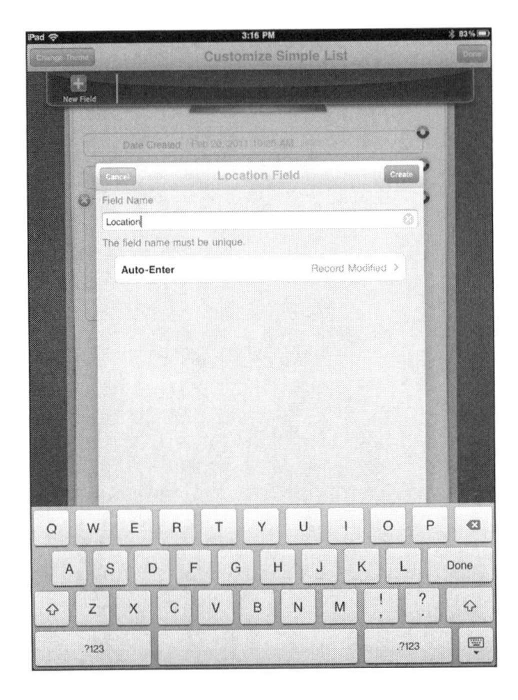

Figure 12.10
Name the field and supply optional additional information.

Modifying an Existing Field

When you tap the pencil at the right of the toolbar to start editing fields (as shown previously in Figure 12.8), you see a list of the fields in the current form and, in the new toolbar at the top, all the fields that are not in the form. Together with the New Field button at the left of the toolbar, you have access to all of your library's fields, both present and future.

To modify an existing field, tap the Info button (i) at the right. This launches you into the same sequence of fields that you saw beginning in Figure 12.9 except that you cannot change the field type for an existing field.

The dot at the far right (just past the Info button) removes that field from the form and places it back in the toolbar. (Remember that between the toolbar and the form's list of fields, you see all of your fields.)

Deleting a Field

At the left, the Delete button (x) lets you delete a field. As you see in Figure 12.11, you are asked to confirm the delete. This is a delete from the library as well as from the form. The field and its data will be gone.

Editing Fields on iPhone

With Bento for iPhone, the processes are much the same, but because the screen is smaller, there are a few other steps. In addition, with Bento for iPhone, there is no concept of a form; all fields are displayed automatically. However, you can change the order of the fields.

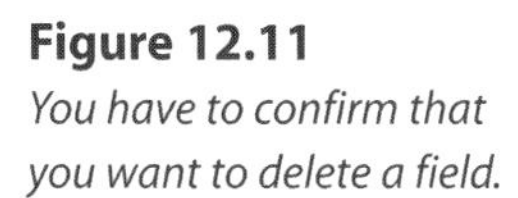

Figure 12.11
You have to confirm that you want to delete a field.

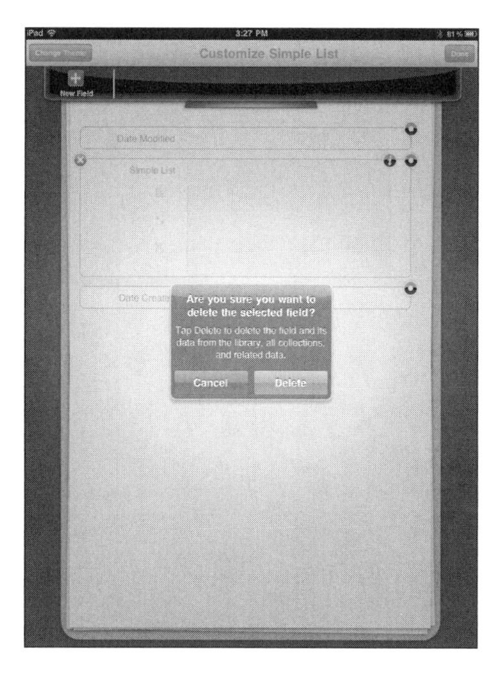

How to Add a New Field with Bento for iPhone

Here is the process to add a new field:

1. **From the Bento Home screen, tap the library you want to edit.**
2. **Tap a record.** It does not matter which record you go to. If you have no records, add one by tapping + in the lower right of the records list, as shown in Figure 12.12.

Figure 12.12
Select a record (any record).

3. When the record is displayed as shown in Figure 12.13a, tap Edit Fields in the upper right to show the editing screen shown in Figure 12.13b.

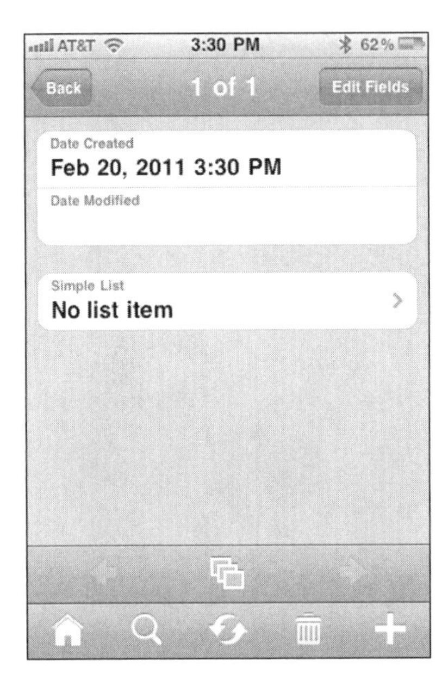

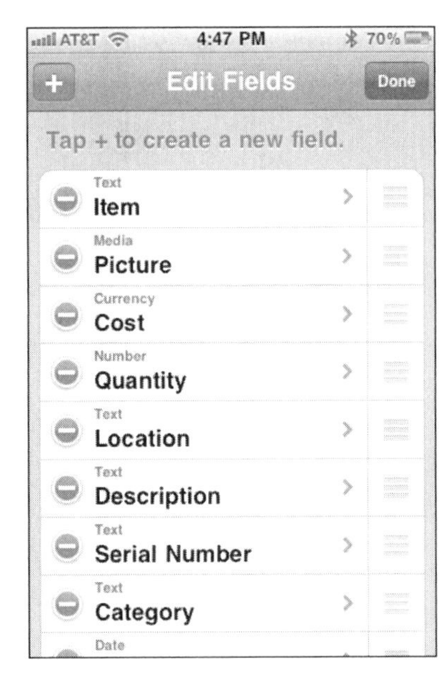

Figure 12.13
Display a record and tap Edit Fields.

4. Reorder fields by dragging them up or down.
5. Use + in the upper left to add a field.
6. Choose a field type as shown in Figure 12.14.
7. Tap Create to continue.
8. Enter the field name and any additional options, as shown in Figure 12.15.
9. Tap Create.
10. Tap Done to return to the records, shown previously in Figure 12.13.

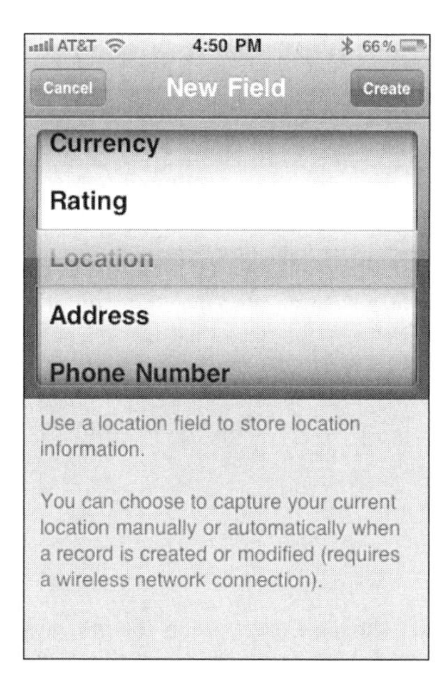

Figure 12.14
Select a field type.

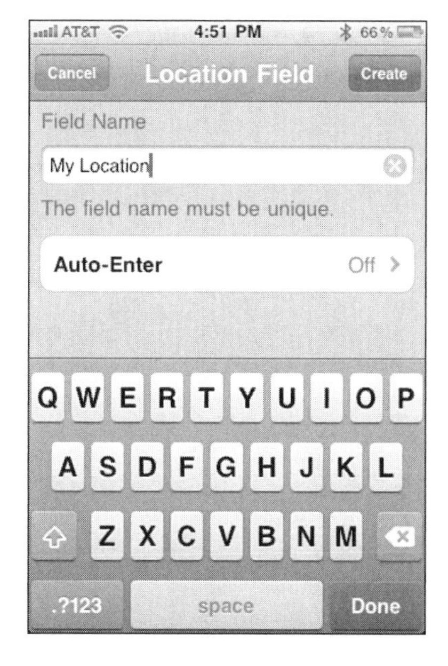

Figure 12.15
Enter a field name and any other options.

Modifying an Existing Field

The process for modifying a field is the same as described in the previous how-to about adding a new field, except that after Step 3, tap a field name and continue with Step 8.

Introducing the View Types on Mac OS X

The fields represent the structure of the database. What you care about most of the time is your data. Here is where Bento really shines in the world of mobile devices. On Mac OS X, there are three types of Bento views: table, grid, and form views. The table and grid views are created automatically for you based on the structure of your database. You design your own form views, and you can customize their appearance.

On Bento for iPhone and Bento for iPad, the automatically created table and grid views are not used, but as you will see, some of their features are automatically integrated into the interface. Form views for the iOS devices are generated automatically by Bento.

The automatic displays mean that you do not have to worry about the issues of fonts and font sizes as you move from a Mac to an iPhone and to an iPad; Bento does it for you. Bento makes your interface look right for each device.

Using Table Views

Table views display your data in a spreadsheet-like table of rows and columns, as you can see in Figure 12.16. This display shows each record in a row, and it shows fields as columns. You always have one—and only one—table view in your library. You can show your table view using the table view button at the left of the bottom row of the navigation bar, as shown in Figure 12.3, or by using View, Table View or ⌘+1. As you see later in this section, you can customize the table view by selecting which fields are shown in it, as well as their width and their left-to-right order in the table view.

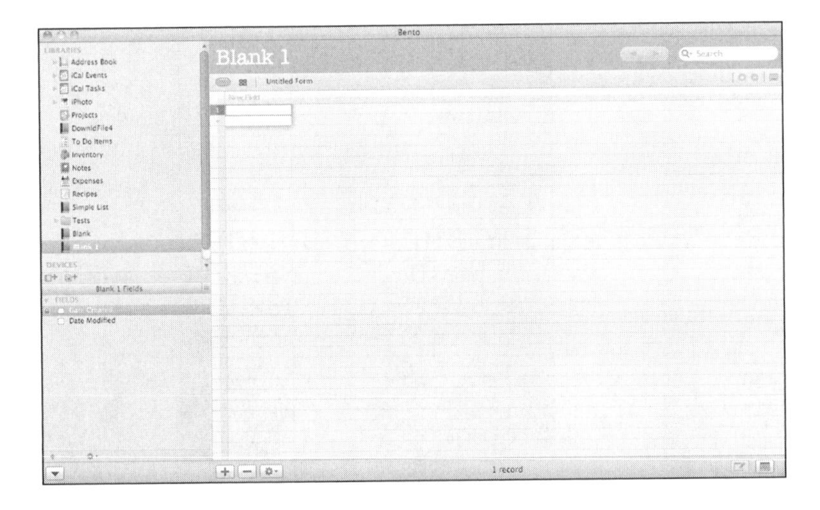

Figure 12.16
Work with a table view.

Bento always displays an added row and an added column so that you can easily modify the data and the fields in your database. The added column is labeled New Field, as you see at the right in Figure 12.16. The expansion row is labeled + in the left margin. Rows above the expansion row are numbered just as in a spreadsheet, and columns to the left of the expansion column (if they exist) have names that you can provide.

If you double-click the title of the first column (New Field), three things happen:

1. You create an actual field (not just an expansion column). The field appears in the Fields pane.
2. The field name is set to Field 1. If you change its name in the table or in the Fields pane, its name changes in the other place.
3. In addition, now that you have a field, the expansion column (labeled New Field) moves to the right.

This is shown in Figure 12.17.

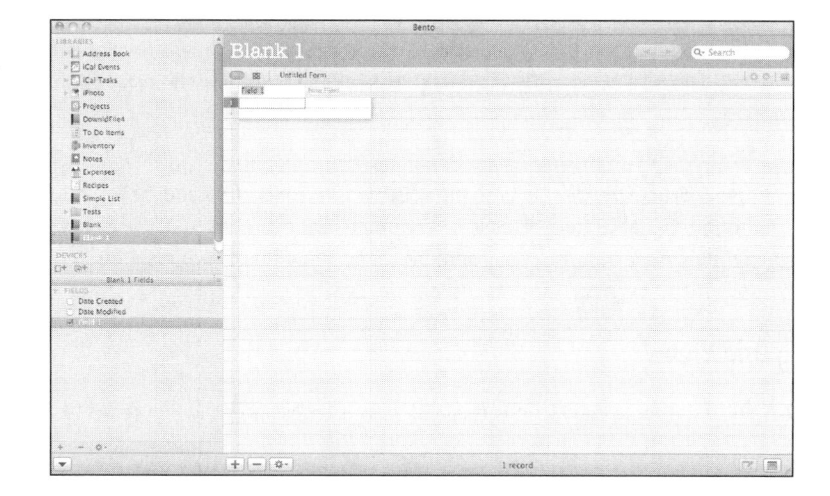

Figure 12.17
Add a field in table view.

The small arrow to the right of the field name in the title row of the table view brings up the shortcut menu shown in Figure 12.18.

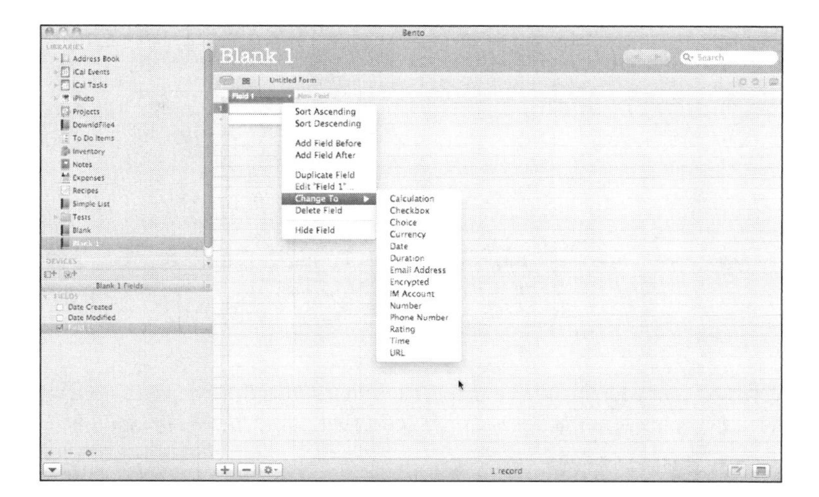

Figure 12.18
Use the shortcut menu in a column's title row to modify a column in table view.

You can sort the column's data, as well as add, edit, or delete the field. If you hide the field, it is removed from the table, but it remains in the Fields pane.

To add an unshown field to the table view from the Fields pane, click the checkbox to the left of the field name. A field can only appear in one column, so the checkbox is either on or off. You can rearrange columns in the table view by dragging their titles to the right or left.

Because each library has one and only one table view, and because you can select the fields in it with the checkboxes to the left of their names in the Fields pane, Bento can rely on the fields selected for the table view for other purposes. For example, when you are exporting data from Bento (see Chapter 14, "Importing and Exporting Bento and FileMaker Data"), you can choose to select only some fields for export. The choice is to use all fields for export or just to use the fields selected for the table view for exporting.

Some people (particularly people who are used to the world of spreadsheets) might work totally in table views. Although the table view display is very much like a spreadsheet, the Fields pane and its associated logic give you a much more powerful environment than the world of spreadsheets.

Using Grid Views

To the right of the Table View icon is the Grid View icon, as shown in Figure 12.19.

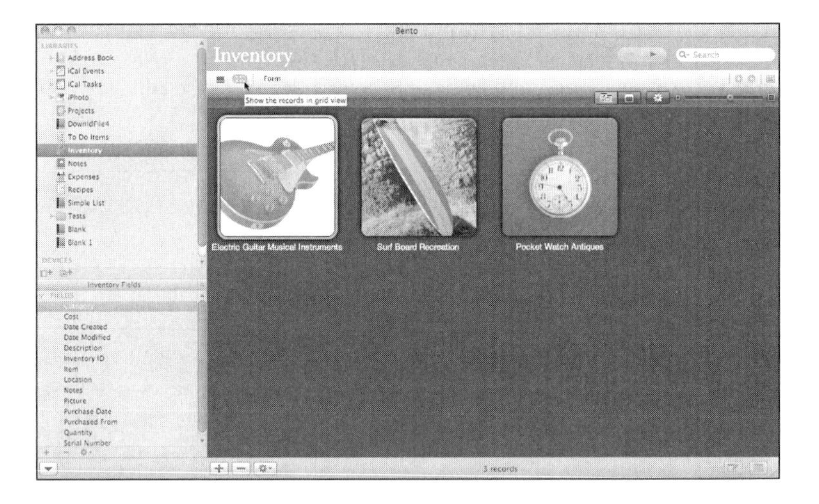

Figure 12.19
View the library in a grid view.

Grid views present your library in a graphical format with a thumbnail for each record. You control the grid view with the controls at the top right of the grid view itself. There are four buttons. From left to right, they provide four types of customization:

- **Image thumbnails.** If you have media fields in your library, you can choose to use images in your thumbnails. If there is more than one image field in each record, the images change as you move your mouse across the thumbnail.

- **Form thumbnails.** The thumbnail shows small images of your forms. As you move the mouse across the thumbnail, the form displayed changes if you have more than one form (you always have at least one).

  Image and form thumbnails are mutually exclusive. Pick the type you want.

- **Grid settings.** The third button (the gear) controls grid settings. These settings apply both image and form thumbnails, whichever you have selected. Grid settings let you select a title and subtitle. Pop-up menus display text fields in your library; you can place up to two text fields on the title line and up to two on the subtitle line.

  In addition, the small down-pointing arrow at the bottom right of the image in Grid Settings lets you choose which form or which image is the starting image or form. As you move your mouse over the thumbnail, you browse through the other images or forms.

- **Thumbnail size.** Finally , the slider lets you control the size of the thumbnails. Bento automatically rearranges them as the size changes.

Using Form Views

At the bottom of the navigation bar are the controls for the table view and the grid view. Beyond the vertical line are your form views. You always have at least one form view, and it starts out with a newly-created library containing the two default fields (creation and modification date). Figure 12.20 shows a form view right after creation.

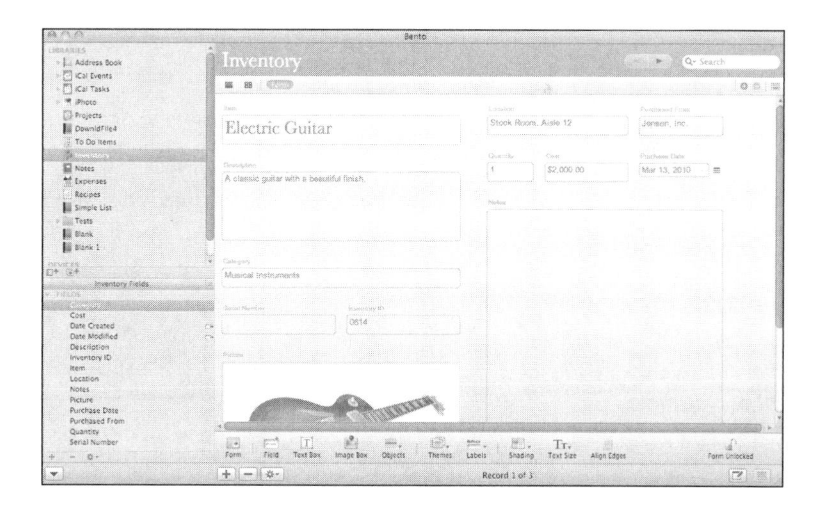

Figure 12.20
You always have a form view.

Create a new form with Forms, New Form or with the create form button at the far right of the navigation bar. You can also show the form tools as described in the following section and use the Form tool at the left.

Delete the current form with Forms, Delete Form or the delete form button at the far right of the navigation bar. Both the button and the menu command enforce the rule that you must always have at least one form. You cannot delete the last form in your library.

Adding Fields to Forms

In table views, you add fields to the table view with the checkbox to the left of a field's name. This works because there is only one table view. Because there can be more than one form view, you drag the fields you want into the current form. Note the dragging symbol to the right of the field names in the Fields pane in Figure 12.20. (You cannot place a field in two places on the form.) A field can only appear once in a form, so the dragging symbols are not shown in the Fields pane for fields that are already in the form.

Fields on forms display the data that has been entered. This is what you would expect. However, certain Bento fields contain sub-fields, and they have special functionality and appearance on forms.

For example, Figure 12.21 shows a Bento address field.

Figure 12.21
Display an address field in a form.

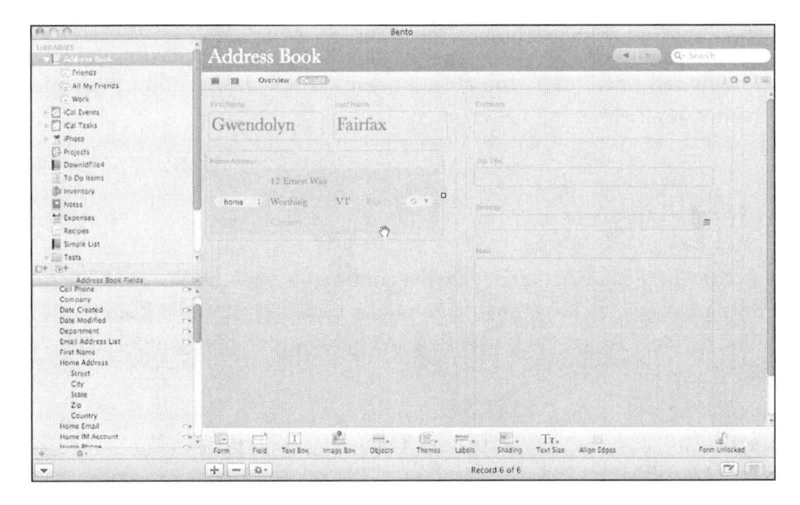

Notice the Street, City, Zip, and Country sub-fields. You do not have to add them individually. (In addition, you cannot remove any of the sub-fields; the address field is a single entity.)

To the left, the pop-up menu lets you choose the type of address that is displayed. As you can see in Figure 12.22, you can also change the format.

Figure 12.22
You can change information about the address with the pop-up menu.

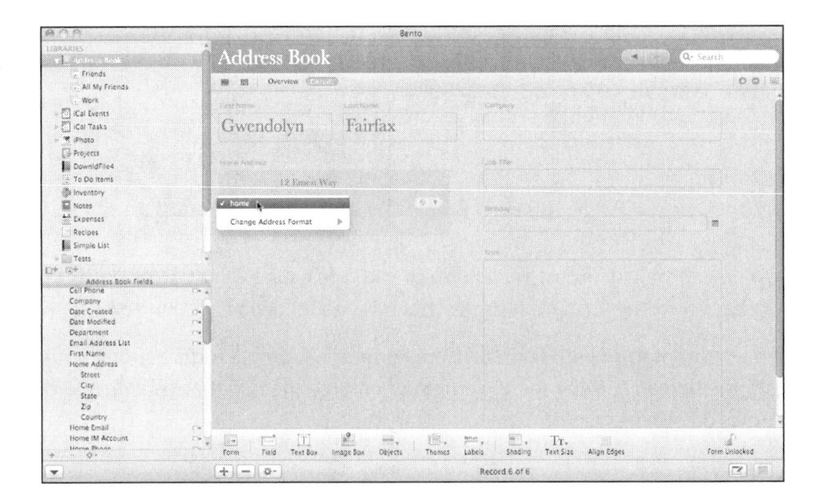

To the right of the address are two buttons. The one on the right (the down-pointing arrow) lists various commands available for the address, as you see in Figure 12.23.

Figure 12.23
Bento provides functionality for an address field.

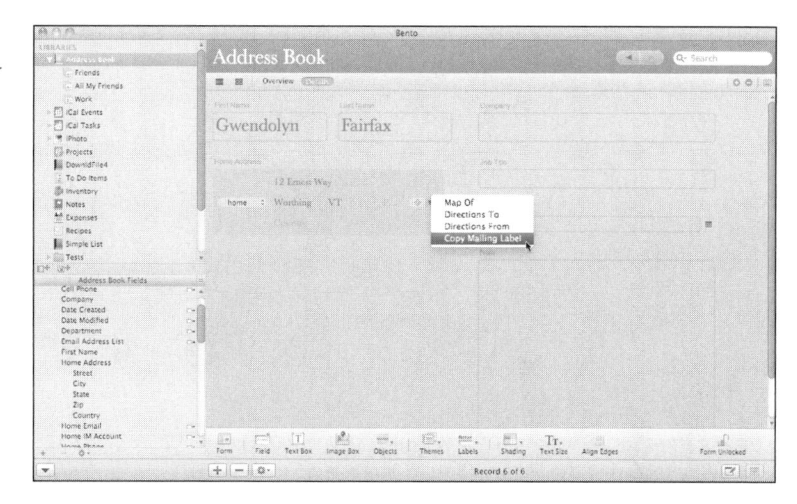

This enables you to map an address, find directions to or from the location, and to create a mailing label.

To the left of the down-pointing arrow, a button lets you map the address.

This interface structure is available for all of the composite fields in Bento—address, email and URL lists, email addresses, and locations. In every case, the down-pointing arrow lists all the options, and the button to its left triggers the first item in that list (which is the most common choice).

➔ This functionality is implemented with a different interface on Bento for iPad and Bento for iPhone, as you see later in this chapter.

Adding Elements to Forms

In addition to dragging fields from the Fields pane into a form, you can add other elements to the form. You access these by using the form tools shown optionally at the bottom of a form. Show them using View, Show Form Tools or by using the form tools button at the right of the bottom window frame. Figure 12.24 shows the form tools in action.

Click on a form tool to add it to the form. After it is added, you can move it around or resize it. The other form elements move aside to accommodate the moved or resized element. These are the form tools you can add with a mouse click to the form tools:

- **Form.** At the left, you can create a new form from the form tools.
- **Field.** You can add a field from the form tools; it is automatically added to the form. (This is the same behavior as the + at the bottom of the Fields pane or Insert, New Field.)
- **Text box.** This inserts a box into which you can type text. This is the same text for every record in the library. It is part of the form.

Figure 12.24
Use form tools.

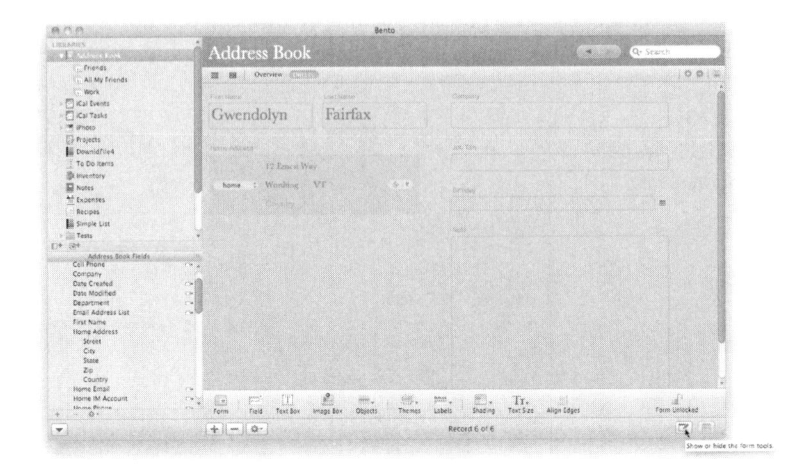

- **Image box.** This inserts a box into which you can paste an image. Like a text box, it is the same image for every record in the library. (You might use it for a logo.)
- **Objects.** Choose from this pop-up menu to insert graphical or spacing elements:
    - **Horizontal separator.** This inserts a horizontal line. You can resize it.
    - **Column divider.** Use this to split your form into columns. You can move fields from one column to another by dragging them.
    - **Spacer.** This is an invisible element that simply takes up space. Because Bento flows elements from top to bottom, you can insert a spacer at the top of a column to push down subsequent elements and leave a space.
- **Themes.** Bento's themes combine color schemes and graphics. Clicking Themes opens a theme picker, as shown in Figure 12.25. Experiment: You can always change them back with a mouse click.

Figure 12.25
Change Bento themes.

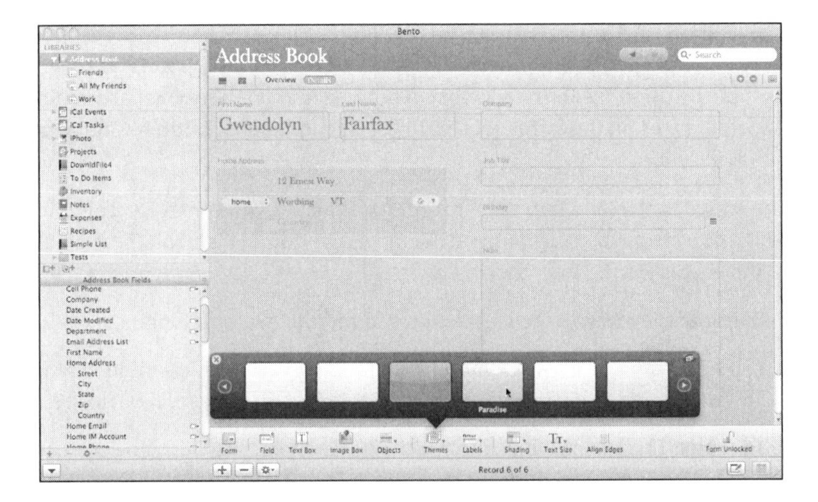

- **Labels.** You can choose to place the automatically produced labels above or beside their fields, and you can make them small, medium, or large. The settings apply to all labels and fields shown in a form.
- **Shading.** You can control shading (none, light, or dark), for the selected fields in a form.
- **Text size.** Select one or more fields and set its text size: smallest, small, medium, large, or largest. Note that this is the size of the text in the field; the size of the label's text is set in Labels.
- **Align edges.** With two or more objects selected in the form, you can choose align edges to make them the same width.

You can select fields in a form to move and resize by clicking on them; the label is the safest place to click because clicking on the interior of a field can send you into the mode in which you edit the data.

Using Split Views

Choose to use a split view from View, Split View or from the button at the far right of the bottom of the navigation bar, as shown in Figure 12.26.

Figure 12.26
Use a split view.

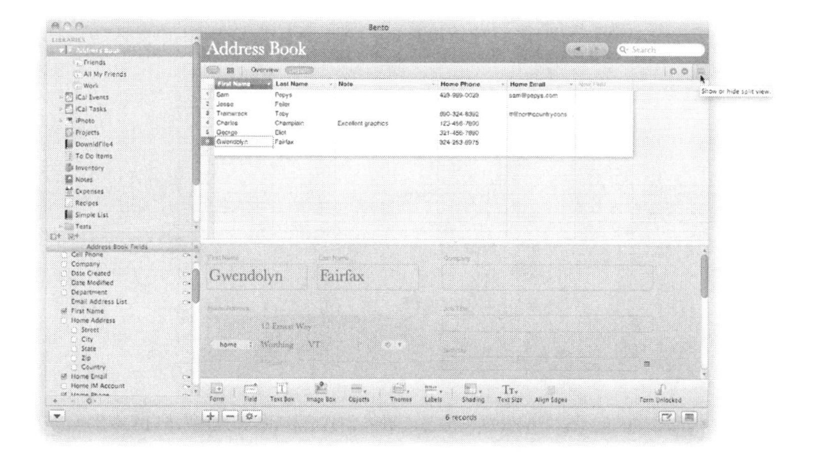

A split view has either a grid view or a table view on the top; a form view is shown below. You can drag the divider up or down so that either is larger or smaller. As you switch from one form to another or from table to grid view, the split view reflects your new choices.

4.1.1

Split view **has two meanings in the world of Bento for iPad. A split view on iPad is a view that contains two views. When the iPad is in landscape mode, one view is shown at the left, and the main view is in the center and right. (You see this later in Figure 12.35.) When you rotate the iPad to portrait orientation, the left view disappears; it can be reopened as a popover using a button at the left of the toolbar. In Bento for Mac, split view refers to the composite view shown in Figure 12.23, with a table or grid view at the top and a form view at the bottom. In practice, there is little confusion because these are very separate uses of the phrase.**

Working with Views on Bento for iPhone

With both Bento for iPhone and Bento for iPad, the views are all automatically generated. You do not have the forms that you can customize on Mac OS X. (Obviously, the same mechanisms that enable Bento for Mac to automatically create table and grid views are at play behind the scenes on the iOS devices.)

With Bento for iPhone, you start on the Home screen as shown in Figure 12.27. (Like most iOS apps, Bento generally returns to wherever you were the last time you used it, so you might not start at the Home screen each time you use Bento for iPhone. You can always get to the Home screen with the Home icon at the bottom of most screens.)

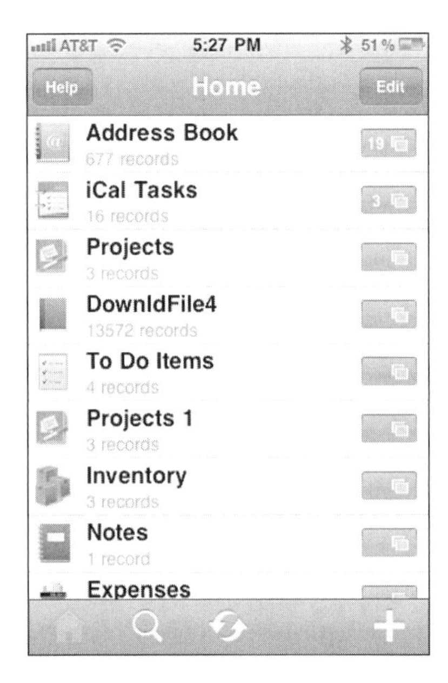

Figure 12.27
Start from the Home screen.

Tap the library you want to work with to get the list of records shown in Figure 12.28.

At the bottom right, you can change the sort order and determine what data is shown in each record's row. Tap the sort order (the stacked items second from the right in the bottom toolbar) to go to the screen shown in Figure 12.29.

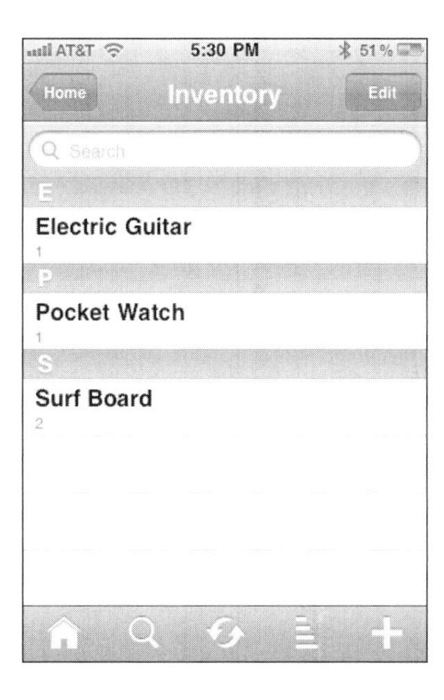

Figure 12.28
Browse the list of records.

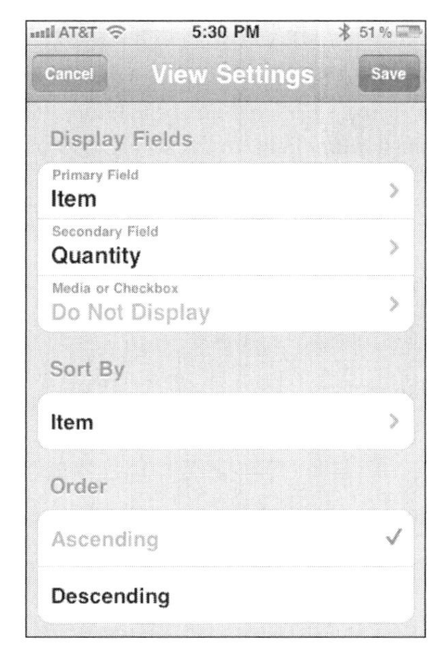

Figure 12.29
Set the sort order.

As you can see, this controls not only the sort order, but also the second line of text below the item name. Tap any of the fields in Figure 12.29, and you can choose a new field to use for sorting or display, as you see in Figure 12.30.

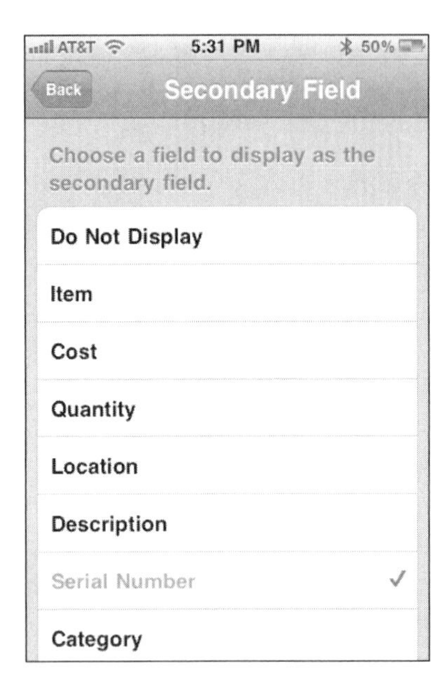

Figure 12.30
Select fields to display and sort.

The fields you can display and sort by are text fields or fields that can be converted into text (dates and times, for example). Bento for iPad does handle media fields, but you cannot sort by images.

From the list of records (shown previously in Figure 12.28), you can search for data. Note that this searches only the data displayed on the record list; data in other fields in the library are not searched.

When you tap a record in the records list, you see its fields, as shown in Figure 12.31.

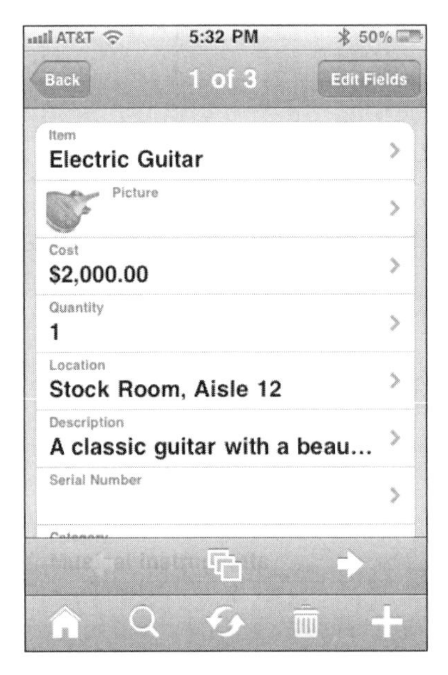

Figure 12.31
Edit fields.

The Edit Fields button lets you edit a field name or change the order of fields in the display. If there is a disclosure triangle at the right of the field in the records list, you can tap it to edit data within the field, as you see in Figure 12.32. The appropriate editor appears based on the field type: in this case, it is a phone number.

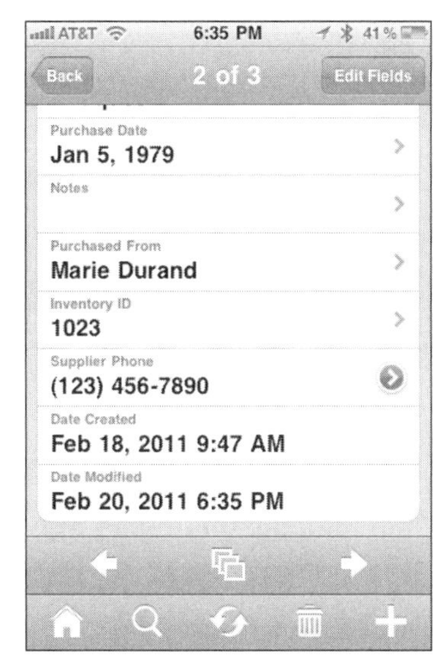

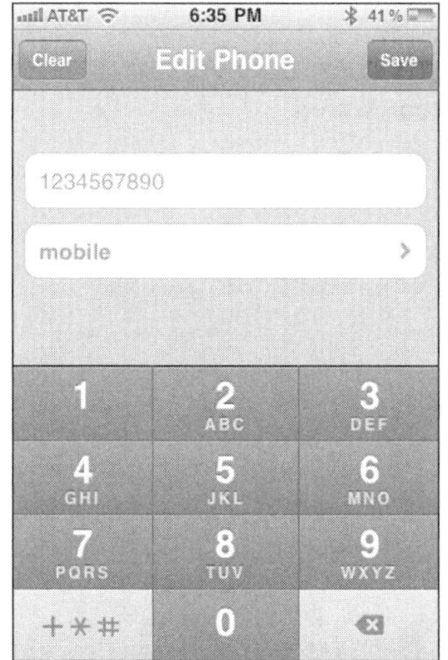

Figure 12.32
View (a), edit (b), and use (c) a field's data.

At the bottom of the screen, you can go to the next or previous record, add the current record to a collection, or delete the record. The only one of these steps that has its own interface is the one to add to a collection (the third from the left).

Tap it, and you see a list of collections in this library, as shown in Figure 12.33.

Note that at the bottom of the screen, you can create a new collection.

Finally, if you want to add a new record use the + in the lower right of the screen as shown in Figure 12.28.

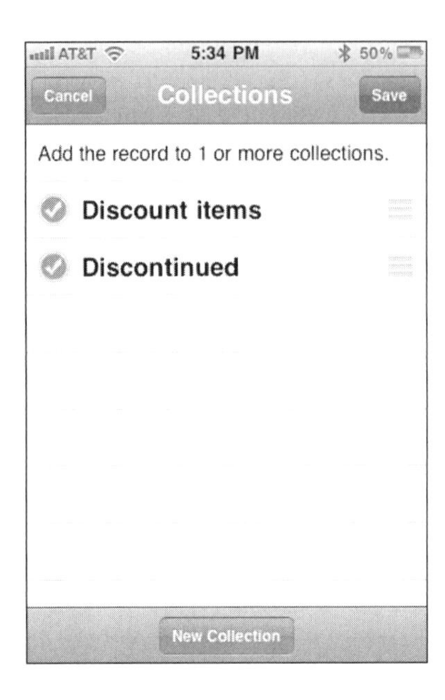

Figure 12.33
Add the current record to a collection.

Working with Views on Bento for iPad

Bento for iPad provides some additional functionality. Most of the features work the same way as they do with Bento for iPhone, but you have additional options, such as the ability to put pictures into the records list.

The basic Bento for iPad window is shown in the landscape orientation in Figure 12.34 and in the portrait orientation in Figure 12.35.

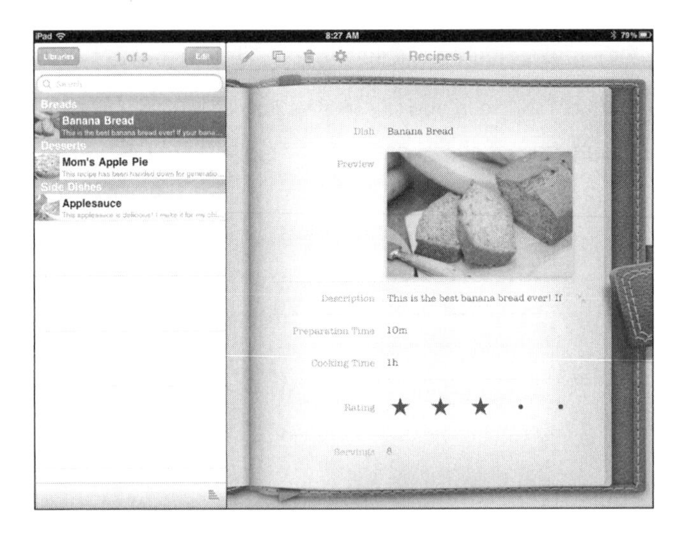

Figure 12.34
Use Bento for iPad in landscape orientation.

The Libraries button at the left of the toolbar lets you show the Libraries popover, as you see in Figure 12.36.

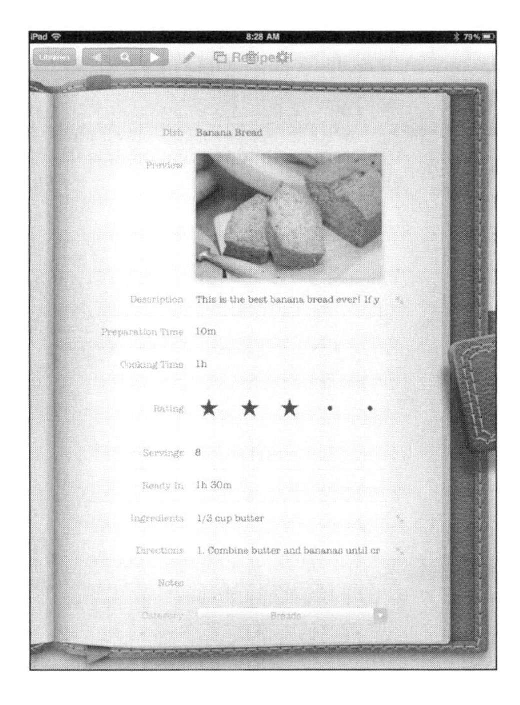

Figure 12.35
Use Bento for iPad in portrait orientation.

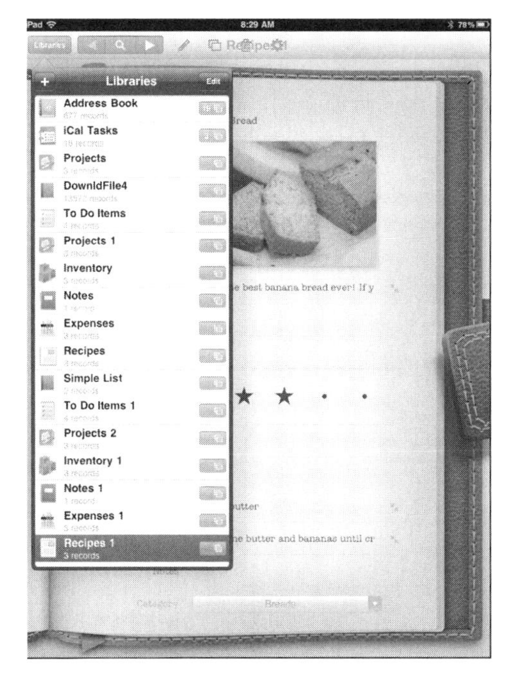

Figure 12.36
Show the Libraries popover with the Libraries button at the left of the toolbar.

This aspect of the interface is slightly different from some other interfaces you might see on iPad. In landscape orientation, the left-hand side of the split view consists of the search field and a list of the records. That data is shown in a popover that you access from the search button (the magnifying glass) in the left side of the toolbar in portrait orientation. Thus, as shown in Figure 12.37, you can show the popover on top of the list at the left-hand side of the screen in horizontal orientation.

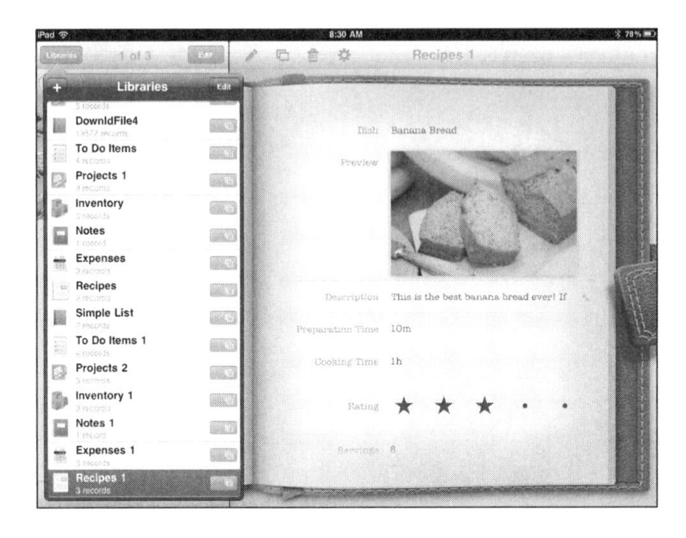

Figure 12.37
The popover can appear in landscape orientation.

The layout and behavior of the records list in Bento for iPad is exactly the same as it is in Bento for iPhone, with one exception: On Bento for iPhone, you add a new record with + at the bottom of the records list, as shown previously in Figure 12.28. With Bento for iPad, the + to add a record is toward the right of the toolbar. This means that it is available in more circumstances.

The view settings shown on Bento for iPhone in Figure 12.29 are almost the same in Bento for iPad. When you tap Media or Checkbox in view settings, you can choose the field you want to use, as shown in Figure 12.38.

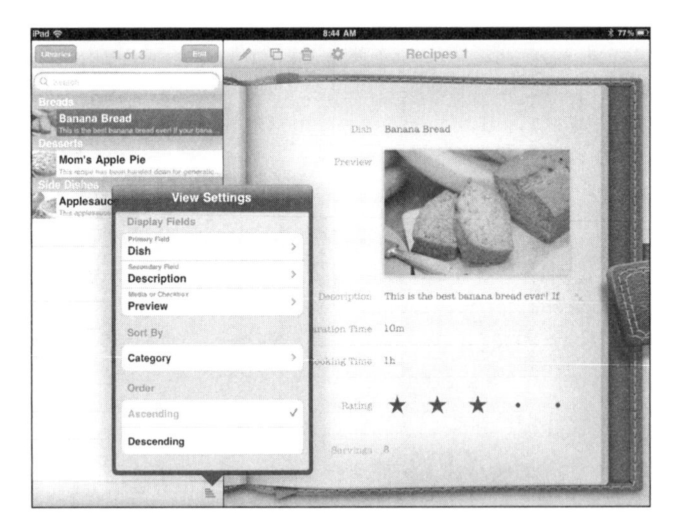

Figure 12.38
Add a media or checkbox field to the records list.

You can change the Bento for iPad theme for a library. These themes differ from the Bento for Mac themes. To change a theme, open the library you want to change, tap the pencil at the right of the toolbar, and then tap Change Theme at the left of the toolbar, as you see in Figure 12.39.

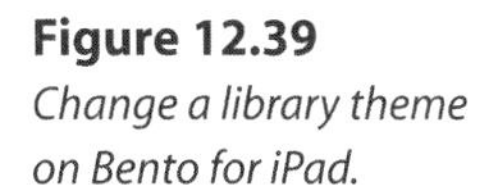

Figure 12.39
Change a library theme on Bento for iPad.

On iPad, individual records are displayed in the automatically generated form view. This means that for composite field types such as addresses, the display looks much like the version on Mac OS X, with a pair of buttons providing additional functionality such as mapping. Figure 12.40 shows an address on Bento for iPad.

Figure 12.40
Use Bento tools for composite fields on iPad.

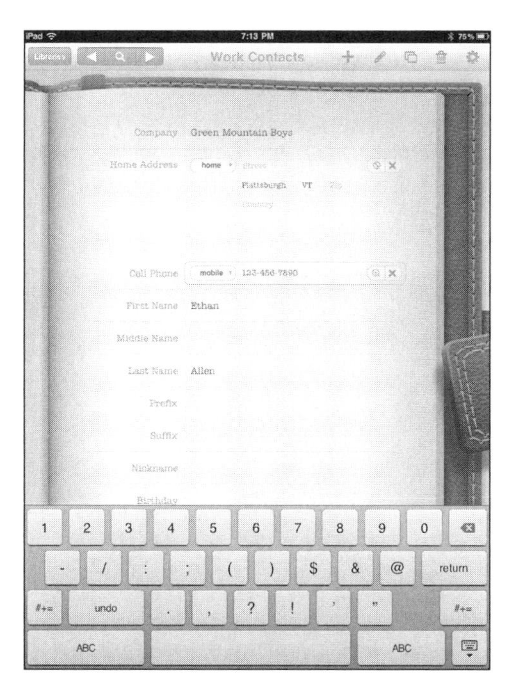

Working with Related Data

Like FileMaker, Bento implements the concept of *relational data*. Relational data is data in two separate libraries that are joined by a *relationship*. For example, you might have a library that contains data for a class. Figure 12.41 shows such a library in Bento for Mac (it is one of the templates). This library is related to another library, "Assignments & Project Due Dates." Related libraries are shown in their own section of the Fields pane, as you see at the lower left of Figure 12.41.

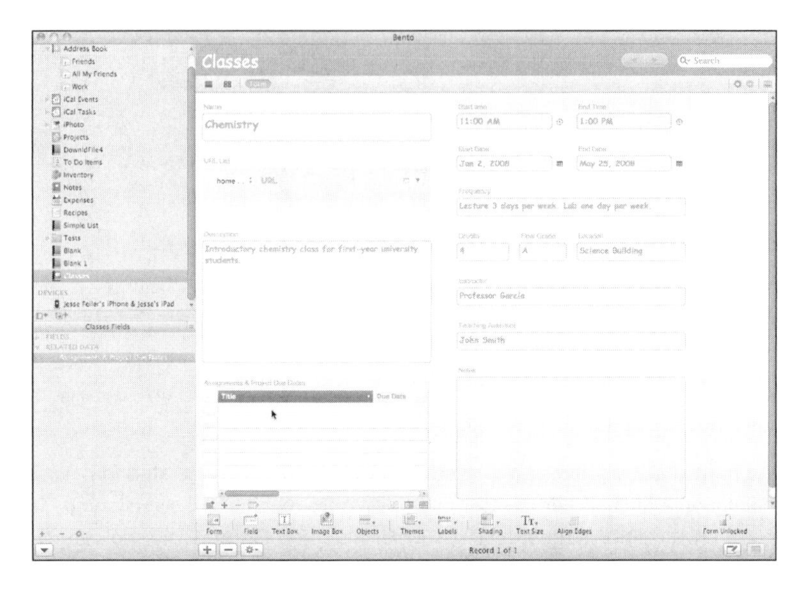

Figure 12.41
Create a Classes library from the built-in template.

As you can see in Figure 12.41, records from the related library can be shown on the form. In this case, the Classes library is designed from the standpoint of an individual student (note at the middle right of the form, there is space for the student's final grade). The related library is the iCal tasks library. On Bento for Mac, you can add a record to the Assignments & Project Due Dates related library by clicking the + at the bottom left of the related library on the form. Type in a distinctive name such as My Related Record, as you see in Figure 12.42.

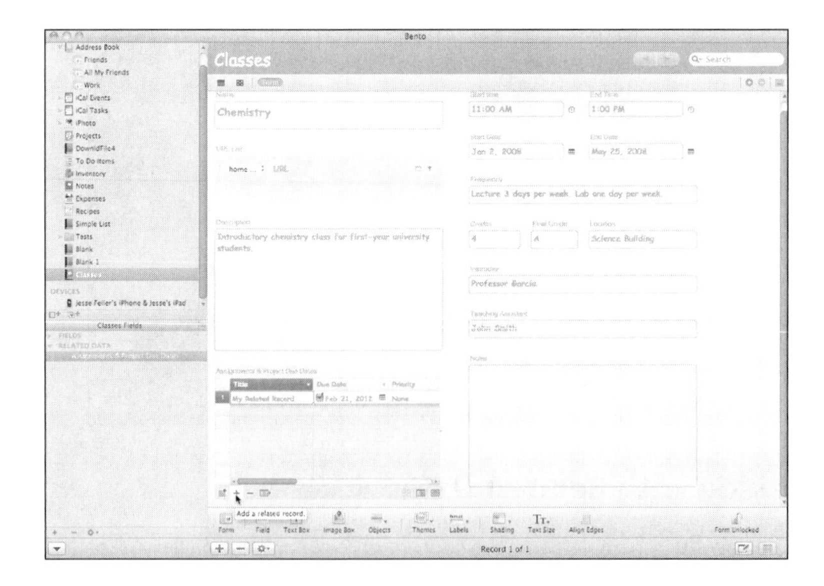

Figure 12.42
Add a related record.

Now go to the iCal Tasks library and look for that record; you will find it. Instead of searching yourself, you can let Bento do the work. With the My Related Record selected, use the link arrow at the lower left of the related records field to go directly to the related record. Figure 12.43 shows how you go to the related record, and Figure 12.44 shows the result after you have clicked. Note the left-pointing arrow in the navigation bar; it returns you to the main record shown previously in Figure 12.43.

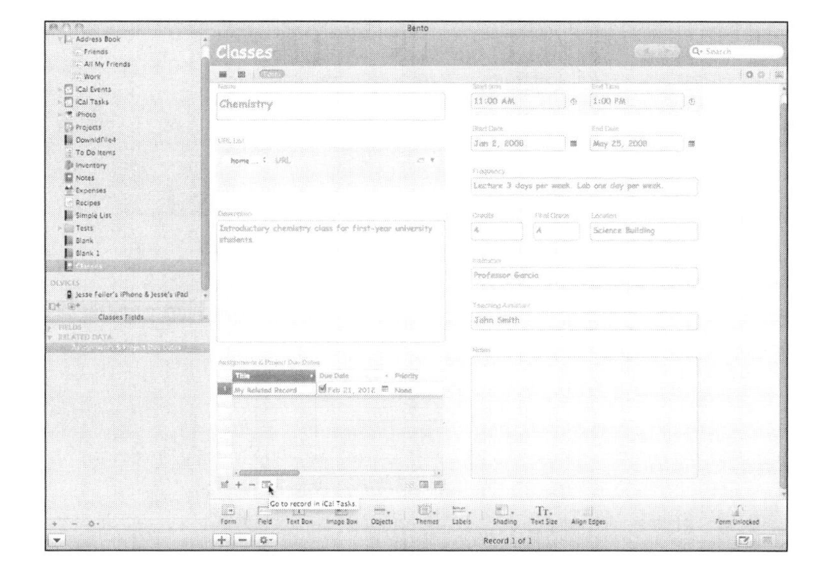

Figure 12.43

You can go to a related record.

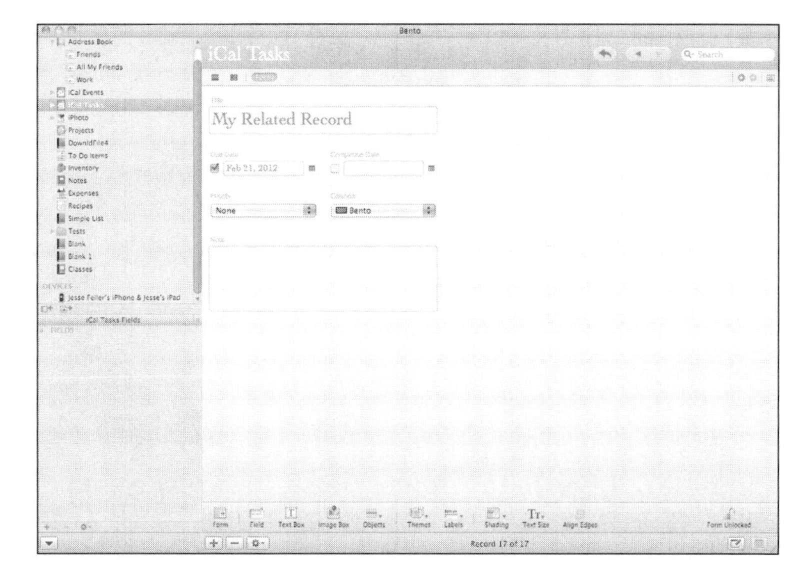

Figure 12.44

The related record is displayed in its own library.

When you add a library in the related records field, it shows up in the related library. You can also work in reverse. Use the button at the bottom left of the related records field to browse the existing records in the related library, as shown in Figure 12.45. Choose the records you want to relate to the main library (in this case, the Classes library). Bento takes care of forging the links for you.

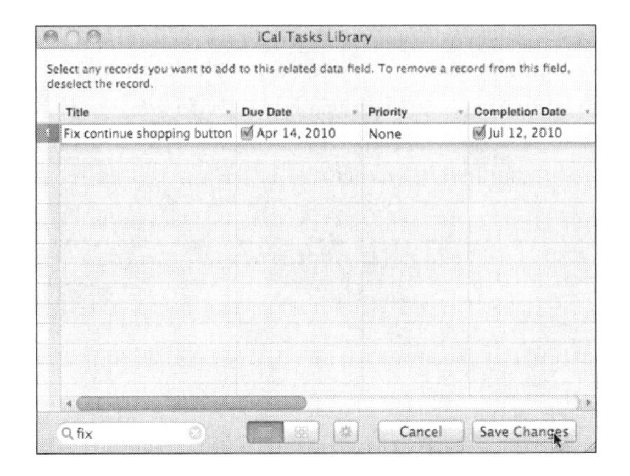

Figure 12.45
You can add records from a related library.

When you use Bento's sync functionality, you can create and edit libraries and library data on various devices and share it without a hassle. In the section that follows, you see two libraries built on Mac OS X and then modified with both their structure and with related record data on iPad.

The Reservations example shown in Chapter 1, "Making Data Mobile," provides an example of using related records. As you can see in Figure 12.46, the primary table is events.

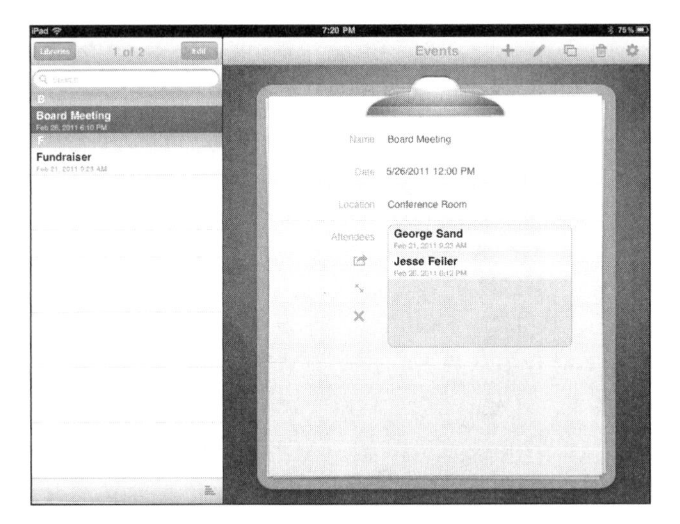

Figure 12.46
Events are at the heart of the Reservations example.

You can add records to the Attendees related records field, as you see in Figure 12.47.

Figure 12.47
Attendees are related to Events.

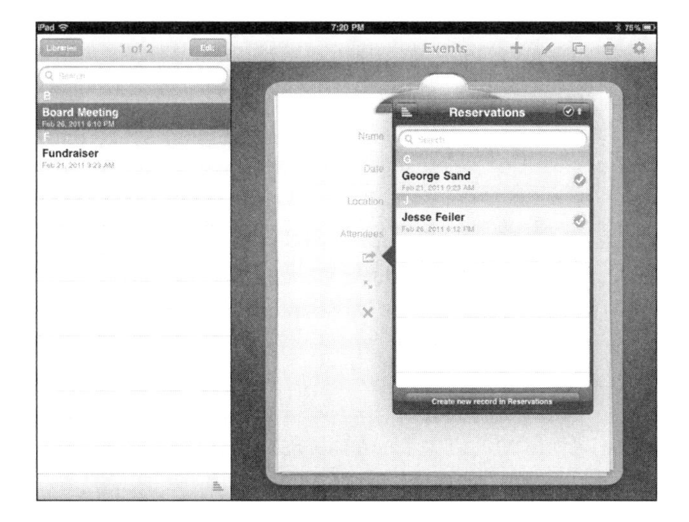

In many cases, it makes as much sense to look at a relationship from one side as from the other. As you can see in Figure 12.48 using Bento for iPhone, you can select an attendee and see the events related to the attendee.

Figure 12.48
View the records list of attendees (left) and then the events for that person (right).

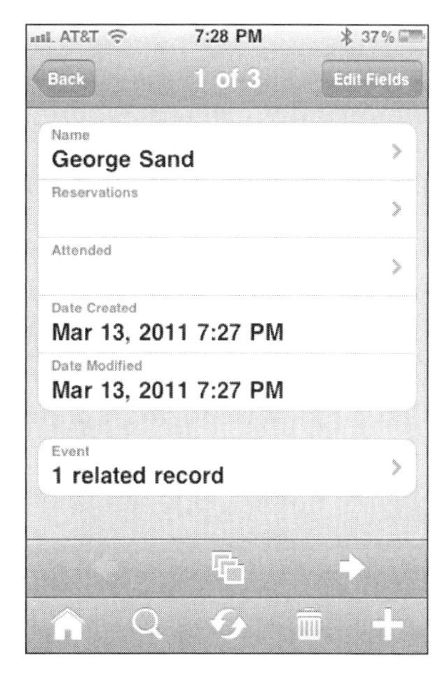

Further Steps

To get a good understanding of how relationships work in Bento, try this experiment. Create a new library from the Classes template. Instead of using it as is (that is, from the student's perspective), change it to work from the teacher's perspective. First, delete the Final Grade field and the Assignments & Due Dates relationship field.

In the place of that relationship field, create a new related records field called Students. Use your Address Book library as the related library. Experiment by adding students in bulk, as shown previously in Figure 12.37, as well as by adding them one at a time from the Classes related records field.

13

IN THIS CHAPTER

- Managing Location Services
- Using Photos and Other Media

Working with Location and Media Fields

As you have seen in Chapter 11, "Using Bento and Bento Libraries," and Chapter 12, "Using Bento Records, Fields, Forms, and Tables," some structural modifications of your libraries are done only with Bento for Mac.

That is similar in some ways to FileMaker Pro. You cannot modify the database structure with FileMaker Go, but you can work with the data that is entered into a structure that has been built on FileMaker Pro. (Note that these are general similarities between FileMaker and Bento. With FileMaker, you cannot make any structural changes on the mobile devices, but with Bento, you can add fields. This difference reflects the different internal architectures of Bento and FileMaker.)

You might be thinking that Bento for mobile devices is in some ways less functional than Bento for Mac, and this chapter should disabuse you of that thought. The two sets of functionality are different in the different environments, and one is not better than the other. In this chapter, you see two features of Bento on iPhone and Bento on iPad that you do not find on Bento for Mac. True, you can accomplish the same results on the Mac, but they are not nearly as easy as they are on the mobile devices.

And what are these mobile-only features? They are mobile-only and mobile-centric location fields, with the ability to automatically enter your location into your Bento library as well as media fields that you can fill with a photo at the tap of a button. Then, with your next sync (see Chapter 3, "Managing Data on the Move"), you can move that data to your Mac, and from there to FileMaker Pro or other applications.

Introducing Location Fields

One of the major features of Bento 4 is the addition of location fields. These fields store latitude and longitude in a composite data structure similar to the structure used to store addresses. You can access each component of the location separately, just as you can access the state or the postal code for an address. And just as is the case with addresses, you need not provide all of the components. You can have a location field into which you type only latitude or another into which you type only longitude.

But typing into these fields is error-prone and messy in some cases. As is so often the case when you encounter something in the world of Bento that is a bit tricky, you have the option to let Bento handle it for you—and that is the heart of Bento's location implementation.

Reviewing Geolocation Basics

Location fields store location values, but most of the time they are used for geolocation purposes. The difference is important. Latitude and longitude let you pinpoint a location (often called a *position*) on the earth's surface. *Geolocation* is the process by which a position is associated with a geographic element, such as a street address. If you use mapping software such as Google Maps, what happens behind the scenes is often that a street address is converted to a position, and it is that position that is mapped. These conversions are visible to you if you are writing mapping mashups and using the Google Maps API, but most of the time you use higher-level software that does the work for you. Nevertheless, it is the latitude and longitude that are at the heart of mapping. In part, this is because the position (the latitude and longitude location) is more precise and less ambiguous than a street address.

→ For more information about the Google Maps API family, see http://code.google.com/apis/maps/index.html.

Manhattan's Sixth Avenue was first mapped out in 1811. In 1945, the New York City Council officially changed its name to Avenue of the Americas. The name was never widely used despite the replacement of street signs. Today, Sixth Avenue is the more common usage. Throughout this entire episode, the latitude and longitude positions of the street have not changed. That is why they are better to use for computational purposes whereas the street address is better for non-computational purposes. (For example in most parts of the world, latitude and longitude are not often used for directions to taxi drivers.)

There are two formats used for latitude and longitude. Both consist of two values separated by a comma. In the traditional degree/minutes/seconds notations (DMS), the earth is divided into degrees, which in turn are divided into minutes and seconds. (There are 360 degrees in a circle, 60 minutes in a degree, and 60 seconds in a minute.) With the advent of computers, decimal degrees (DD) are frequently used so that they can easily be used in computations.

In fact, each component (latitude and longitude) identifies a specific point on a hemisphere—half of the world. For longitude (the east/west positioning), the position that is identified is either east or west of the Prime Meridian (or International Meridian or Greenwich[England]Meridian). The numeric value followed by E or W identifies a longitude location.

For latitude (the north/south positioning), the numeric value followed by N or S locates a point north or south of the equator.

This enables any point on the planet to be identified with remarkable accuracy. Because Bento can identify your location, you do not have to worry about the numbers, but if you are using location fields with numbers that you type in, you have to be aware of what you are dealing with. With DMS notation, the numbers are of the form 42°45′27″ N, 73°53′4″ W. This can pose transcription and typing issues as people look for the special characters.

With DD notation, the same location is 42.757583° N, 73.884577° W. These numbers are easier to type in, but beware of the decimal places. Bento usually uses six decimal places, and that degree of precision means that the last decimal place is approximately 4.37 inches or 1.11 centimeters.

Of course, this summary approaches the matter from the computational side. Whether you are using a GPS system or are typing in a location from a printed source, you might not have the degree of precision that can support locations down to the centimeter. In fact, if your iPad or iPhone is in Airplane mode, you might not have any position information at all, although iOS uses a variety of techniques to try to come up with a location even in the absence of true GPS.

Here are the tips for working with latitude and longitude on computers today:

- Whenever possible, use values that are calculated by the computer. Use decimal degrees, and avoid rounding them. If, as with Bento, you have six positions of precision to the right of the decimal point, keep them intact.
- When importing and exporting data, watch to make certain that the numbers are not rounded.
- Do not lose the E/W or N/S indicators that follow the numbers (this can happen if you import or export the data with software that assumes it is numeric and therefore ignores letters).
- Whether you are using Bento or any other mapping software, do some reality checks. Using Wikipedia or another reference, look up several latitude/longitude positions in various places around the world—in both northern and southern hemispheres as well as east and west. Test these values with the mapping software.

Finding Your Location: The Operating System Side

On both Mac OS X and on iOS, Bento asks the operating system for the current location. It then uses the location as necessary. The operating systems also have built-in location manipulation routines, and Bento uses those, too, as necessary.

There is a fundamental difference in the way in which the two operating systems find your location. iPhone, as well as iPad with 3G, has the ability to find the location of the nearest cell tower (that is necessary for connecting to the telephone network). iOS can use nearby cell tower locations to triangulate the current position. It also can use built-in GPS if available as well as location information from Wi-Fi networks. For most Macs, the location is calculated using information from Wi-Fi networks as well as from the location of your IP address.

Note that these location tools might not function when Airplane Mode is turned on for a mobile device; the loss of an Internet connection can likewise compromise location services on a desktop computer. By relying on the built-in location services of the operating system, Bento does not have to reinvent the wheel. However, it is only as good as the data that the operating system can pass on.

Creating Location Fields

You create location fields the same way you do any other fields. They are toward the bottom of the field type list, as you see in Figure 13.1.

Figure 13.1

Create a location field.

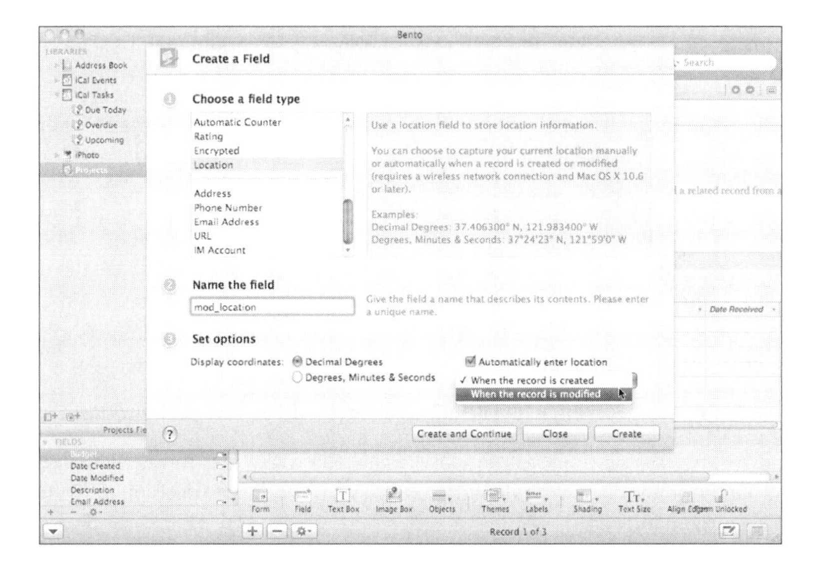

As always, provide the field name, and then specify which format you want to use for the latitude and longitude (it must be the same for both values—it would be very peculiar to use decimal values for latitude and degree/minute/second values for longitude).

At the lower right, you can choose to enter the location automatically. You can choose to enter it when the record is created or when the record is modified.

If you specify automatic entry, note that you cannot override it with a manual value. After you have selected automatic entry, that choice cannot be modified.

Location fields are available in Bento for Mac when you are using Bento 4. You cannot create new location fields on Bento for iPad or Bento for iPhone until after you have done your first sync with Bento for Mac. Thereafter, the fields show up in the fields list on your mobile device and you can add them wherever you want.

Using Location Fields

Using the automatic entry option for a location field means that record creation or modification triggers the location entry (just as you can use a timestamp automatically in the same way). When you are using a mobile device that is location-aware (that is, not in Airplane mode or otherwise unable to find a location), everything happens automatically. This enables you to set up a Bento library that can simply record locations without your worrying about it.

After the library is created and filled with automatically entered location data, you can synchronize it with your other Bento libraries so that the location data is shared. It is not changeable if you have used the automatic entry option, but it can be a good reference tool. In fact, knowing that automatically entered location data cannot be changed after it has been entered is a benefit in many cases.

For example, you can use your iPhone or iPad with Bento to identify locations that are relevant to local businesses, locations of facilities for team games in your region, and other similar projects.

➔ See Chapter 14, "Importing and Exporting Bento and FileMaker Data," for examples and a case study.

Working with Location Fields in Forms, Tables, and Grid Views on Mac OS X

After your location data is in your library, you can use it in forms, tables, and grid views. In form views, you drag the location field from the Fields pane into your form, as you see in Figure 13.2. As always, when you have a form open while you add a new field, that field is added automatically to the form. Of course, you can move it around or remove it and then you can add it back in as shown in Figure 13.2.

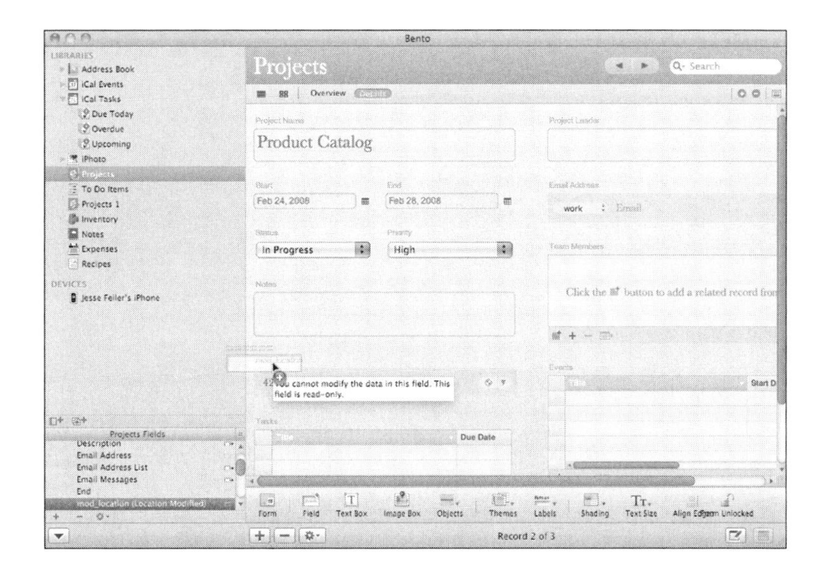

Figure 13.2
Add a location field to a form.

Note that you add the location field with its two subfields (latitude and longitude); you cannot add only one of the subfields. However, if you want to use the latitude or longitude sub-field in a form, there is a way. Create a new calculation field, as you see in Figure 13.3. Inside the calculation, you can access either the latitude or longitude subfield. You can use that data in a calculation, or you can just leave the calculation to show latitude or longitude, as shown in Figure 13.3.

When the location field is added to a form, two controls appear at the right just as they do with address, email, list, and other composite fields. In all cases, you can map the location and find directions to and from it (relative to your current position). If you have not used the automatic entry options, you can also set the field to your current location. You can see this in Figure 13.4.

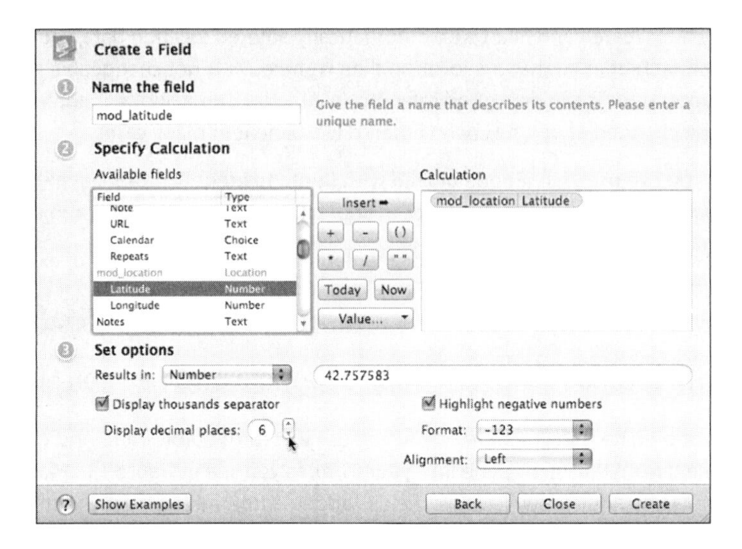

Figure 13.3
Access latitude or longitude in a calculation.

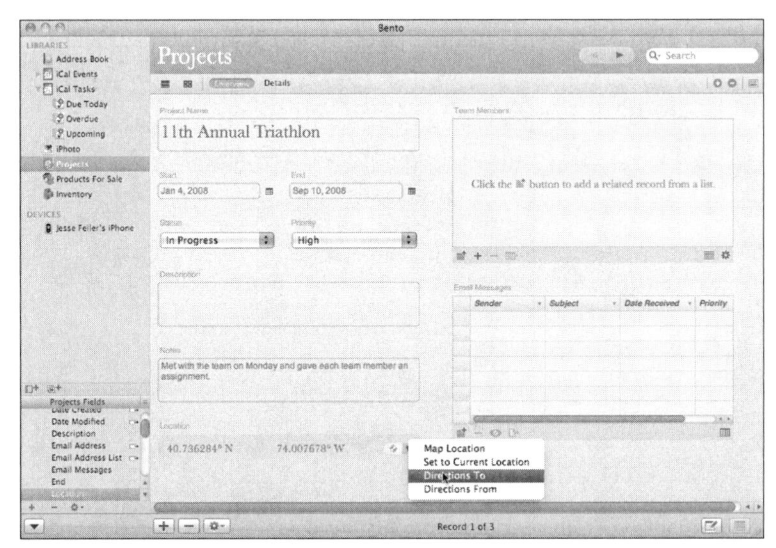

Figure 13.4
Work with a location field in a form.

In table views, a small button in the grid lets you map the location, as you see in Figure 13.5.

The mapping is handed off to Google Maps, so when you click the button in the table view, your browser automatically opens with the address. Because you are now on a standard Google Maps web page, you have all the features for finding directions and distances, as shown in Figure 13.6.

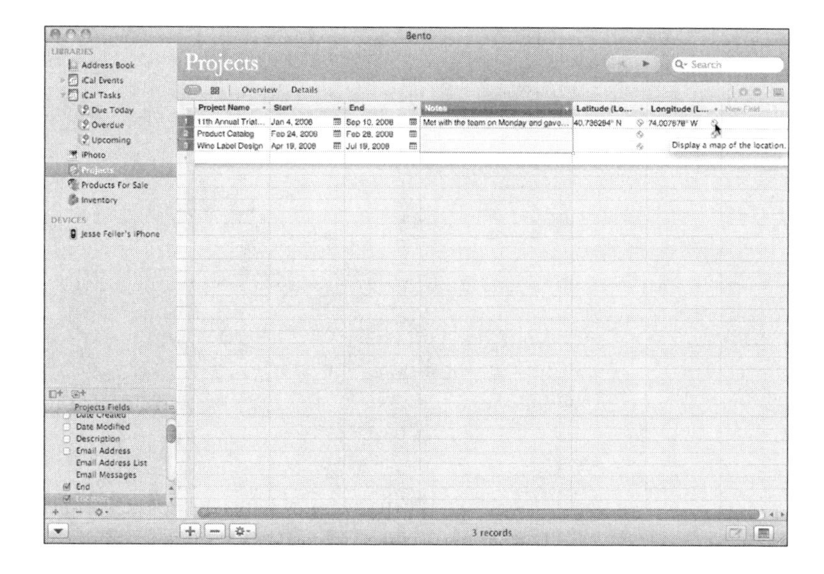

Figure 13.5

Map a location field in a table view.

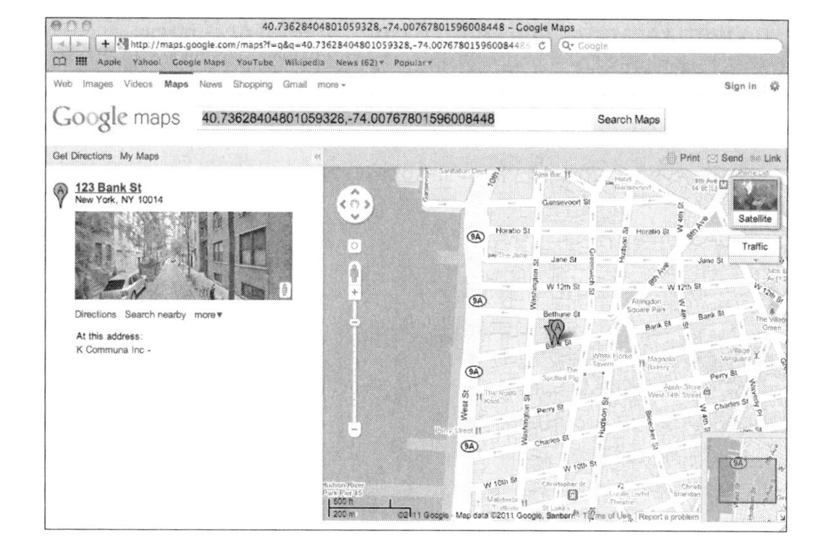

Figure 13.6

Bento launches Google Maps in your browser.

Grid views provide yet another way to use location data in Bento for Mac. As displayed in Figure 13.7, Grid Settings lets you use the individual subfields in the title or subtitle of your thumbnail.

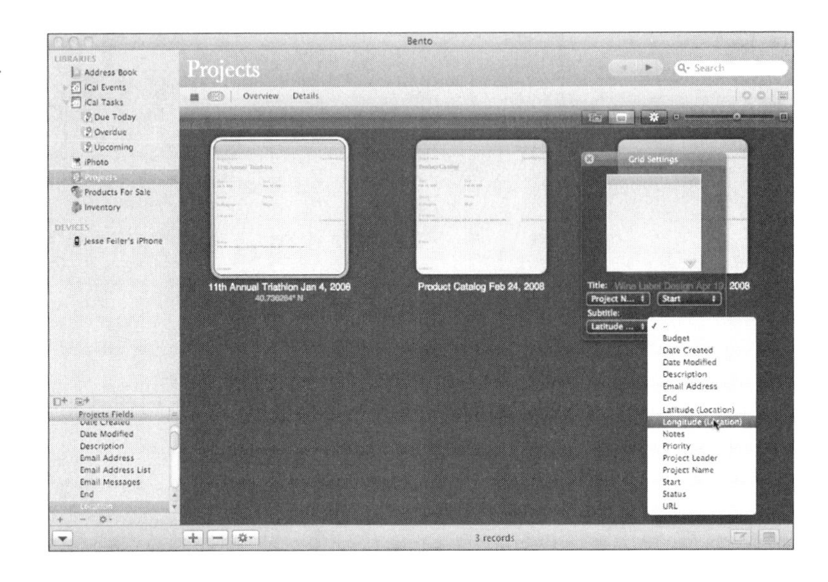

Figure 13.7
Use location data in your grid settings.

Working with Location Fields with Bento for iPad or iPhone

After you have a location field in a Bento library, you can use the data just as you would with Bento for Mac. The interface is different, but the functionality is the same.

Figure 13.8 shows how a location field can appear on iPhone. Note how the position found on the Mac in Figure 13.4 automatically can be coverted to a geolocation (a street address).

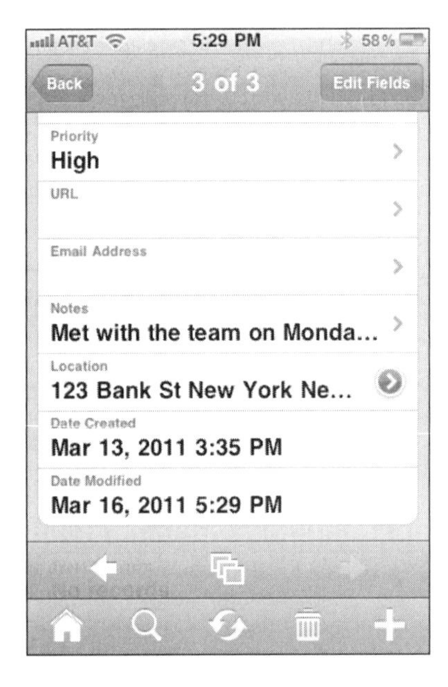

Figure 13.8
You can interact with location fields on iOS devices.

Tapping the detail disclosure button at the right takes you to a map where the title is automatically set to the appropriate value from Bento and the location, as you see in Figure 13.9. You can also interact with the map as usual to find directions or open it in Maps as you always do on an iOS mobile device.

Figure 13.9
Set the current location.

Looking at Media Fields

Media fields can store photos or movies. With Bento for Mac, media fields appear as shown in Figure 13.10. When you click in the field, the small control appears and enables you to choose photos, movies, PDF files, or audio.

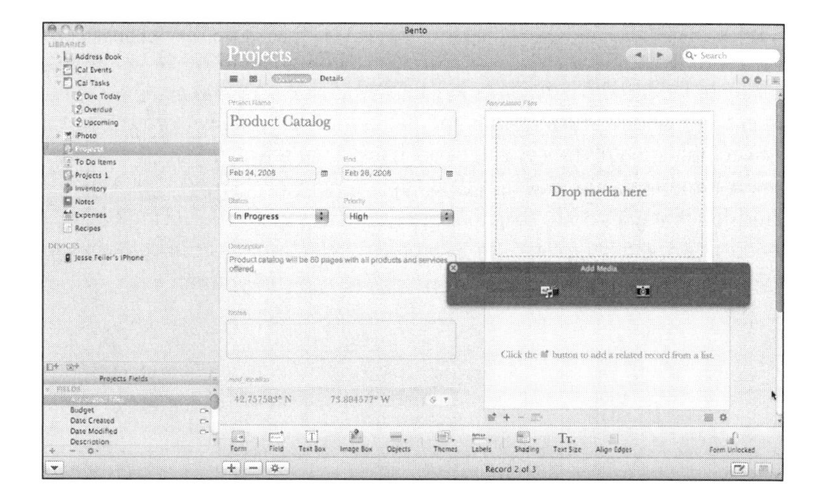

Figure 13.10
Add media to your Bento form.

If your Mac has a camera, clicking the photo icon enables you to take a photo right then and there. The photo is inserted into the media field.

Exactly the same thing happens with Bento for iPhone, although the interface is different. In Figure 13.11, you see an empty media field on the left; tapping the action button in the lower left of the screen enables you to add content as shown at the right of Figure 13.11.

Figure 13.11
Work with a media field with Bento for iPhone.

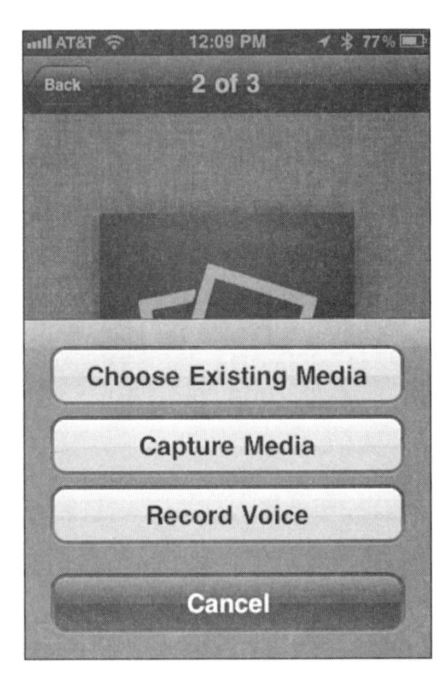

The process on iPad is basically the same as for iPhone.

Further Steps

Create a Bento library with a media field, a location field, and a text field for the name of the location. Experiment with photos and locations and then sync the library with your Mac and on to your iPad (or with whatever combination of devices you have). Try to get a sense for how the photos look on each device, and much more important, how much precision you have with location fields. Although theoretically you might have precision that goes down to inches or centimeters, in practice, using GPS technology at the small scale is not a good idea. Try to get a sense for how precise your locations can be. Because this depends in part on where you are, you should be able to see what level of detail you can use.

14

IN THIS CHAPTER

- Importing and Exporting Text-Based Data
- Creating and Using Bento Templates

Importing and Exporting Bento and FileMaker Data

Bento and FileMaker Pro along with many other software products let you import and export data using common formats. This reflects the reality that, in today's world, you often need to deal with data in a variety of formats that is stored in a variety of products. If your data is locked in a proprietary fortress to which you do not have access, it is not really *your* data; it is more like a hostage.

With the advent of mobile devices, many people thought of them as extensions of databases that resided on servers; the mobile devices were just new ways of getting into the corporate database. Over time, however, people have come to realize that mobile devices are not just handy ways to gain access to the main databases, they are also an important way of providing data to the database. The computer on your desk in the office is not really a great tool to use for building a database of street signs that need replacing. True, you can take photos and move them to your computer's disk, but with a mobile device, you can take the photo and store it in Bento or FileMaker Go without bothering with memory cards or cables.

Sometimes, such as with photos, the sharing is most easily accomplished by using Bento template sharing. Other times, you are just sharing text data, and that can be done with any of these formats. Remember that when it comes to mobile devices and Bento, you have access to some very important textual data that can be tremendously important when moved to FileMaker databases and even spreadsheets or word processing documents. Those two coordinates of location are text data, but it is the mobile device that has the ability to capture those numbers. (So, too, can a desktop Mac, but dragging it around to various locations so that you can ascertain a location requires an awfully long power cord.)

This chapter provides an overview of how you can cross the boundaries of formats and software products to take charge of your data no matter where it is.

Sharing Data with FileMaker and Bento

There are three models of data sharing that are supported in this environment (these three models are separate from the synchronization processes described in Chapter 3, "Managing Data on the Move"):

- **Importing and exporting records.** Most database and spreadsheet applications today (along with many word processing products) let you import and export records in common formats. Both FileMaker and Bento let you do this. The import and export works with records in a single table, so importing and exporting does not work for related tables (if you have visible keys to provide a link between the tables, you can import and export each table separately).

  The import/export process copies data so that at the end of the process, you have two identical sets of data. (In some cases, you deliberately delete the data that you have used for your import because it is now in FileMaker or Bento.)

  Importing and exporting to and from FileMaker Pro and Bento is summarized in this chapter.

- **Using data sources.** FileMaker lets you define a *data source* that can become part of your solution's relationships graph. A data source can be another FileMaker table, or it can be an SQL data source that might be running in another type of database manager. Data sources can be part of your FileMaker layouts, and your users might never need to know which fields in a layout are in a local FileMaker database and which are located elsewhere and in another type of database.

  What is important is that data sources let you access data in real time. You have a single point of access to your data through FileMaker's relationships graph, but the data is stored in whatever database manager is defined to the relationships graph. There is only one copy of the data. In other words, the link is not from file to file but from application to application. If either the sending or receiving application is not running, the data is inaccessible. But if they both are running, you have a powerful and dynamic data-sharing tool.

  FileMaker data sources are beyond the scope of this book. For more information on the topic, see www.filemaker.com and search for "data source."

- **Bento template sharing.** Bento and its templates do something that is fairly rare in the database world: You can share a database and its data—and that includes related data.

  Bento template sharing is described further in this chapter.

Importing and Exporting Records in Bento and FileMaker

Importing and exporting records normally means taking a single table (rows and columns, like a spreadsheet) and copying it into a file that might or might not use the same format. As noted previously, at the end you wind up with two copies of the data.

The import/export process can be done over time; it is not a dynamic real-time process such as is used with data sources. It is common to export data from one database to a file, store or transmit that file, and then import it into another database. This process can happen over a span of minutes, or it can be interrupted. The exported file may be saved for months or years before it is imported into a database.

The formats used for importing and exporting data in this way are not new. In the early days of computers, data was typically entered into fields of fixed width using punched cards or magnetic tape. The location of a particular digit or letter made it clear to what data field it belonged. This method wasted a lot of space (a field that could hold 10 characters might only need 2), and it was prone to typing errors that could throw off data entry. (Keypunch machines had programs that could in some ways mitigate these problems.) A common solution was to use a special symbol to *delimit* each field. Common symbols were (and are) commas, semicolons, and tabs. Of course, if any of these symbols are used in the data itself, their value as delimiters is compromised; as a result, many delimited fields are not only delimited but are also surrounded by quotes.

Despite the challenges of delimited data entry, it works most of the time and has been supported since at least the 1960s. (The biggest objection to using delimited data entry in the 1960s was that because the data needed to be scanned programmatically, it used more machine processing power than fixed-width data entry.)

This discussion of delimiters and the "waste" of computer cycles for scanning data might seem quaint, and indeed delimited files are a remnant of a long-gone computer age. Nevertheless, they are still used because they (mostly) work, and because so many software programs support them for import and export. As a result, they are the primary data-sharing technologies for many people. Both FileMaker Pro and Bento support them.

There is one important point you might notice: These are all text-based technologies. Digital data needs to be converted to or from text and the appropriate delimiters added or inserted. Bento and FileMaker along with many, many other hardware and software products are equally adept at manipulating text as they are with digital data, such as three-dimensional animations, video, and still images. How do you import these with a text-based format?

The answer is two-fold. On the one hand, you do not put non-text data into these files; on the other, you can use a URL of a file that contains the digital data and use that text-based URL for conversion. (FileMaker Pro makes this particularly easy.)

These are the general issues that arise with importing and exporting data using the standard formats. There is more information on the topic in *FileMaker Pro 10 in Depth*, as well as on the FileMaker website. What is most important to realize is that the powerful importing and exporting mechanisms in FileMaker Pro, Bento, and many other software products generally work flawlessly with text-based data. As soon as you are thinking about importing and exporting non-text data, you need to do a little more research, because that might be where problems arise.

Table 14.1 shows the formats that are supported for import and export by FileMaker Pro and Bento.

Table 14.1 IMPORT/EXPORT FORMATS FOR FILEMAKER PRO AND BENTO

	FileMaker		Bento	
Format	Import	Export	Import	Export
Tab-separated text (.tab .txt .tsv)	X	X	X	X
*Comma-separated text (.csv .txt)	X	X	X	X
DBF				
Merge (.mer)	X	X		
HTML table (.htm)	X	X		
FileMaker Pro (.fp7)	X	X		
XML (.xml)	X	X		
Excel 95–2004 (.xsl)	X	X	X	
Excel Workbook (.xslx)	X	X	X	X
FileMaker (.fp7)	X			
Numbers (.numbers)			X	X
Bento template			X	X
Bento data source	X	X		
XML data source	X	X		
ODBC data source	X	X		

**Bento lets you choose commas or semicolons as the delimiter in CSV exports.*

The following sections provide step-by-step guides showing you how to important and export data.

How To Import Data into Bento

The first step in importing data into Bento is to generate the import file by exporting it from Bento, FileMaker or another program such as Numbers, Excel or any other program that can generate a file for Bento to import.

1. **In Bento, use File, Import, Data to open the dialog shown in Figure 14.1.**

2. **Using the button at the top, choose the file to import.** If you cannot see it, chances are it has not been exported into a format that Bento can recognize. Check the file extension for starters.

3.a. **If you are importing from a Numbers or Excel spreadsheet, you can choose the specific worksheet, as shown in Figure 14.2.**

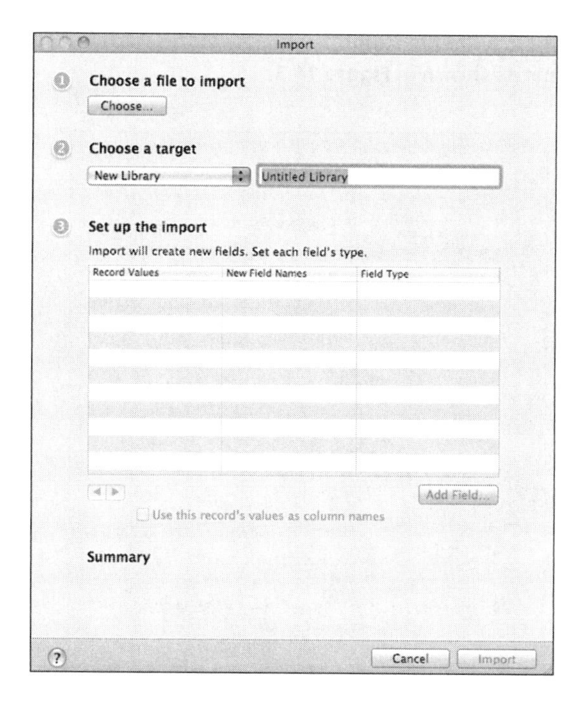

Figure 14.1
Import data into Bento.

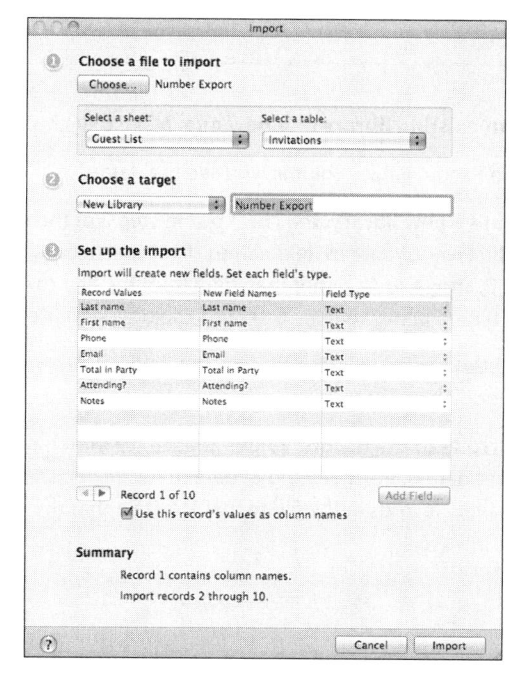

Figure 14.2
Choose a worksheet to use for importing.

3.b. If you are importing from a flat file (comma- or semicolon-separated values or tab-delimited) you can choose your format as shown in Figure 14.3.

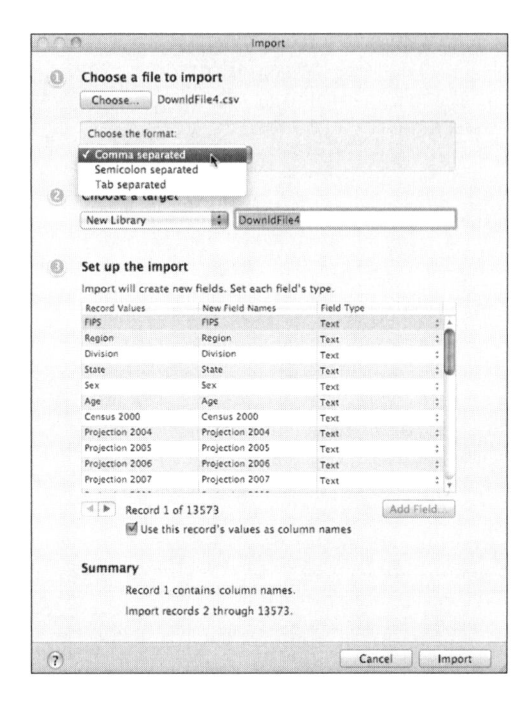

Figure 14.3
Choose an input format.

4. **You can add the data to an existing library or a new one.** Make that choice on the dialog.

5. **Bento reads the file.** In the Record Values column, you see the data.

   If you have chosen to create a new library, the checkbox in step 3 of the dialog lets you choose whether or not to use the first row of data as field names. If you uncheck it, the New Field Names column contains Field 1, Field 2, and so forth rather than data from the first row of data. Either way, you can modify the field names by typing in new names. Select the field type to use.

   If you have chosen to add data to an existing library, the field name and field type are taken from that library. They are not modifiable.

6. **Use the double-headed arrows in the Record Values column to match data to a specific field.** If the exported data is correctly formatted, the matching of data to field is the same for each record. Use the arrows to browse through a few records in the file to make certain that the data is in the right place.

7. **Click Import.**

Exporting Data from Bento for Mac

You can export data using a variety of formats, as you see in Table 14.1.

How to Export Data from Bento

Exporting data from Bento means that you are creating a file that can be imported. You can use this symmetry to test your import and export knowledge: Start from a Bento table, export it, and then import it. You should wind up with the same data.

1. **Open the Bento library you want to export.**
2. **If you only want to export some records, open a smart collection based on that library or perform a search.** Otherwise, make certain all records are shown.
3. **Choose File, Export to open the dialog shown in Figure 14.4.**

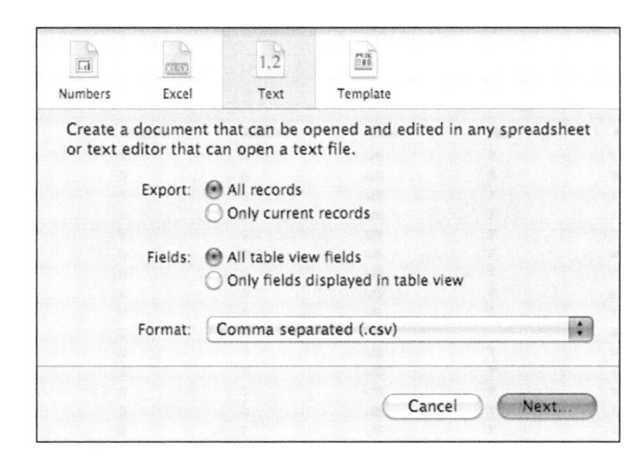

Figure 14.4
Export data from Bento.

4. **Choose the format you want to use: Excel, Numbers, or Text.** Bento templates are described in the following part of this chapter.

   Choose the checkbox to export selected or all records.

   Use the next checkbox to select whether all fields are exported or only the fields you have chosen for the table view in your library.

   For text exports, use the pop-up menu to choose your delimiter: comma, semicolon, or tab. Click Next.
5. **You will be asked to name the file.**
6. **Click Save.**

Importing Data into FileMaker Pro

How To Import Data into FileMaker Pro

In part because FileMaker has many more features than Bento, importing has a few additional steps on FileMaker Pro.

➔ This is a summary of importing features in FileMaker Pro. For more details, see *FileMaker Pro 10 in Depth*.

1. **Begin by creating or locating the export field that you want to import into FileMaker Pro.**
2. **Open or create a FileMaker file to receive the data.** You can create a new database file and not add any fields to it as shown in the accompanying figures. FileMaker Pro creates the table from the data just as Bento can do. You can also import into an existing FileMaker table.
3. **Choose File, Import Records, File to open the dialog shown in Figure 14.5.**

Figure 14.5
Choose the import file.

4. **At the bottom, FileMaker enables you to set this up as a recurring import (for example, of nightly export data).** If you choose to make it a recurring import, after you click Open, you fill in the data shown in Figure 14.6.

   Note that what happens is that a script is created. You can run it manually or from a button that is added to the layout that is created. This gives you a one-click way to update data from an external source.

   This is the end of the process for repeated imports. One-time imports require that mouse click and then let you configure additional options.

Figure 14.6
Provide the recurring import data.

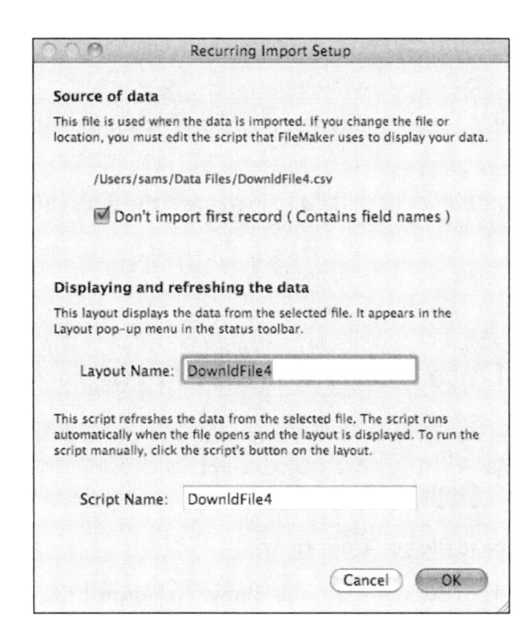

5. **For a one-time import, the Import dialog opens, as shown in Figure 14.7.**

Figure 14.7
Configure the import.

6. **Use the pop-up menu at the top right to import into an existing table or to create a new one.** (Creating a new table is what Bento does on its imports.)

7. **Browse through records with the back and forward arrows in the lower left.** The data in the field appear at the left in the Source Fields column. You can choose to skip the first record, just as you do in

Bento. The target fields in the right column can consist of the first record's data or the default f1, f2, f3... field names.

8. **Use the two-headed arrows to move the target field up or down to match them with the appropriate data.**

9. **Click the Field Mapping symbol for each field to indicate if the data should be imported or not.**

10. **Click Import.**

How to Export Data from FileMaker Pro

As you will see, the FileMaker Pro exporting process is the same as it is for Bento. In fact, the export process is basically the same for Bento, FileMaker Pro, Microsoft Excel, Numbers, and any other program that is going to export data in a common format.

1. **Open a layout in the database file you want to export.**

2. **Choose File, Export Records to open the window shown in Figure 14.8.**

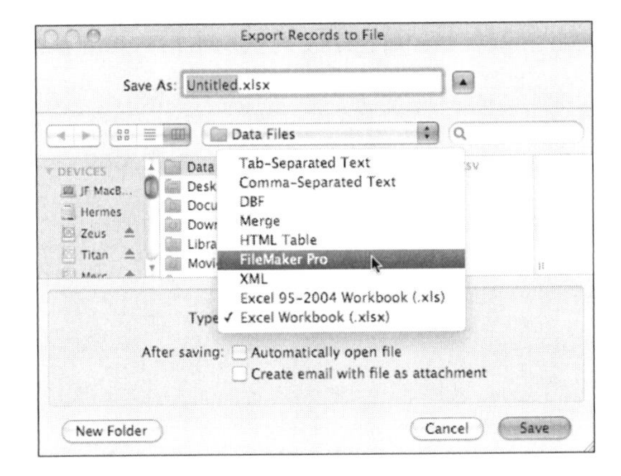

Figure 14.8
Set up the export.

3. **Enter the file name and choose the export format.**

4. **Specify Field Order for Export opens, as shown in Figure 14.9.**

5. **Select the fields to export.** The pop-up menu at the top left lets you choose fields from the current layout, the current table, and for related or unrelated tables. You can use any combination of these fields.

6. **Double-click the fields you want to export or select them and click Move (or Move All).**

7. **The fields are moved to the Field Export Order list.**

8. **Drag the fields up or down to produce the order you want.** Note that if you control both importing and exporting, you can specify the order either in importing or exporting. The flexibility is needed for cases in which you have to use an export order that someone else has implemented.

9. **Click Export.**

Figure 14.9
Select the export fields.

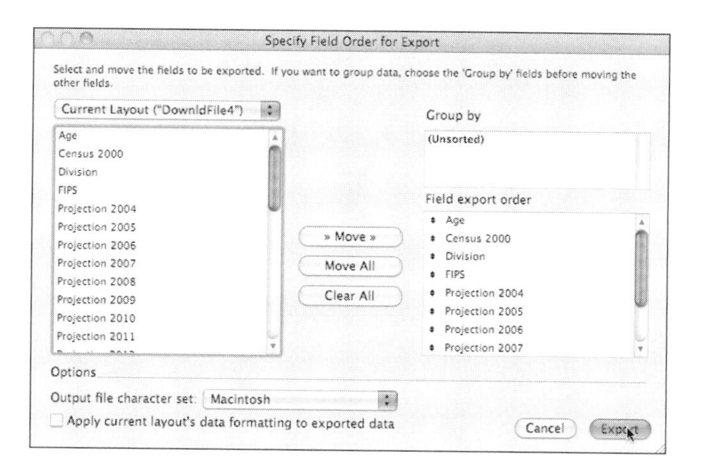

Importing and Exporting Data with Bento Templates with Bento for Mac

When you are moving data into and out of Bento, you are not limited by the text-based formats described in the previous section. You can create a Bento template that includes the library design including fields and forms as well as the data in the library. Because it is not a text-based format, you can export and import graphics and media files without worrying.

This is the simplest way of moving data around with Bento. For example, you can create a Bento library as a tour guide with records for important stops on a walking tour. Take a photo at each stop and store the location so that users can use it to find directions if they get lost. Then export the library with its data, and perhaps post it on your town's website.

This sort of scenario is very different from synchronizing and sharing Bento data over your local area network. It is a way to publish Bento data.

The Bento-to-Bento export process is very simple. Start by opening the library you want to share. Choose File, Export to open the dialog shown in Figure 14.10.

Click the Template tab. You have two checkboxes you can use:

- **Related records.** You can choose to include related records and libraries or not.
- **Data.** You can choose to export blank libraries or to include the data.

If you include the data, you might want to make a copy of your actual Bento library and clean it up. Perhaps you want to remove some of the records so that the distributed template has only a few records to demonstrate how to use it.

Click Next, enter a file name, and you have created your template.

You can take the template file and post it on the web or send it via email.

To use a template, choose File, Import, Template and select the template you want to use. (New Library from Template is for the built-in templates, not the ones you create.)

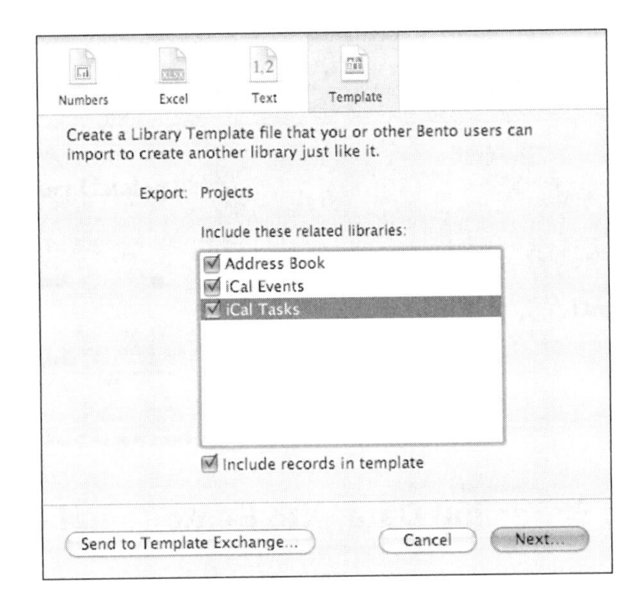

Figure 14.10
Export a Bento template.

Further Steps

Take some time to experiment with importing and exporting; it's a good idea to try the process before you need to do it for real. You might have some problems with downloaded files that you want to import, so get used to the kinds of repairs you might have to make. The most frequent issue that you find with downloading spreadsheets and flat files is that they might have been "cleaned up" and "improved." You have to take out these clean up and improvements.

Note that both Bento and FileMaker Pro provide the option to use the first line as column/field headings or not. Many people go beyond this and create three or four title rows. You have to take out all of them or all except one; otherwise, those additional title rows show up as data.

Likewise, the basic rows-by-columns format of a spreadsheet is often improved on, as in some places that are in fact several tables on a single sheet. There, too, you need to take out the "improvements."

FileMaker Pro handles quite a few database formats that are more restrictive than the decades-old text formats; they might be your best alternative.

15

IN THIS CHAPTER

Deploying FileMaker/IWP with FileMaker Server Advanced

An Overview of Instant Web Publishing

Instant Web Publishing (IWP) lets you publish your FileMaker databases to a web server with remarkably little effort.

The IWP Story So Far

Broadly speaking, Instant Web Publishing (IWP) is one of the two primary options for sharing data from a FileMaker database to the web. The other major option is using PHP: Hypertext Preprocessor with the Web Publishing Engine.

→ See Chapter 16, "Deploying FileMaker/CWP with FileMaker Server," **p. 347**.

The goal of IWP is to translate to a web browser as much of the appearance and functionality of a FileMaker Pro database as possible, without requiring that a developer do any additional programming. FileMaker layouts are rendered in the user's browser almost exactly as they appear to users of the FileMaker Pro desktop application.

IWP is more, though, than simply rendering your layouts as web pages. IWP users have much, if not all, of the same application functionality as do FileMaker Pro users. They can run scripts and view, create, edit, and delete data just like traditional FileMaker Pro users.

Almost all the differences between IWP and FileMaker access to databases have to do with the fact that in IWP, the database displays in a web browser, and in FileMaker, it displays in a FileMaker window. The consequence of this is that the menus at the top of the window or display belong to the browser, not to FileMaker. Therefore, all FileMaker menu commands available in IWP have to be provided as buttons or other controls in the IWP controls area at the top of the window or the layout itself.

Because Safari is built into the iOS devices, nothing prevents you from directing Safari to an IWP page: It renders it as accurately as possible, preserving the FileMaker look and feel through the browser interface. This is the easiest way to get your FileMaker databases onto a mobile device. This chapter shows you how to do that. (Mobile Safari is the browser for the iOS devices, but other mobile devices have their own browsers. FileMaker/IWP works with most up-to-date browsers, so it can certainly be worth experimenting to see how wide a net you can cast with your FileMaker database after it is IWP-enabled.)

But there is much more you can do. You have to go back to one of the first decisions you must make about FileMaker databases on mobile devices: Do you want the mobile result to look like FileMaker or like iOS? You have seen a variety of techniques that you can use to accommodate the smaller screen of a mobile device, and the issues that you confront when typing is a bit more complicated than it is on a desktop-based computer. You might want to use those techniques in your FileMaker layouts so that when they are translated onto mobile Safari, they look as good as can be.

For many people, the mobile Safari presentations of IWP databases are so important and used by so many people that it is worthwhile to create at least a few special-purpose layouts specifically designed for IWP on mobile devices. What do those special-purpose layouts need to incorporate? The answers are the same as always: larger controls and font sizes and more space between elements. And before you sign off on the deployment, test with all of the mobile devices you might be using.

For starters, though, if you have an existing FileMaker database that you want to publish to a mobile device, you might consider just turning on IWP and to see what it looks like without alternate layouts of tweaking.

See Chapter 16 to see how to use PHP for more control of your web-published databases. If you want to publish to iPhone, FileMaker/CWP is usually a better choice than FileMaker/IWP.

Getting Started with IWP

After you decide that IWP is something you want to try, there isn't too much you'll need to do to get started. There are two ways to deploy IWP. You can use the regular FileMaker Pro desktop application, in which case you're limited to publishing a maximum of 10 database files to at most five concurrent users. Alternatively, you can use FileMaker Server Advanced, which allows for significantly more files and users. The configurations for these options are covered in detail in the next section.

The host machine—whether running FileMaker Pro or FileMaker Server Advanced—of course needs to have an Internet (or intranet) connection. Ideally, it will be a persistent connection (for example, cable, T1, or DSL).

The host machine also must have a static IP address. If you don't have a static IP address on the host machine, remote users can have a difficult time accessing your solution. Finally, any databases you want users to access via IWP need to be open on the host machine.

A static IP address can be provided by your ISP; it will enable people outside your organization to connect to the IWP databases. If you have a local area network that communicates with the Internet through a single connection, each machine on the network has a separate IP address. They can be created dynamically, or they can be assigned within the network. (They typically start with 192.168 or with 10.0, depending on the size of the network.) Your IWP host machine can have a static IP address within the network that you assign even if the entire network's IP address (visible from the outside) changes. This static IP address will enable everyone inside the network to consistently connect to the IWP host, but outsiders will not be able to do so.

To access your IWP-enabled files, remote users need to have an Internet connection and a compatible browser. Compatible browsers must support cascading style sheets (CSS), JavaScript, and other modern web tools. Fortunately, most browsers on mobile devices are up-to-date; you are not likely to run into an ancient copy of a pre-CSS browser on a mobile phone or tablet.

Enabling and Configuring IWP

To publish databases to the web via IWP, you must enable and configure IWP on the host machine, and you have to set up one or more database files to allow IWP access. Each of these topics is covered in detail in the sections that follow.

Configuring FileMaker Pro for IWP

Using FileMaker Pro, you can share up to 10 databases with up to five users. To share more files or share with more users, you need to use FileMaker Server Advanced as your IWP host. FileMaker Pro can serve only files that it opens as a host—that is, it's not possible for FileMaker Pro to open a file as a guest of FileMaker Server Advanced and to further share it to IWP users.

You cannot share files with IWP with both FileMaker Server and FileMaker Pro on the same computer at the same time. Only one can manage the sharing software.

Figure 15.1 shows the Instant Web Publishing setup screen in FileMaker Pro. In Windows, you get to this screen by choosing Edit, Sharing, Instant Web Publishing. On Mac, choose FileMaker Pro, Sharing, Instant Web Publishing. The top half of the Instant Web Publishing dialog box relates to the status of IWP at the application level; the bottom half details the sharing status of any currently open database files. The two halves function independently of one another and are discussed separately here. For now, we're just concerned with getting IWP working at the application level and therefore limit our discussion to the options on the top half of the Instant Web Publishing dialog.

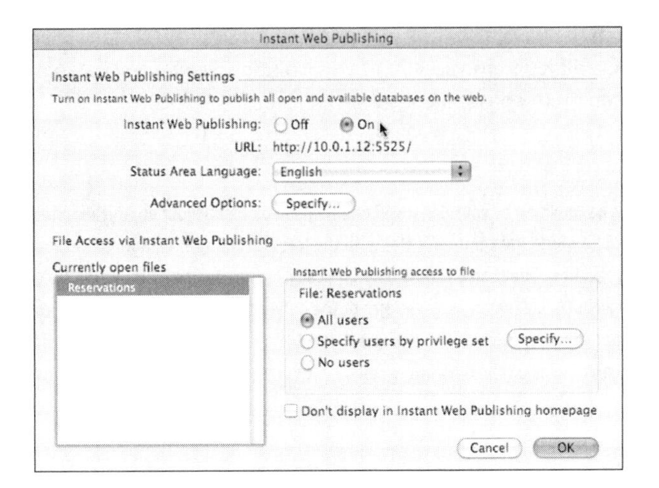

Figure 15.1
To enable Instant Web Publishing in FileMaker Pro, simply select On from the IWP configuration screen.

Turning Instant Web Publishing on and off is as simple as toggling the Off/On selection. Selecting On enables this particular copy of FileMaker Pro to act as an IWP host. You can choose the language that will be used on the IWP Database home page and in the IWP controls at the top of the window. You can also configure a handful of advanced options, as shown in Figure 15.2.

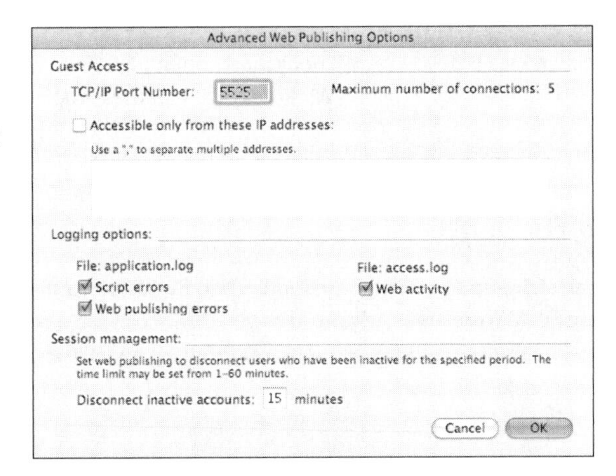

Figure 15.2
On the Advanced Web Publishing Options dialog, you can configure the port number, logging options, IP restrictions, and session disconnect time.

Port Number

By default, IWP is configured to use port 80 on the host machine. If another application, such as a web server, is already using that port, you see an error message and are asked to specify a different port to use. FileMaker, Inc., has registered port 591 with the Internet Assigned Numbers Authority (IANA), so that's the recommended alternative port number. The only downside of using a port other than 80 is that users need to explicitly specify the port as part of the URL to access IWP. For instance, instead of typing `127.0.0.1`, your users would have to type `127.0.0.1:591` (or whatever port number you specified).

If you are using Mac OS X, you might be asked to type your computer's passphrase if you attempt to change the port number when configuring IWP within the FileMaker client.

Security

If you know the IP addresses of the machines your IWP users will use when accessing your solution, you can greatly increase your solution's security by restricting access to only those addresses. You can enter multiple IP addresses as a comma-separated list. You can use an asterisk (`*`) as a wildcard in place of any part of the IP address (except for the first part). That is, entering `192.168.101.*` causes any IP address from 192.168.101.0 to 192.168.101.255 to be accepted. Entering `192.*` allows access to any user whose IP address begins with 192.

If you don't set IP restrictions, anyone in the world who knows the IP address of your host machine and has network access to it can see at least the IWP Database home page (which lists IWP-enabled files). And if you've enabled the Instant Web Publishing extended privilege on the Guest privilege set, remote users could open the files as well. This is, of course, exactly the behavior you would want when IWP is used as part of a publicly accessible website.

Logging

You can enable two activity logs for tracking and monitoring your IWP solution: the application log and the access log. The application log tracks script errors and web publishing errors:

- **Script errors.** These errors occur when a web user runs a script that contains non-web-compatible script steps. See the section "Scripting for IWP," later in this chapter, for more information about what particular steps are not web compatible. A script error can also occur if a user attempts to do something (via a script) that's not permitted by that user's privilege set. Logging script errors—especially as you're testing an existing solution for IWP friendliness—is a great way to troubleshoot potential problems.
- **Web publishing errors.** These errors include more generic errors, such as "page not found" errors. The log entry generated by one of these generic errors is very sparse and might not be terribly helpful for troubleshooting purposes.

The access log records all IWP activity at a granular level: Every hit is recorded, just as you'd find with any web server. As a result, the access log can grow quite large very quickly, and there are no mechanisms that allow for automatic purging of the logs. Be sure to check the size of the logs periodically and to prune them as necessary to keep them from eating up disk space. (A knowledgeable system administrator can configure both Windows and Mac OS X to periodically trim or rotate logs to prevent uncontrolled log growth.)

Each of the two logs can be read with any text editor, but you might find it helpful to build a FileMaker database into which you can import log data. It will be much easier to read and search that way.

Ending a Session

The final option on the Advanced Web Publishing Options dialog is the setting for the session disconnect time. As mentioned previously, IWP establishes a unique database session for each web user. This means that as a user interacts with the system, things such as global values, the current layout, and the active found set are remembered. Rather than just treating requests from the web as discrete and unrelated events, as was the case in previous incarnations of IWP, the host maintains session data on each IWP user.

Because only five sessions can be active at any given time when using FileMaker Pro as an IWP host, it's important that sessions be ended at some point. A session can be ended in several ways:

- A user can click the Log Out button in the IWP controls.
- The `Exit Application` script step ends an IWP session and returns the user to the Database home page.
- You can terminate a session after a certain amount of inactivity. The default is 15 minutes, but you can set it to anything from 1 to 60 minutes.

Clicking the home icon in the IWP controls to return to the Database home page does not end a session. If a user reenters the file from the Database home page without ending his session, he returns to exactly the same place he left, even if a startup script or default layout is specified for the file.

Closing the browser window or quitting the browser application does not end a session, so be sure to train your users to click the Log Out button (or an equivalent button that you provide). One of the problems you could run into is that an IWP user might quit his browser but still have a record lock. Until the session times out, no other user can modify that record. If you experience this problem, try reducing the session timeout setting to something like five minutes.

Configuring FileMaker Server Advanced for IWP

One of the best features of the FileMaker product line is the capability to do Instant Web Publishing directly from files hosted by FileMaker Server Advanced. Using FileMaker Pro as an IWP host works well for development, testing, and some limited deployment situations, but for many business applications, you'll find that you want the added power and stability that come from using FileMaker Server Advanced for this purpose.

Using FileMaker Server Advanced as your IWP host provides several significant benefits. The first is simply that it scales better. With FileMaker Pro, you are limited to 5 concurrent IWP sessions; with FileMaker Server Advanced, you can have up to 100 IWP sessions. FileMaker Server Advanced can also host up to 125 files, compared to FileMaker Pro's 10. Even more important, you have the option to use SSL for data encryption when using FileMaker Server Advanced as the web host. FileMaker Server Advanced is a more reliable web host as well. It is more likely that the shared files will always be available for web users, that they'll be backed up on a regular basis, and that the site's IP address won't change when you use FileMaker Server. (Even in organizations that use dynamic addressing for desktop machines, servers are typically assigned static IP addresses.)

FileMaker Server Admin Console is a Java configuration tool that enables you to attach a Web Publishing Engine to a FileMaker Server and configure it. As shown in Figure 15.3, you turn on Instant Web Publishing for FileMaker Server simply by checking the box on the Instant Web Publishing pane of the Web Publishing screen.

You can see a list of the databases accessible via IWP on the server by going to the Databases page, shown in Figure 15.4. For a database to be IWP-accessible, one or more privilege sets needs to have the fmiwp extended privilege enabled in the database itself, as described in the following section. There's no configuration or setup that you need to do in FileMaker Server Admin Console nor to the files themselves before hosting them with FileMaker Server Advanced. In fact, even while a file is being hosted by FileMaker Server, a user with the privilege to manage extended privileges can use FileMaker Pro to open the file remotely and edit the privilege sets so that the file is or isn't IWP accessible.

Figure 15.3
Use the FileMaker Server Admin Console to allow FileMaker Server Advanced to host IWP-enabled databases.

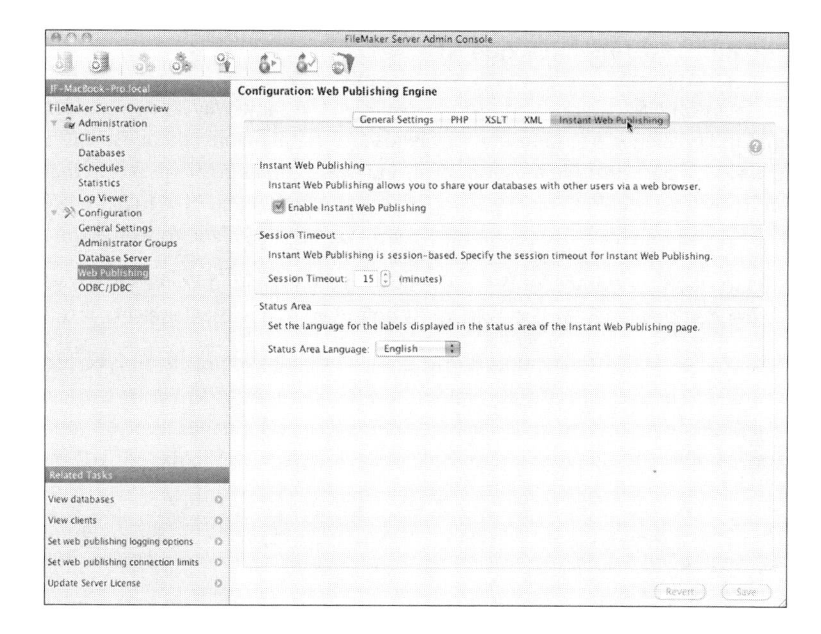

Figure 15.4
FileMaker Server Admin Console lists all the web-accessible databases on the server, but you don't need to do any configuration here at the file level to allow something to be shared to IWP.

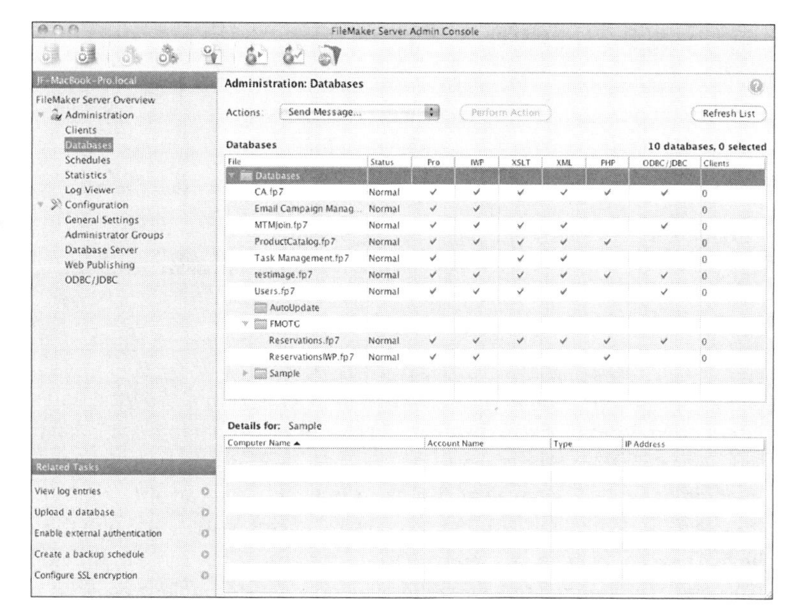

If you want a file to be accessible via IWP but not to show up on the Database home page, you have to open the file with FileMaker Pro (open it directly—that is, not simply as a client of FileMaker Server) and go into the Instant Web Publishing configuration screen. After you are there, select the file and then check the Don't Display in Instant Web Publishing home page checkbox. You do not need to actually enable IWP or add any extended privileges to privilege sets to have access to this setting.

Sharing and Securing Files via IWP

Security for Instant Web Publishing users is managed the same way it's managed for FileMaker Pro users: via accounts and privileges. Accounts and privileges also dictate which database files are accessible via IWP. To be shared via IWP, a particular file must be open, and one or more privilege sets in that file must have the fmiwp extended privilege enabled. This is true regardless of whether you plan to use FileMaker Pro or FileMaker Server Advanced as the web host. You assign the fmiwp extended privilege to a privilege set in any of three ways:

- Go to File, Manage, Accounts & Privileges. On the Extended Privileges tab, you'll see a list of the various extended privileges and be able to assign fmiwp to any privilege sets you want.
- Also in File, Manage, Accounts & Privileges, on the Privilege Sets tab, you can select fmiwp as an extended privilege for the currently active privilege set, as shown at the lower left in Figure 15.5. The FileMaker Mobile (fmmobile) privilege refers to a no-longer-supported product for mobile devices. It does not need to be enabled to use mobile devices with FileMaker/CWP, FileMaker/IWP, or FileMaker Go.

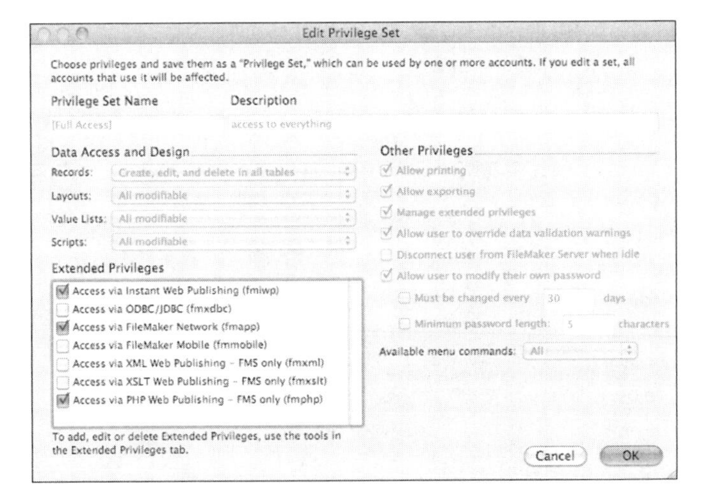

Figure 15.5
Make certain that fmiwp is an extended privilege for the relevant privilege sets.

- On the Instant Web Publishing setup screen (refer to Figure 15.1), the bottom half of the screen shows a list of open database files. When you select a particular database, you can manage the fmiwp extended privilege right from this screen. If you select All Users or No Users, the fmiwp extended privilege is granted or removed from all privilege sets in the file. You can also select Specify Users by Privilege Set to select those privilege sets that should have access to IWP. Although the words *extended privilege* and *fmiwp* never appear on this screen, it functions exactly the same as the Extended Privilege detail screen. This screen is intended to be more user-friendly and convenient, especially when you are working with multiple files.

To assign extended privileges in any of these ways, a user must be logged in with a password that grants rights to Manage Extended Privileges.

The other sharing option you can configure on the Instant Web Publishing setup screen is whether the database name appears on the Database home page. In a multifile solution, you might want to have only a single file appear so that users are forced to enter the system through a single, controlled point of entry.

Any changes made in the sharing settings and privileges for a file take effect immediately; you do not need to restart FileMaker or close the file.

When users type the IP address (or domain name) of the IWP host in their browsers, the first thing they'll see is the IWP Database home page, an example of which is shown in Figure 15.6. The Database home page lists, in alphabetical order, all files on the host machine that have at least some privilege sets with the fmiwp extended privilege enabled. The Database home page cannot be suppressed, although it can be customized or replaced, as explained later in this chapter.

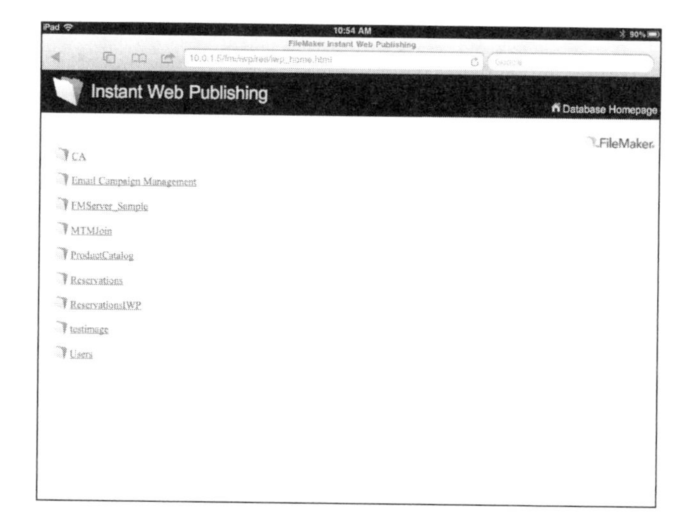

Figure 15.6
The Database home page provides users with a list of accessible files.

Users aren't prompted for a password on their way to the Database home page. The password prompt occurs (unless you are logged in as a guest, as described in the following bulleted list) when users first try to interact with a database. IWP now uses an HTML forms-based interface for entering a username and password, as shown in Figure 15.7. To be authenticated, users must enter an active, valid username and password, and their accounts must be associated with a privilege set that has the fmiwp extended privilege enabled.

You should know a number of things about how accounts and privileges are authenticated under IWP:

- As in regular FileMaker authentication, the password is case sensitive (although the account name is not).
- IWP ignores any default login account information that has been set up under File Options.

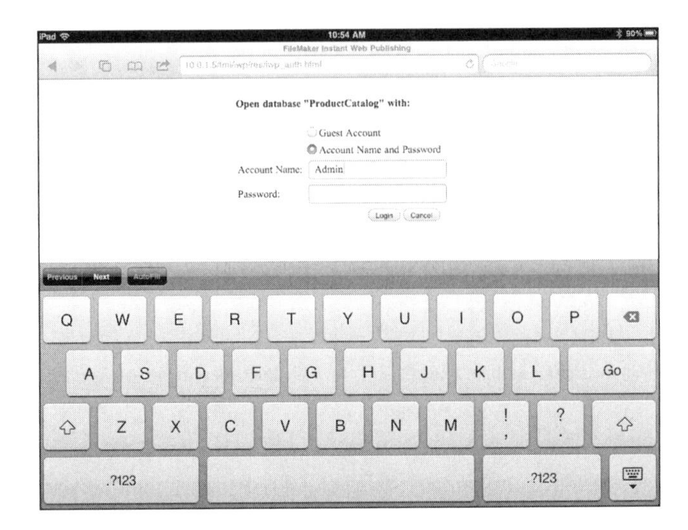

Figure 15.7
You need to log in to the database to access it.

- IWP does not support the Account option to require users to change their passwords after the next successful login. Changing passwords is not a feature supported by IWP. If this option has been set, a web user who tries to log in with that username and password receives an Error 211, `Password Has Expired`, and cannot enter the system.
- If the Guest account has been activated and given the fmiwp extended privilege, users might not be prompted for a username/password to access the database. For the users to skip the login screen, though, it's necessary that the fmiwp extended privilege be assigned only to the [Read-Only Access] privilege set (the privilege set used by the Guest account). Anyone automatically logged in in this fashion will have the privileges of the Guest account. Such a configuration would typically be used only for websites that need to be accessed by the general public.

You can create a script that uses the account management script steps to create your own customized login routine. Users would use Guest privileges to get to your login screen, and then your script would use the re-login step to reauthenticate them as different users.

After a user is authenticated as a valid user of the file, that user's privilege set then controls which actions can be performed, just as it does for users of the FileMaker Pro desktop application. Field and layout restrictions, record level access, creation and deletion of records—all of these are managed exactly the same for IWP users as for FileMaker Pro users. The capability to make use of this unified security model is truly one of the best features of FileMaker IWP and makes it much simpler to deploy robust and secure IWP solutions.

Designing for IWP Deployment on iPad

The preceding sections discussed how to enable IWP at the application level and how to set a file so that users can access it via IWP. Although this is enough for IWP to function, there are usability issues to consider as well. Not all layouts and scripts translate well to the web, and some FileMaker features simply don't work via IWP.

Beyond those issues, iPad poses additional problems. Although you can create iPhone layouts for IWP, in most cases it is easier to use PHP as described in Chapter 16. It is almost always the case that you cannot take existing FileMaker layouts and use them for IWP without any modifications.

The following sections discuss the constraints that you, as a developer, must be aware of when deploying an IWP solution. There is also a discussion of several development techniques that can make an IWP solution feel more like a typical web application.

Constraints of IWP

Most of the core functionality of FileMaker Pro is available to IWP users. This includes being able to view layouts, find and edit data, and perform scripts attached to buttons. However, a number of FileMaker features are not available to IWP users. (It might not surprise you to notice that some of these IWP constraints are also constraints for FileMaker Go.) It's important to keep these points in mind, especially when trying to port an already existing solution to the web:

- IWP users have no database development tools. This means IWP users can't create new files; define tables, fields, and relationships; alter layouts; manage user privileges; or edit scripts.
- IWP users can't use any of the FileMaker Pro keyboard shortcuts. Be sure that you leave the IWP controls visible or provide your users with ample scripted routines for tasks such as executing finds and committing records.
- There is no capability to import or export data from an IWP session. In general, any action that interacts with another application, the file system, or the operating system is not possible via IWP.
- IWP has no Preview mode. This means that sliding and multicolumn layouts, all of which require being in Preview mode to view, are not available to IWP users. Similarly, printing is not supported. IWP users can choose to print the contents of the browser window as they would any other web page, but the results will not be the same as printing from FileMaker Pro. (That is, headers and footers won't appear on each page, page setups will not be honored, and so on.)
- There are a few data-entry differences for IWP users. For instance, web users can't edit rich text formatting in fields. That is, they can't change the style, font, or size of text in a field. They can generally, however, see rich text formatting that has already been applied to a field.
- Most window manipulation tools and techniques do not translate well to IWP. The user's browser can show only the contents of the currently active window in the virtual FileMaker environment. That environment can maintain multiple virtual windows and switch between them, but a user can't have multiple visible windows in the browser, and cannot resize or move windows except to the extent allowed by the browser. In other words, the users can manually resize their browser windows, but precision movement and placement of windows using script steps such as Move Window is not supported in IWP.
- Spell-checking is not available via IWP.
- Many graphical layout elements are rendered differently, or not at all, on the web. This includes diagonal lines, rounded rectangles, rotated objects, ovals, and fill patterns. The sections that follow discuss this topic in greater detail.
- IWP users can't edit value lists through a web browser.

- There is no built-in way for users to change their passwords via IWP, even if they have the privilege to do so. If you need this sort of functionality, you have to use the account management script steps and come up with your own scripted routine.

Scripting for IWP

One of the greatest recent advances in Instant Web Publishing is script support. In versions of FileMaker before version 7, only a handful of script steps could be executed from the web, and scripts could be no more than three steps long. Under those severe restrictions, it was quite difficult to build anything but the most basic web applications.

IWP supports more than 70 script steps, and scripts can be of any length and complexity. Also, because IWP is now session-based, scripts executed from the web operate within what might be thought of as a virtual FileMaker environment. This means that changes to the environment (active layout, found set, and so on) are persistent and affect the browser experience, which is a good thing.

Even though IWP script support has come a long way, there are still some behaviors, constraints, and techniques you should be aware of.

Unsupported Script Steps

ScriptMaker itself has an option that makes identifying unsupported script steps quite easy. When you use the Web Compatibility pop-up menu, all the unsupported script steps are dimmed. This affects both the list of script steps and the steps in whatever script you're viewing. Figure 15.8 shows an example of what script step dimming looks like. The Show Compatibility pop-up menu has no effect other than showing you which steps are not supported; how you choose to use that information is up to you (although unsupported script steps are dimmed out, you can still add them to a script). Additionally, its status is not tied to any particular script. That is, it is either turned on or off for the entire file, and it remains that way until a developer changes it. This is pointed out explicitly because the checkbox right next to it, Run Script with Full Access Privileges, is a script-specific setting.

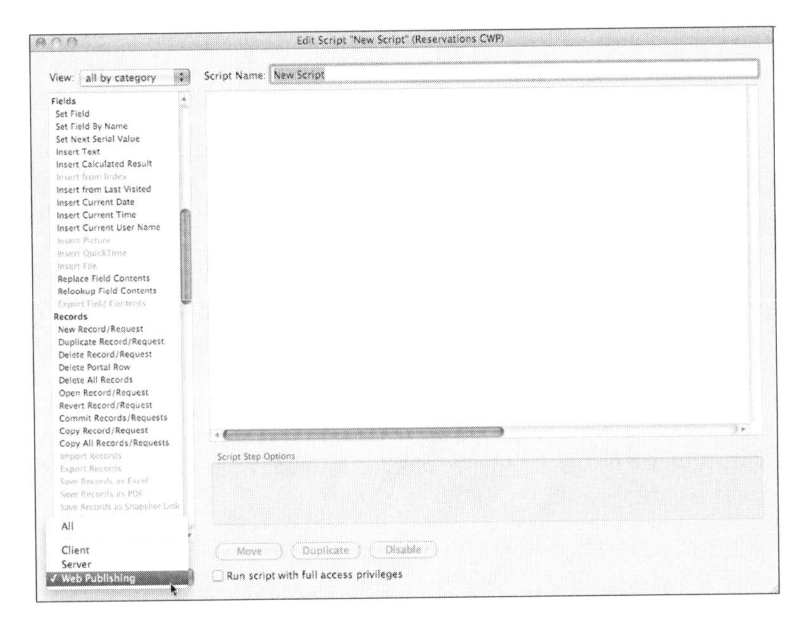

Figure 15.8

When you are writing scripts that will be used via IWP, turn on the Indicate Web Compatibility checkbox to dim out incompatible script steps.

Additionally, the option to perform with a dialog is not supported in a number of supported script steps. They include Delete Record/Request, Replace Field Contents, Omit Multiple Records, and Sort Records. These steps are always performed without a dialog via IWP, regardless of which dialog option has been selected in ScriptMaker.

Error Capture

The outcome of running a script (from the web) that contains unsupported script steps depends on whether the Allow User Abort setting has been turned on or off. If it's not explicitly specified, a script executes on the web as if Allow User Abort had been turned on. So, not specifying any setting is the same as explicitly turning it on.

If user abort is on (or not set at all), script execution halts when an unsupported step is encountered. Steps before the offending script step are performed as normal. If you've chosen to log script errors, the offending step is logged as an error in the application log. The user does not see any error message or have any knowledge that anything is amiss.

If user abort has been turned off, a script simply bypasses any unsupported scripts and attempts to perform subsequent steps. It's performed as if the offending step was simply not there. No error is logged to the application log when this occurs.

Script steps with the unsupported "perform with dialog" options discussed earlier are not affected at all by the error capture setting. These script steps will always be performed as if Perform Without Dialog had been checked, regardless of error capture.

Committing Records

If a script run via IWP causes a record to be altered in any way (such as using a `Set Field` script step), be sure that you explicitly save the change by using the `Commit Record/Request` step sometime before the end of the script. If you don't, your web user will be left in Edit mode and, provided that the IWP controls are visible, will have the option to submit or cancel the changes, which is likely not an option you want to offer at that point. Canceling would undo any changes made by the script.

Startup and Shutdown Scripts

If you have specified a startup script for a file, it is performed for IWP users when the session is initiated. Similarly, IWP also switches to a particular layout on startup if you've selected that option. The shutdown script is performed when the user logs out, even if the logout is the result of timing out.

The startup script executes only once per session, when the user navigates there from the Database home page or follows an equivalent link from another web page. The startup script is not run if a file is activated through the performance of an external script.

Performing Subscripts in Other Files

A script can call a subscript in another file, but that file has to be open and enabled for IWP for the subscript to execute. Calling a subscript does not force open an external file, as happens in the FileMaker Pro desktop application.

If your subscript activates a window in the external file, the IWP user sees that window in the browser. Unless you provide navigation back to the first file, a user has no way of returning, except by logging out and logging back in. You should make sure that any record changes are fully committed before navigating to a window in another file. It's possible that the record will remain in an uncommitted, locked state, even though the IWP user has no idea this has occurred.

Testing for IWP Execution Within a Script

If you have a solution that will be accessed by both FileMaker Pro desktop users and IWP users, chances are that they'll use some of the same scripts. If those scripts contain unsupported script steps, you might want to add conditional logic to them so that they behave differently for IWP users than they do for FileMaker users. You can do this by using the `Get (ApplicationVersion)` function. If the words `Web Publishing` are found within the string returned by this function, it means the person executing the script is a web user. It's not possible to discriminate between an IWP user and a CWP user with this function; you simply know you have a web user. The actual syntax for performing the test is as follows:

```
PatternCount (Get (ApplicationVersion); "Web Publishing")
```

Layout Design

Most layouts you design in FileMaker Pro will be rendered almost perfectly in a web browser via Instant Web Publishing. IWP does this by using the absolute positioning capability of cascading style sheets, Level 2. FileMaker helps out with autosizing positioning that works with Safari and the iOS accelerometers so that layout objects move and size appropriately. (This is yet another of the many reasons for making certain that you set those options properly when you design new layouts; you never know when a layout is going to be displayed on a mobile device.)

Graphic Elements

Rounded rectangles, ovals, diagonal lines, rotated objects, and fill patterns are not rendered properly in the web browser and should be avoided. In some cases, IWP displays altered versions of the objects; in other cases, the objects simply do not show up.

Test with your mobile device in portrait and landscape orientation as you go along. In general, the rules for using larger font sizes and more spaces around objects are your starting point. If your layout looks good on a desktop-based computer, then congratulations! But that means next to nothing when you are viewing it on a mobile device.

"View As" Options

Web users have the same ability that FileMaker desktop users have to switch between View As Form, View As List, and View As Table on a given layout, unless you restrict that ability at the layout level. To do so, go into the

Layout Setup options, shown in Figure 15.9, and simply uncheck any inappropriate views. The additional Table View options that can be specified all translate well to IWP, except for resizable and reorderable columns.

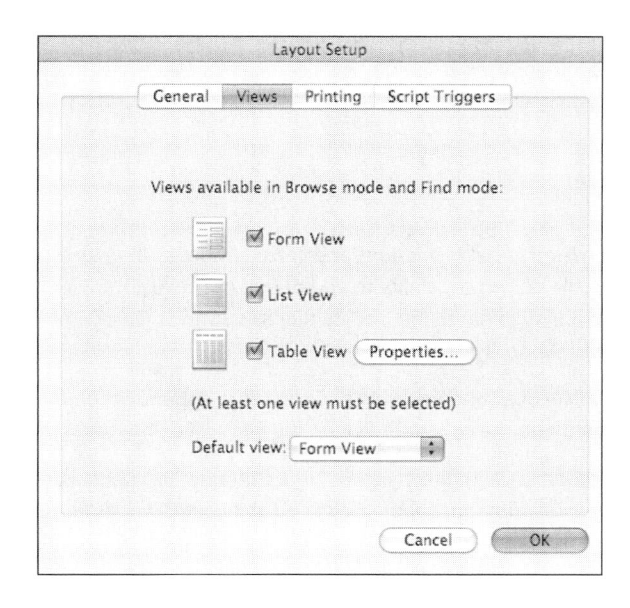

Figure 15.9
In Layout Setup, you can specify the types of views to which a user should be able to switch for a given layout.

You should be aware of a few special characteristics of List and Table views in IWP. By default, View As List shows a set of at most 5 records, and View As Table shows a set of at most 20 records. You cannot change these settings. Also, while in List or Table view, whenever a user clicks on a record to edit it, the active record jumps to the top of the set. This response can be slightly disconcerting for users habituated to working with lists of records in FileMaker. For instance, if a user is viewing records 6–10 of a set as a list and clicks on record 8, record 8 jumps to the top, and the screen then displays records 8–12.

Layout Parts

IWP can render any and all parts that compose a layout. There are a few differences, however, between how and when parts display in IWP and how and when they display in the FileMaker desktop application.

First of all, in Form view, the vertical size of a part displayed via IWP is the size that the part was defined to be. It doesn't stretch to fill the vertical space. This is different from how FileMaker Pro behaves. In FileMaker Pro, the last visible part expands to fill any remaining vertical space. Say, for instance, that you have a layout that consists of only a single, colored body part. Via IWP, if a user resizes a browser window so that it's larger (vertically) than the body part, the space between the bottom of the part and the bottom of the browser is a white void. This also means that if your layout has a footer part, it won't necessarily (indeed, won't likely) be displayed at the bottom of the browser window.

View As List in a browser also has some differences from its FileMaker counterpart. In FileMaker, a header or footer part is locked on the screen at the top or bottom. The area in between displays as many body records as space permits. In FileMaker, leading and trailing grand summary parts display in List view, but title header, title footer, and subsummary parts do not (in Browse mode).

In a browser, List view always contains 25 records (except, of course, when the found set is fewer than five or if the active record is one of the last four of the found set). The header and footer are not fixed elements as they are in FileMaker. If the 24 records of the list take up less than the full browser window, the footer simply shows up in the middle of the screen; if they take up more than the full window, a user would need to scroll to see the footer. Another major difference is that title header, title footer, and subsummary parts are all visible in the browser at all times (in List view).

Container Fields

You should know about a few special restrictions and considerations when using container fields in an IWP solution. Most important, there is no capability to add or edit data in a container field via IWP; these fields are strictly view-only. The capability to enter and update pictures, sounds, QuickTime movies, files, and objects is available only to regular FileMaker Pro users.

The visibility and accessibility of a container field's contents are dependent on the types of objects they are and how they were entered into the container field in the first place:

- Graphic images that have been directly stored in a container field (that is, not stored as a reference) are visible through a browser. Images should be stored as pictures, not as files.
- Graphic images that have been stored as a reference are visible to IWP users only if the images are stored in the web folder of the FileMaker Pro application (if FileMaker Pro is the IWP host) or if they are stored in the root folder of the web server (if FileMaker Server is the IWP host).
- QuickTime movies can't be accessed directly from the web browser. If you insert them as files rather than as QuickTime, however, a user can play or download them.
- Files stored directly in a container field render to an IWP user as a hyperlink. Clicking the link begins a download of the file. No icon or other graphic representation of the file is visible to web users.
- Sounds that have been stored directly in container fields cannot be played via IWP.

Application Flow

You've seen many of the technical limitations and details of how various FileMaker features translate to the web. We turn now to more practical development matters. Certain routines and development habits that work well in the FileMaker desktop application don't work as well from a web browser. The following sections discuss how the constraints of IWP will influence how you develop solutions.

If you're designing a new solution and you know that you'll have IWP users, you might consider thinking about how you would develop the solution if it were a web application. Because there are more constraints placed on designing for the browser, anything you build for the browser should work well for FileMaker users also.

Explicit Record Commits

HTTP—the underlying protocol of the web—is a stateless protocol. This means that every request a browser makes to a web server is separate and independent from every other request. Put differently, the web server doesn't maintain a persistent connection to the web client. After it has processed a request from someone's browser, it simply stands by, waiting for the next request to come in. To make HTTP connections appear to be persistent, web programmers need to add information to each request from a single client and then let some piece of web server middleware keep track of which client is which, based on this extra request data. This technique is referred to as *session management*.

The client/server connection between FileMaker Pro and FileMaker Server is persistent. The two are constantly talking back and forth, exchanging information and making sure that the other is still there. FileMaker Server is actively aware of all the client sessions. When FileMaker Server receives new record data from any client on the network, it immediately broadcasts that information to all the other clients. And when a user clicks into a field and starts editing data, FileMaker Server immediately knows to consider that record as locked and to prevent other users from modifying the record.

The fact that IWP is now capable of performing session management means that FileMaker maintains information about what's happening on the web in a virtual FileMaker environment. Even though this doesn't change the fact that HTTP is stateless, using sessions gives IWP a semblance of persistence. Essentially, the server stores a bunch of information about each IWP user; each request from a user includes certain session identifiers that enable the server to recognize the IWP guest and to know the context by which to evaluate the request. One of the benefits of this session model is that IWP users can lock records, and they are notified if they try to edit a record that a regular FileMaker Pro user has locked.

Still, the statelessness of the web makes the application flow for something even as basic as editing a record much different in IWP than it is in FileMaker. In FileMaker, of course, a user just clicks into a field, makes some changes, and then clicks out of the field to commit (save) the change. On the web, editing a record involves two distinct transactions. First, by clicking an editable field or using the Edit Record button in the IWP controls, the user generates a request to the server to return an edit form for that record and to mark the record as locked. As we discussed earlier in this chapter, Edit mode in the browser is distinctly different from Browse mode.

The second transaction occurs when the user clicks the Commit button in the IWP controls (or clicks a similar button you've provided for this purpose). No actual data is modified in the database until and unless the record is committed explicitly.

This transaction model for data entry might feel alien to users who are accustomed to working with a FileMaker interface. As you evaluate the web-friendliness of existing layouts or build new layouts for IWP users, try to make the application flow work well as a series of discrete and independent transactions. One common way to do this is by having tightly controlled routines that users follow to accomplish certain tasks. For instance, instead of letting users just create new records anywhere they want, create a "new record" routine that walks users through a series of screens where they enter data and are required to click a Next Screen or Submit button to move forward through the routine.

Hiding the IWP Controls

As when designing a solution for FileMaker users, you have the option to leave the IWP controls visible for your IWP users or to hide it from them. And as with regular FileMaker, unless you lock it open or closed, users can toggle it themselves.

By default, the IWP controls are visible for your IWP users. The script step `Show/Hide Status Area` enables you to programmatically control the visibility of the IWP controls. Typically, if you want to hide the IWP controls, you do so as part of a startup script.

The script step Show/Hide Status Area shows or hides the status area in pre-FileMaker Pro 10 software and shows or hides the Status toolbar in FileMaker Pro 10. The name of the script step was kept the same for version compatibility.

There are certainly benefits to having the IWP controls visible. Most important, the IWP controls provide a wealth of functionality for the IWP user. Navigation, complex searching, and a host of record manipulation tools are features that come at no charge in the IWP controls.

There are also reasons that developers want to hide the IWP controls from users. The first is simply to constrain users' activities by forcing them to use just the tools you give them. This is generally why developers hide the Status toolbar for FileMaker desktop deployments as well. Hiding the IWP controls also makes your application more web-like. If you are using IWP alongside an existing website or plan to have the general public access your site, you'll probably want to hide the IWP controls. Public users are more likely to expect a web experience than a FileMaker experience.

If you do decide to hide the IWP controls, you must provide buttons in your interface for every user action you want or need to allow, including committing records, submitting Find requests, and logging out of the application. Because users have no keyboard shortcuts—including using the Enter key to submit Find requests and continue paused scripts, for example—and no pull-down menus, you'll probably need even more buttons than you would when designing without the Status toolbar for FileMaker users.

Portals

IWP does an astonishingly good job of displaying portals in a browser, complete with scrollbars, alternating row colors, and the capability to add data through the last line of a portal (providing, of course, that the underlying relationship allows it). Another nice feature of portals in IWP is that you can edit multiple portal rows at once and submit them together as a batch.

When you are designing an IWP application that requires displaying search results as a list, consider whether you can use a portal instead of a List view. A portal gives you flexibility as far as the number of records that display, and you can use the space to the left or right of it for other purposes.

The best way we've found to make a portal display an ad-hoc set of records, such as those returned by a user search, is to place all the record keys of the found set into a return-delimited global text field by using the Copy All Records script step and then to establish a relationship between that field and the file's primary key. Because you can let the portal scroll, you don't strictly need to create Next and Previous links, but it would make your application more web-like if you did. One option to do this is to take the return-delimited list of record IDs and extract the subset that corresponds to a given page worth of IDs. The `MiddleValues` function comes in handy for this task. You would simply need to have a global field that kept track of the current page number. Then, the function

```
MiddleValues (gRecordKeys; (gPageNumber-1) * 8 + 1; 8)
```

would return the eight record IDs on that page. Substitute a different number of records per page in place of 8, of course, if you want to have a hitlist with some other number of records on it. The scripts to navigate to the next and previous pages then simply need to set the page number appropriately and refresh the screen.

Creating Links to IWP from Other Web Pages

The IWP Database home page provides a convenient access point for entering web-enabled databases. It's possible also to create your own links into a file from a separate HTML page, which is perhaps more desirable for publicly accessible sites. To do this, you simply create a URL link with the following syntax:

`http://`*`ip address:port number`*`/fmi/iwp/cgi?-db=`*`databasename`*`&-loadframes`

This syntax is different than it was in versions of FileMaker prior to version 7, so be sure to update external links if you're upgrading an IWP solution from FileMaker 6 or earlier. You might think that this is an unlikely situation, but it is far more common than you guess. One of the great features of FileMaker is that after the databases are set up and IWP is enabled, things can chug along for years and years. The changes are to the data in the database, but the infrastructure needs remarkably little attention over time.

If you are using FileMaker Pro itself as your IWP host (as opposed to FileMaker Server Advanced), you can place static HTML files and any images that need to be accessible to IWP users in the web folder inside the FileMaker Pro folder. The web folder is considered the root level when FileMaker Pro acts as a web server. If you had, for example, an HTML page called `foo.html` in the web folder, the URL to access that page would be the following:

`http://`*`ip address:port number`*`/foo.html`

If you develop a solution that uses FileMaker Pro as the host and later decide to migrate to FileMaker Server Advanced, you should move the entire contents of the web folder (if you've put any documents or images there) to the root folder of your web server.

Creating a Custom Home Page

You can override the default page with a page of your own devising. The new file must be called `iwp_home.html`. It can be used when serving files via IWP either from FileMaker Pro (in which case it belongs in the web directory inside the FileMaker Pro application folder) or from FileMaker Server Advanced (in which case it belongs in the FileMaker Server/Web Publishing/iwp folder).

There are several approaches to creating such a file. You could devise your own file from scratch, creating your own look and feel, and populate that file with hard-coded links to specific databases, as described in the preceding section. Or, if you want a file that dynamically assembles a list of all available databases, the way the default home page does, you'll want to customize the default page. An example of that default page can be found on the FileMaker Pro product CD. (For the curious, it can also be found in the FileMaker application folder:

On Windows, it's found in Extensions/Web Support/Resources/iwpres; on Mac OS, it's found in Extensions/Web Support/FM Web Publishing/Contents/Resources/iwpres. On Mac OS, Extensions/Web Support/FM Web Publishing is an OS X package, not a directory, so you'll have to right-click it and select Show Package Contents to drill deeper.) The default page makes heavy use of JavaScript and, in particular, of JavaScript DOM function calls, so familiarity with those technologies will be desirable if you want to customize the IWP home page.

Using an IWP Solution

The focus of this chapter is on what a developer needs to know to create and share databases to the web using Instant Web Publishing. One crucial piece is an understanding of what IWP looks like and how it functions from the user's perspective.

Browse Mode

If you've hidden the IWP controls from your users, you have complete control over what a user can do and how it's done. When the IWP controls are active, however, a user has access to a great many built-in features, including the capability to perform complex finds, sort records, navigate to other layouts, and manipulate data. Even so, you can still constrain a user's options by placing restrictions on the privilege sets assigned to IWP users.

To avoid a user navigating to a non-IWP-friendly layout, edit the layout options of the IWP-enabled privilege sets. Mark any layouts you want users to avoid as No Access.

Figure 15.10 shows the IWP controls a user sees while in Browse mode. Unless you explicitly lock the IWP controls either open or closed, a user can toggle it open and closed while in any mode.

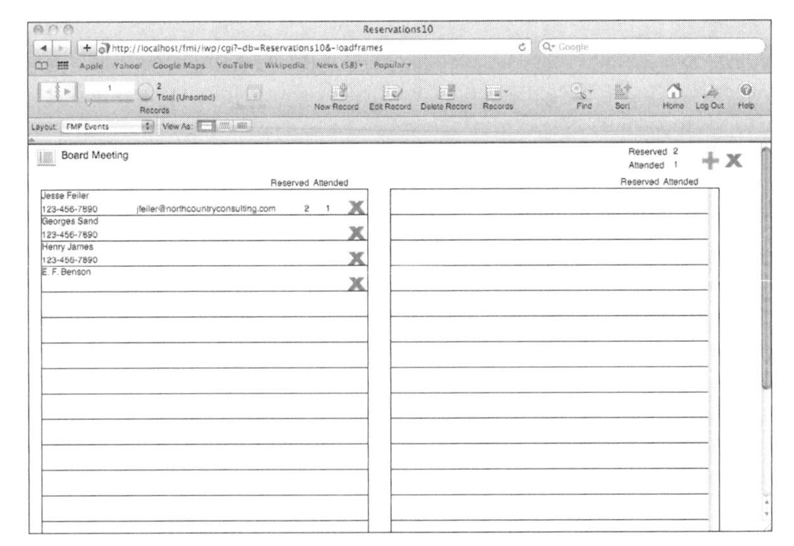

Figure 15.10
The IWP controls in Browse mode contain a number of record manipulation and navigation tools.

While in Browse mode, providing that users have the proper privileges, they can create, edit, duplicate, and delete the current record. They can also sort, find all, omit one record, omit multiple records, and show only the omitted records. Any buttons whose functionality is not permitted by the users' privilege sets are dimmed and inactive.

Edit Mode

As we've discussed, one of the biggest differences between the user experience in IWP versus the FileMaker Pro desktop application is the explicit distinction between being in Browse mode and being in Edit mode. A user can enter Edit mode in a few ways:

- By clicking any field (except container and calculation fields) where the Field Behavior is set to allow entry while in Browse mode
- By running a script that opens a record and doesn't commit it
- By clicking the Edit Record button in the IWP controls

In Edit mode, new buttons at the top appear: Submit, Cancel, or Revert (see Figure 15.11). Any of them ends editing and reverts to Browse mode.

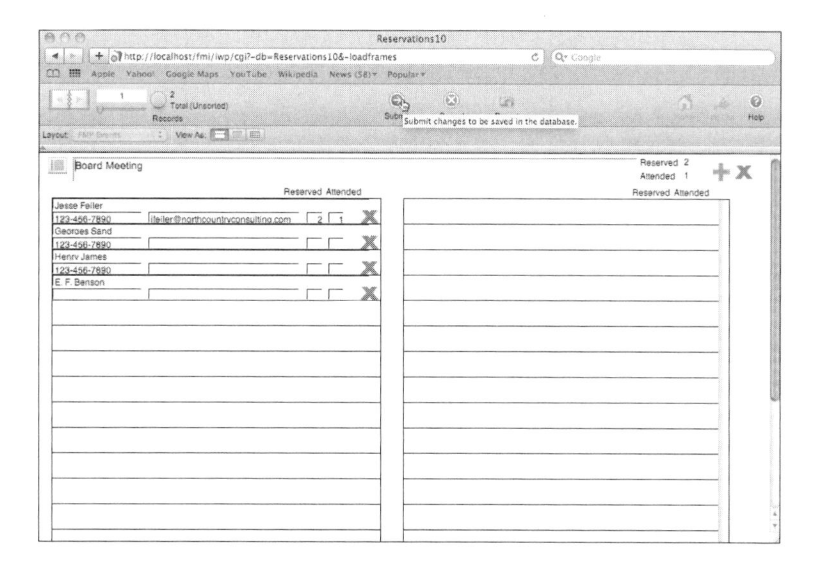

Figure 15.11
Use Submit, Cancel, or Revert to exit Edit mode.

Buttons on a layout are active regardless of whether a user is in Browse, Edit, or Find mode. Executing a script via a button does not change the mode unless a mode-changing step is in the script.

The `Enter Browse Mode` script step does not return an IWP user from Edit mode to Browse mode. Use the `Commit Record/Request` script step for this purpose.

Find Mode

Users can enter Find mode in IWP either by clicking on the magnifying glass icon in the IWP controls or by clicking on a button of your creation that leaves them in Find mode. Figure 15.12 shows the IWP controls as they appear in Find mode. Users can enter their search criteria and execute the find by clicking on the appropriate buttons in the IWP controls. Just as in the FileMaker Pro desktop application, users can create multiple Find requests, pick from a set of find operators, choose to omit found records, and extend or constrain the current found set.

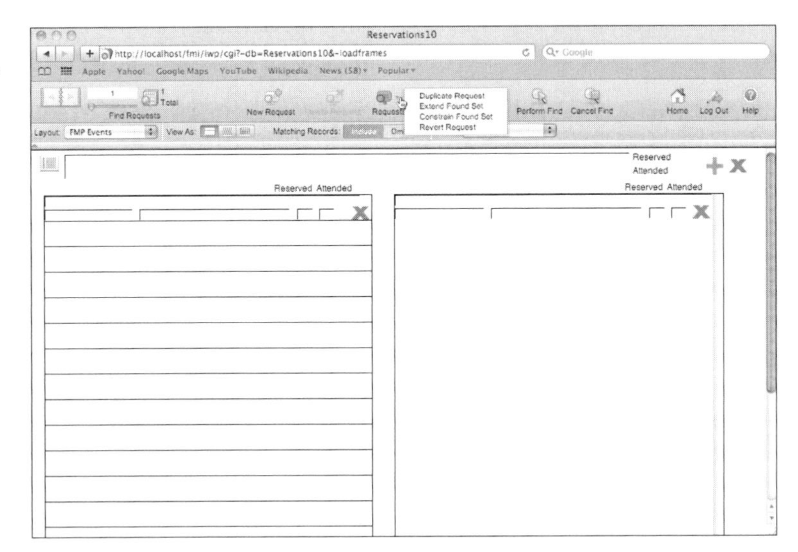

Figure 15.12
In Find mode, the IWP controls contain all the tools necessary for users to create complex ad-hoc searches.

If you are using the `Perform Find`, `Constrain Found Set`, or `Extend Found Set` script steps to execute a find, be aware that if no records are found, IWP does not display an error message to users. You need to trap for that error yourself and take appropriate action, such as navigating to a layout that has a `No records found` message.

Further Steps

Set up IWP for Starter Solutions or some of your own databases to see how it works on various devices. As you explore your databases on PCs, Macs, iPad, and iPhone, you soon see the areas where work needs to be done. You might find some images used in the interface that use transparency that need to be adjusted. You will find scripts that need to be exposed to users through interface elements such as buttons, and you'll also start to get a good idea of which layouts need to be customized for IWP and for mobile devices.

IN THIS CHAPTER

- Build a FileMaker Database Solution for Mobile Safari on iPhone
- Learn How to Modify FileMaker Databases for Custom Web Publishing with PHP
- Explore the PHP Site Assistant
- Deploy Your PHP Database Solution

Deploying FileMaker/CWP with FileMaker Server

Custom Web Publishing (CWP) with PHP gives you the greatest degree of control over the finished product. The goal of Instant Web Publishing (IWP) is to replicate the FileMaker experience as closely as possible in a web browser. With CWP, the goal is to let you design a solution that can look like FileMaker Pro or one that uses another interface model. The most common type of interface, which is very difficult to accomplish with IWP and very easy to accomplish with CWP, is that of iPhone. Among the templates that are built into the PHP Site Assistant is one designed specifically for iPhone. This chapter focuses on that template, but you can use any of the other templates just as easily. The development steps are the same.

Preparing for Custom Web Publishing (CWP) with PHP

CWP with PHP requires that you have FileMaker Server installed alongside a web server. By comparison, IWP requires FileMaker Server Advanced. In addition to that, you have to prepare each database that will use CWP, and you must enable the appropriate CWP technologies on FileMaker Server.

Getting Your Databases Ready for Custom Web Publishing with PHP

Access to a FileMaker database via PHP is handled via the security and privilege system. You can allow or deny PHP access to a file based on whether a user has the appropriate privilege, and you can control that user's rights and privileges down to the record or field level.

Setting the Extended Privilege for PHP

To allow access via PHP, enable the extended privilege with the keyword `fmphp`. Figure 16.1 illustrates the use of these extended privileges.

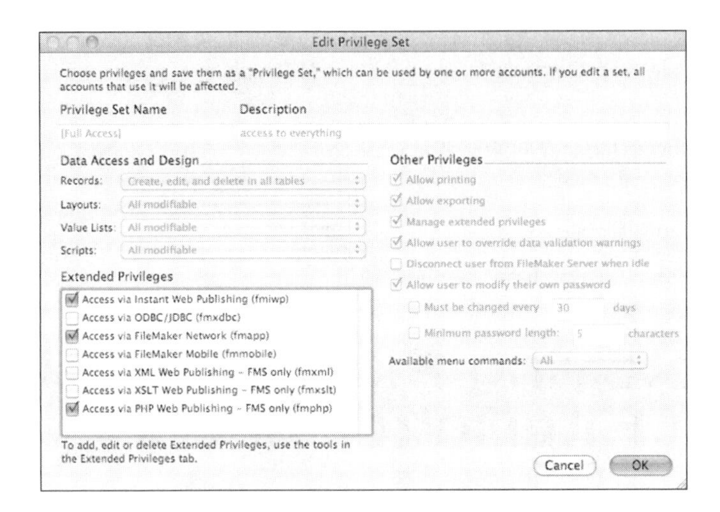

Figure 16.1
You must enable the PHP extended privilege for each database you want to publish with PHP.

In general, it is a good idea to also enable FileMaker Network access. If you do this, you can use the Open Remote command in FileMaker Pro to access the database shared by FileMaker Server. This is an important safety valve for you because it enables you to log on and make changes to accounts and privileges, for example. In general, providing yourself with at least one account that has network access to each shared database is a good idea.

If you expect to see a database served via CWP/PHP and it doesn't appear, check to make sure that the appropriate extended privileges are enabled.

Setting Other Security Measures for the Database

In addition to setting the extended privilege for PHP, you can also take a variety of other steps if you choose. Still in the area of security and access, you might want to review the accounts and privileges that you have set for the database. Even if you are behind a corporate firewall, you often want to limit web access to your database to the absolute minimum. If you are publishing public data in a totally public environment, it is still worth occasionally reviewing your database to make certain that, in its ongoing maintenance and modification, no confidential data has appeared there.

This security issue is particularly relevant to organizations such as schools where the internally public data (class lists, student addresses, and so forth) is generally never shown in public.

Reviewing Layouts

The PHP Site Assistant is based on layouts. A FileMaker layout is based on a single table occurrence in the relationships graph; from that table, you have access to any of the tables that are related to it in the graph, whether they are in the same file or not (and in fact, whether they are FileMaker databases or external data sources using ODBC). You can use existing layouts, or you can create new layouts for your website. You can use PHP to manipulate the data that is presented to web users, but the more manipulation and editing that you can do before PHP comes into play, the more efficient your site will be. This can mean removing from layouts any fields that will never be needed on the web so that you do not have to fuss with them in your PHP code. It also might mean creating some calculation fields that do manipulations you would otherwise do in PHP. Depending on the balance between FileMaker users of your database and web users, you might want to consider whether such changes are worthwhile.

The PHP Site Assistant combines layouts into *layout groups*: All these layouts use a single table occurrence as their primary table occurrence.

A simple way of modifying layouts is to take the existing layouts you are using, duplicate them, and change their names to begin with *Web* or *W*. It can also be a good idea simply to remove unnecessary fields and not to rearrange the remaining fields. The layout might look a little strange with gaps and holes in it, but comparing it to the original layout will be very easy. Remember, for CWP the image of the layout from FileMaker is not shown on the web, unlike IWP.

Getting FileMaker Server Ready for Custom Web Publishing with PHP

You also need to use Admin Console to enable PHP publishing, as shown in Figure 16.2. The other options in this window are described later in this chapter.

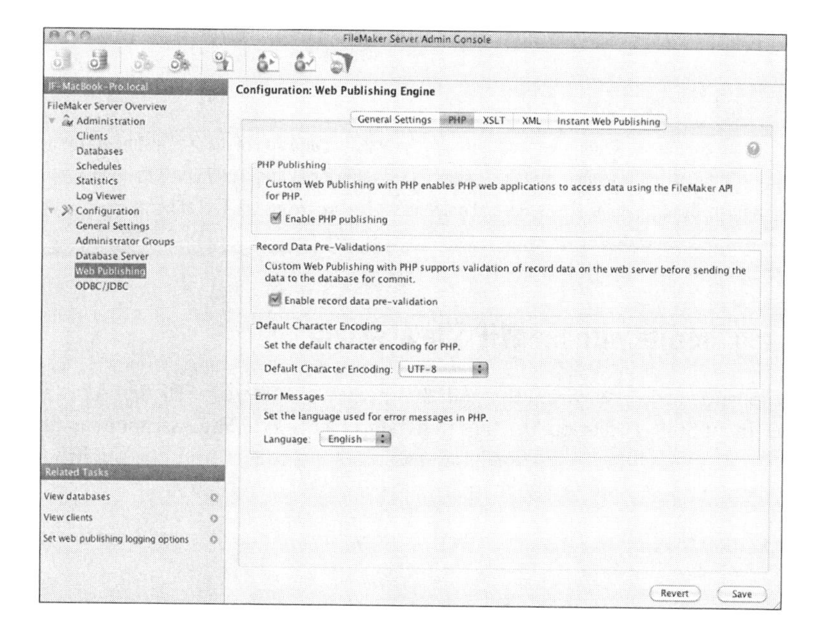

Figure 16.2
Enable PHP Publishing in Admin Console.

Placing Files on the Web Server

There are two types of files you need to worry about placing on the web server: the PHP files themselves and the files that are referred to by reference container fields.

Placing the PHP Files

When you are using Custom Web Publishing with PHP, your PHP files live in the normal web publishing folder on the web server. If you are using multiple machines, the Admin Console setup establishes the link between the web server and the machine running the Web Publishing Engine.

Dealing with Container Fields

If you use container fields that contain references to external files rather than the content of the files themselves, you need to prepare them for the Web Publishing Engine. The first step is to place the files to be inserted inside the Web folder of the FileMaker Pro folder for the version of FileMaker Pro you are using (FileMaker Pro or FileMaker Pro Advanced). Then insert the references as you normally would.

When you are ready to publish the files, move them to the web server's directory. On Windows, this is <drive>:\Inetpub\wwwroot; on Mac OS X, it is /Library/WebServer/Documents.

Using the PHP Site Assistant

After you have set up FileMaker Server and your databases for PHP publishing, you can use the PHP Site Assistant to build a site. It is highly recommended that you start here. You can take the site that is built and modify the graphics to provide your own look and feel. You can also modify the pages to show more or less data. Starting from a functioning site is a big step forward. In fact, many programmers and web developers never start from a blank page: They begin with an existing site and modify it for the task at hand.

If you are building an iPhone site, work through the assistant and do not make adjustments to the look and feel until you have tried it out. One of the biggest mistakes people make when designing for iPhone is to bring another set of interface features with them. Use the iPhone structures and graphics so that your users will feel at home.

Building the Site with the PHP Site Assistant

The following sections walk you through the process of using the PHP Site Assistant to build a site for the Task Management Starter Solution. The files generated by the PHP Site Assistant can be downloaded from the author's website at http://www.northcountryconsulting.com or from the publisher's website at www.informit.com/title/9780789747860.

In the section that follows, you see how to write your own site files and how to modify them.

Launching the PHP Site Assistant

If you have a database open in FileMaker Pro 10 or 11 Advanced, you can choose Tools, Launch PHP Assistant. You might be asked to allow access for the Java app, as shown in Figure 16.3. Click Allow.

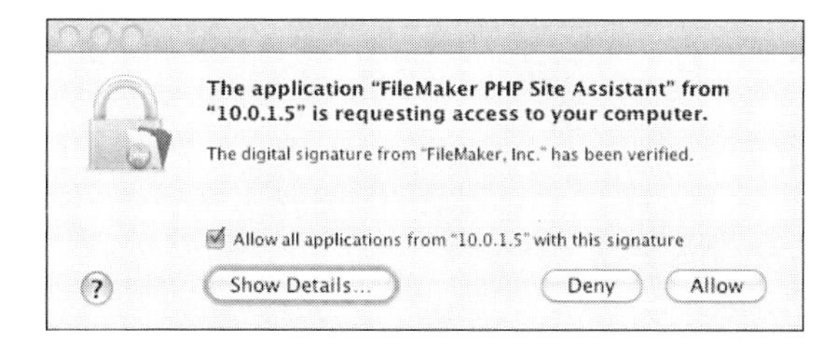

Figure 16.3
Allow access to the PHP Site Assistant.

You can also launch the PHP Site Assistant in other ways:

- There is a link to it at the bottom of the page you can access to connect to Admin Console. The URL is http://10.0.1.2:16000. Replace 10.0.1.2 with the address of your FileMaker Server computer; if you are running on the same computer, you can use 127.0.0.1 or localhost. The link takes you to the tools page from which you can launch either the XML/XSLT Site Assistant or the PHP Site Assistant.
- You can go directly to the tools page at http://10.0.1.2:16000/tools.
- If you have downloaded the PHP Site Assistant, you might have a shortcut on your desktop that you can double-click.

The basic screen opens, as shown in Figure 16.4. You have the choice of creating a new site, opening a site, or viewing help. The PHP Site Assistant process walks you through the construction of a site. At any time, you can save the site; this is the PHP Site Assistant version of the site. At the end of the process, you can then generate the site; that produces PHP files that you can edit if you want to. The ability to save the site separately from the generated PHP files gives you great flexibility.

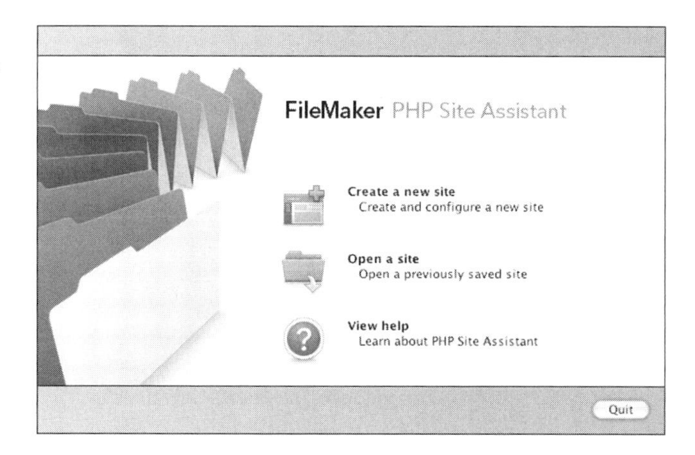

Figure 16.4
Choose what you want to do with the PHP Site Assistant.

Creating or Opening a PHP Site Project

If you are creating a new site, you will be prompted to name it. You will be able to reopen the site and continue with the PHP Site Assistant at another time if you want.

Connecting to the Web Server and Database

In the next step, shown in Figure 16.5, you select the server and connect to it. You should then see the available FileMaker databases. Choose the one you want to use.

Figure 16.5
Connect to FileMaker Server.

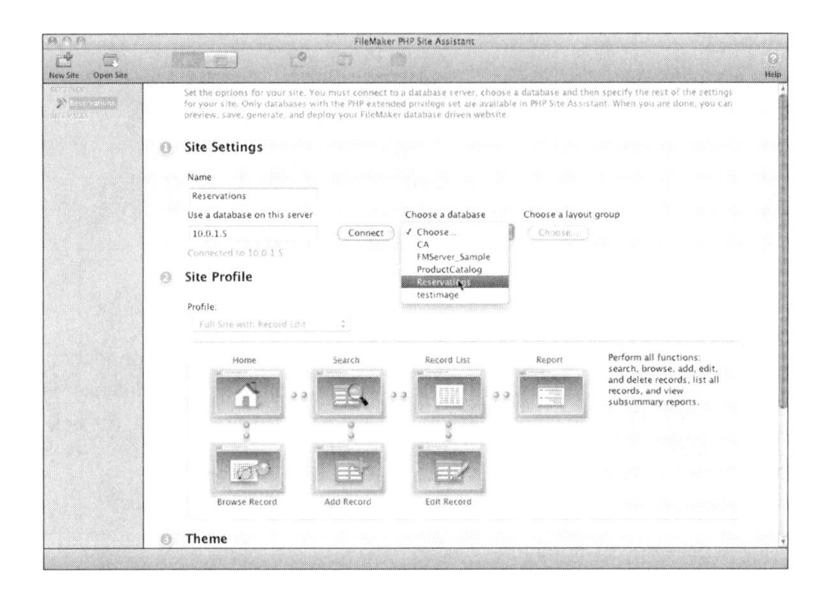

If you do not see the database you expect to see or do not see any databases, try using the Open Remote command in FileMaker Pro or FileMaker Pro Advanced to connect to the same server. If you do not see the databases, use the Admin Console to check that they are open in FileMaker Server.

If you see them in FileMaker Pro with the Open Remote command, check the PHP extended preference for the databases. Use FileMaker Pro and the Open Remote command to do this. Databases without the XSLT extended preference set will not show up in this list.

As you can see in Figure 16.6, you can choose to store the username and password in the PHP site files or to prompt each user for login information. The latter approach is more secure, and it is appropriate for internal systems. If you store a username and password inside the PHP files, anyone with access to the PHP files will be able to log in over the Internet, which might or might not be what you want.

Figure 16.6
Supply the password and choose whether to store it.

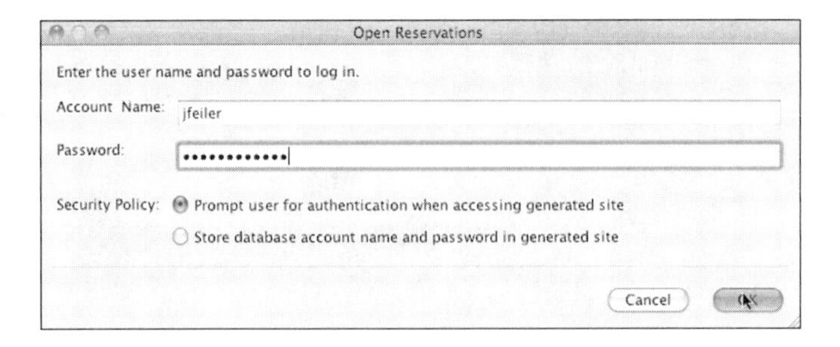

Choosing the Site Profile

The next step is to choose the type of site that you are creating. You can choose some basic sites (such as a simple data entry site) or something as complex as Full Site with Record Edit, as shown in Figure 16.6. Note that because you can build on the PHP pages created, you can create several sites and link the pages together into a single site. Note, too, that now that you have logged in to the database and chosen the type of site, the various pages are shown in the sidebar at the left of the window.

The Full Site with Record Edit is a good site to start with. It includes the following seven pages:

- Home
- Search
- Record List
- Report
- Browse Record
- Add Record
- Edit Record

You can see how they fit together and then come back to modify the site (or start over) with just the pages you need. Get the big picture first.

The pages link to one another as needed. For example, after you perform a search, the records found are presented in a list. You can click an individual record to browse it in more detail; if you want, you can decide to edit it.

Selecting a Layout Group

The next step is to choose a layout group, as shown in Figure 16.7.

Figure 16.7
Choose a layout group.

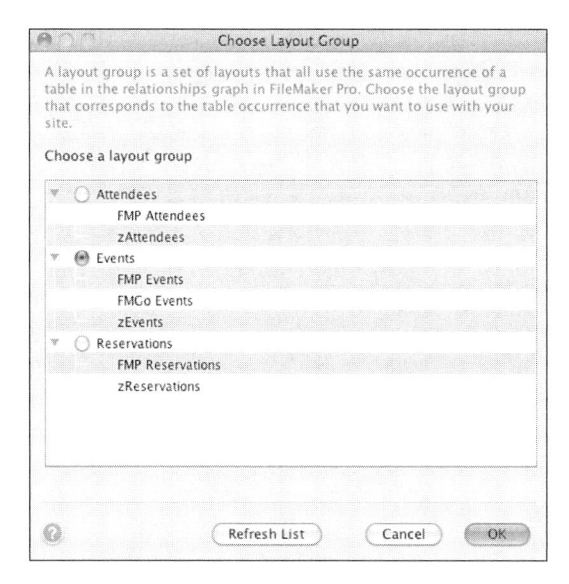

The Relationships Graph can contain multiple instances of a given table. What you are selecting here is basically a single table instance in the Relationships Graph and all the layouts based on it. In a database where the tables have no multiple occurrences, you are simply selecting the table on which to base the PHP site. Although you are choosing a base table, all the fields in the associated layouts—no matter what tables they are in—are available to you. As you can see in Figure 16.7, this database has layouts based on Attendees, Events, and Reservations. The naming conventions help you see that some tables are internal and contain all fields (these start with z), and other tables are purposed to FileMaker Go. As long as they are based on the same table occurrence, you will be able to work with all of their fields.

Selecting a Theme

You now can choose a theme for the site (this is much like iWork and Bento). If you are designing for iPhone, you want the iPhone theme. However, if you are building a generic PHP site, experiment with the various themes. The PHP Site Assistant will provide you with the appropriate styles. You can then use them as you customize the pages if you need to do so.

Specifying Options for the Home Page

After you have chosen the site profile, you can click on each page at the left in the sidebar to customize it.

The home page customization is text based and is shown in Figure 16.8. This is an example of the type of further customization you might want to do. You might want to manually add a logo to the home page in the PHP code that is generated. Alternatively, if you want the PHP Site Assistant to do everything for you, you might want to add a container field in the database for your logo so that you can simply select it as another field on each page in the PHP Site Assistant. Many people find moving a field in PHP easier than adding an image in PHP. It's your choice.

Figure 16.8
Specify home page options.

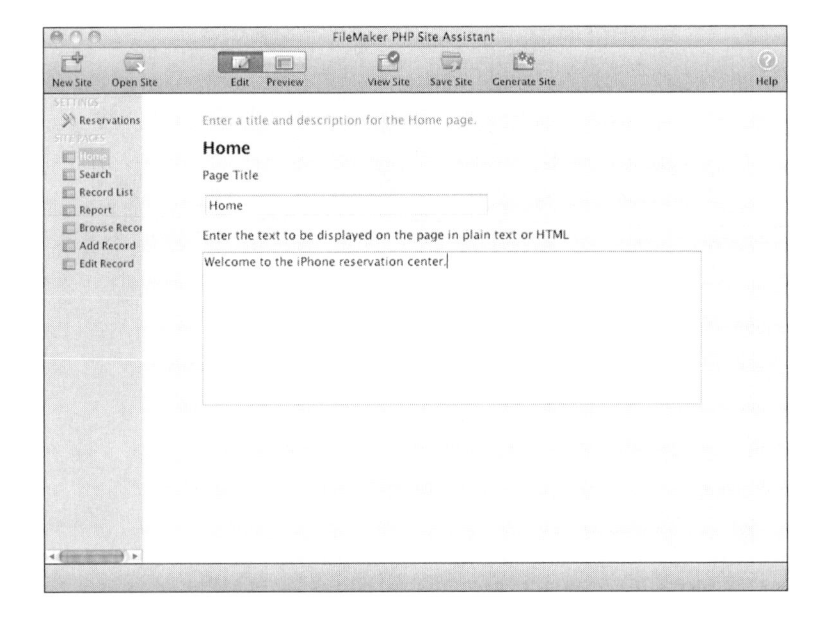

Try Out the Site on Your iPhone

This is the first moment when you can actually try out the site. There is not much on it, but you should confirm that the pieces are all in place. You can use the PHP Site Assistant to preview the site, but that will not tell you what you need to know. Instead of trying to preview the site, click Save Site at the top of the window shown in Figure 16.8.

Next, click Generate Site to actually create and deploy the site. You will be asked for the place to deploy it, as shown in Figure 16.9.

Figure 16.9
Deploy the site.

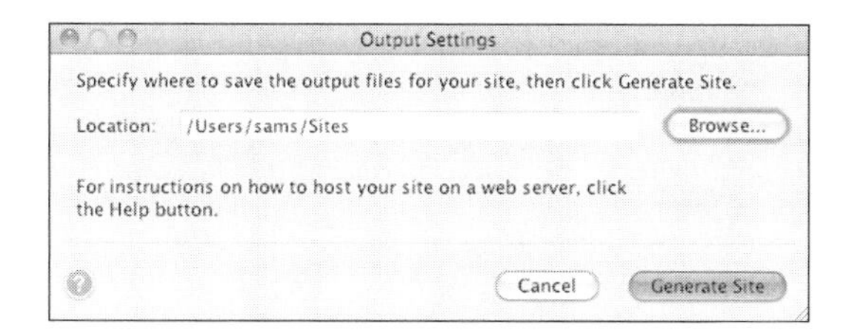

If you have web sharing turned on in System Preferences, you can deploy it in your own Sites folder or in the Sites folder for your computer (that is, for all accounts on the computer). You can also deploy it on any other web server to which you have access. For testing, the best choice is to deploy it to your own Sites folder. This folder is automatically created in your own directory. Its address is /Users/<your account name>/Sites. The site will be deployed in a folder with the name of the site; that folder will contain a number of files. The home page will be home.php. Thus, if your account name is sams, the address will be the following:

/Users/sams/Sites/<project name>

If your computer address is 10.0.1.5, the URL will be as follows:

10.0.1.5//~same/<project name>/home.php.

If you deploy it at the root level of your computer (for all accounts), the URL will be the following:

10.0.1.5/<project name>/home.php

The address on disk will be as follows:

/Webserver/Documents/<project name>/home.php

Try to connect to the site from your iPhone. The first positive sign is shown in Figure 16.10 when you are asked to log in.

Continue to the home page, and you should see the text that you typed in previously in Figure 16.8. Figure 16.11 shows that page on the iPhone.

Figure 16.10
Log in to the site.

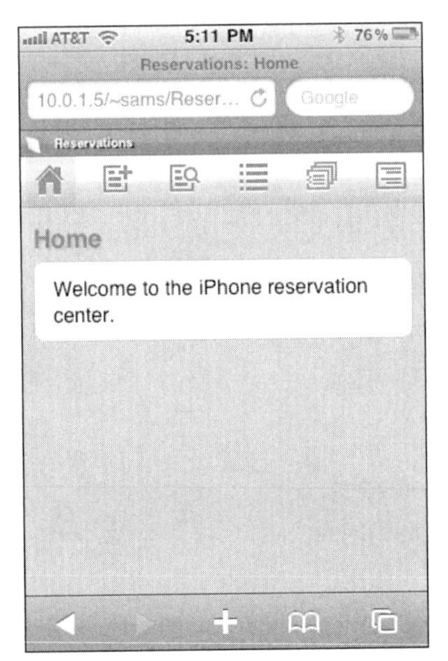

Figure 16.11
View the home page.

Explore the site you have created. It certainly needs work, but you have just moved your FileMaker database to iPhone in record-setting time.

Refine the Pages

Each page has its own customization settings, but they all are similar to the settings shown in Figure 16.12.

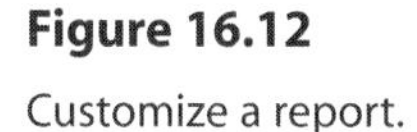

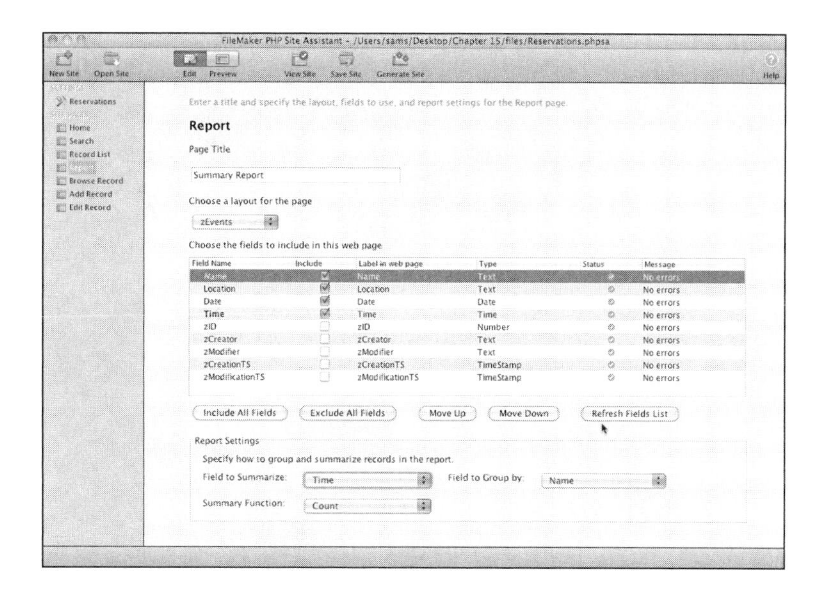

Figure 16.12

Customize a report.

First, you select the page you are interested in from the list of site profile pages at the left of the window. Each page has the same general layout. All the available fields are shown. The quickest way to move ahead at this point is to remove the fields you do not want on the web interface. Just uncheck the box for the internal fields (starting with z) and any other fields irrelevant to that page. If you want, you can change the page title or the layout on which it is based.

Figure 16.13 shows how you create an Add Record page. Compare it with Figure 16.12, and you'll see how the same basic interface is used for each type of page.

You can change field labels here instead of going into the pages after they are created. You can also move fields up or down. Every step that you take here with the graphical user interface is one fewer step to do with the actual PHP code with which you might be less familiar.

If you do not see all the fields you expect, or any fields, check the permissions in the database. In particular, check the permissions for all files involved in the layout. If you have separated the interface from the data in your solution, you will need to expose both files to PHP publishing for the fields to appear. Alternatively, you can publish only from the database file and have nothing to do with the interface because PHP is your interface.

Continue with each of the pages in the site. The same interface is used for the fields in the layout. For each page, you can specify a title, and you can choose the layout to use from among all the layouts in the layout group selected for the site.

Figure 16.13
Specify Add Record page.

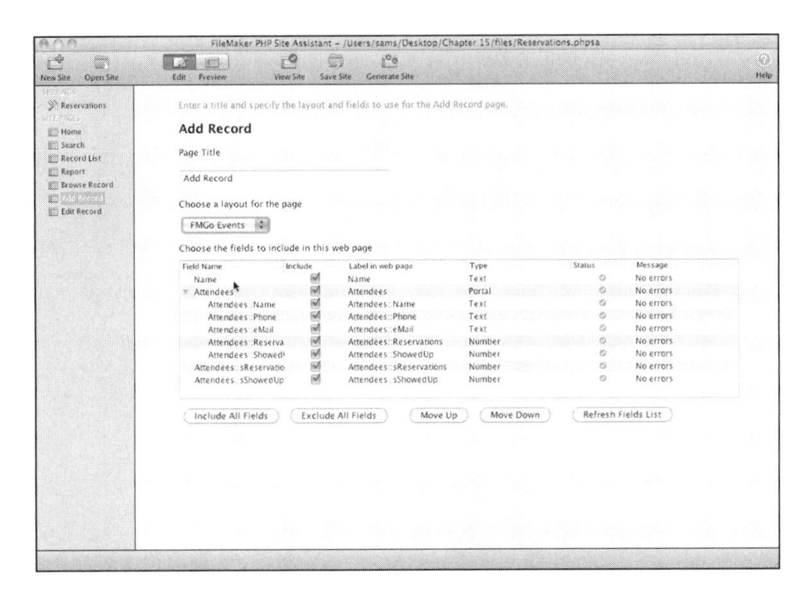

In addition to the general fields settings shown in Figures 16.12 and 16.13, the specific options for types of pages are as follows:

- **Record List.** Enables you to set the sort order and maximum number of rows.
- **Report.** Provides grouping and summarization options.
- **Browse.** Enables you to decide whether to have an edit button.
- **Add Record.** Is available when there are no other options.
- **Edit Record.** Enables you to decide whether to show delete and duplicate buttons.

Further Steps

The immediate next step is clear: experiment. Putting your FileMaker database on the web with the PHP Assistant is so fast that you can turn around various examples and tests very quickly. The best part of it is that you can share your work and get feedback from friends anywhere in the world. Just send them the URL.

The only thing to remember is that the initial setup can sometimes be tricky. If you do not have your iPhone site up and running, go back and follow the steps in this chapter. Check your permissions and security settings.

IN THIS CHAPTER

- Pairing and Connecting Bluetooth Devices
- FileMaker Go External Keyboard Conventions
- Pairing a Bluetooth Keyboard with an iOS Device
- Unpairing a Bluetooth Keyboard from a Compute

Using a Wireless Keyboard

If you are doing significant amounts of data entry on a mobile device, a wireless keyboard can be an enormous help. Bluetooth keyboards are not very expensive, and they can save a lot of time and effort for a skilled typist (or even a slightly skilled typist).

There is nothing special about FileMaker Go when it comes to connecting a wireless keyboard to your device. There are a few issues you should know about with regard to how FileMaker uses certain keys on the external keyboard.

This appendix describes those differences as well as the steps to take to connect a wireless keyboard to your device (this applies to any app on the device). In addition, you see how to disconnect a wireless keyboard from a Mac (the Windows steps are basically the same).

Pairing and Connecting Bluetooth Devices

Bluetooth devices have two aspects to their links: pairing and connections. On Mac OS X, you can see your Bluetooth devices and their status by choosing ⌘, System Preferences and clicking the Bluetooth icon. If Bluetooth status is shown in your menu bar, you can click on the Bluetooth icon there and choose Open Bluetooth Preferences to get to the same place. Either way, the dialog shown in Figure A.1 opens.

Figure A.1
Use Bluetooth preferences on Mac OS X.

To add another device, click the + at the bottom of the list at the left; to delete a device, select it and click –. The devices in this list are *paired* with your computer. Each paired device can be connected or not. The gear wheel lets you connect or disconnect a device and also change its name.

To use a Bluetooth device, it must be both paired and connected. If you use an earphone with your mobile phone, you might have seen this process in action. You normally pair the earphone with your mobile phone and leave the pairing in place. You may disconnect the earphone from the mobile phone (or it might automatically disconnect itself after a period of non-use); this conserves battery power. The process of reconnecting a Bluetooth device that is already paired is simpler than re-pairing it. (One way to notice the distinction is that pairing or re-pairing a device often requires you to type in an ID number. Connecting or re-connecting is usually just the push of a button.) As you will see, moving a Bluetooth keyboard from one device to another (for example, from your computer to your iPad) requires you to unpair it and then re-pair it.

FileMaker Go External Keyboard Conventions

Here are the ways in which FileMaker Go's use of an external keyboard differs from FileMaker Pro.

- **Modifier keys and record navigation keys.** These are not supported in FileMaker Go. Examples are the Option and Alt keys, as well as commands such as Ctrl+↓, which moves you to the next record on FileMaker Pro.
- **Eject key.** Shows or hides the on-screen keyboard (there is no optical disc drive to eject from).
- **Arrow keys.** These do not work for scrolling through value lists as they do on FileMaker Pro.
- **Tab key.** Works for fields that use the on-screen keyboard—that is text fields. With FileMaker Pro, you can tab to fields that are edited with radio buttons, checkboxes, drop-down lists as well as container

fields. When you tab to these fields, they are selected, and you can use the keyboard or arrow keys to move to specific radio buttons or checkboxes as well as to items in a drop-down list. These behaviors are not supported in FileMaker Go.

Other keys work as they do normally.

Pairing a Bluetooth Keyboard with an iOS Device

If you are moving a keyboard from another device such as your computer to your iOS device, first unpair it, as described in the following section.

First, make certain that the keyboard is turned on.

On your iPad, tap Settings and then General, as shown in Figure A.2.

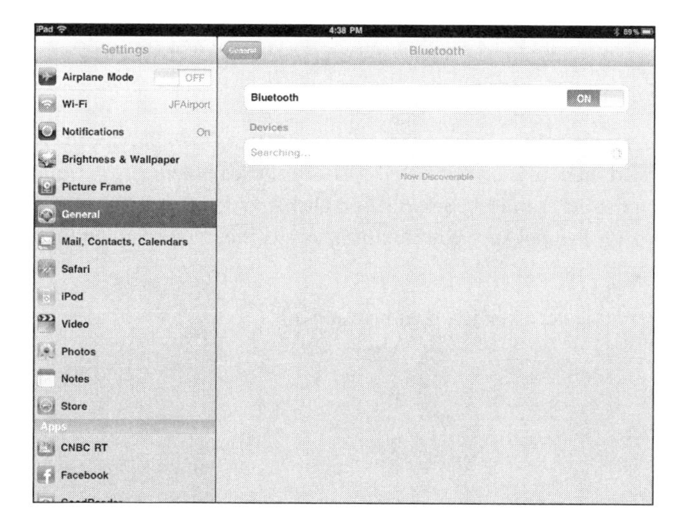

Figure A.2
Turn on Bluetooth.

Make certain that Bluetooth is turned on. Your iPad searches for nearby devices.

When it finds devices, it lists them. Tap the keyboard on the list. Your iPad might find other Bluetooth devices nearby, such as a mouse or an earphone; make certain that you pair with the keyboard. If the keyboard is not found after a short period of time (a minute is plenty), turn it off and turn it on again. Check that the power light is on when you turn it on so that you know the batteries are OK. (The power light goes out after a brief interval to conserve battery power, but the keyboard remains on.)

As you see in Figure A.3, you are given a code to enter on the keyboard. Do so.

You are ready to go.

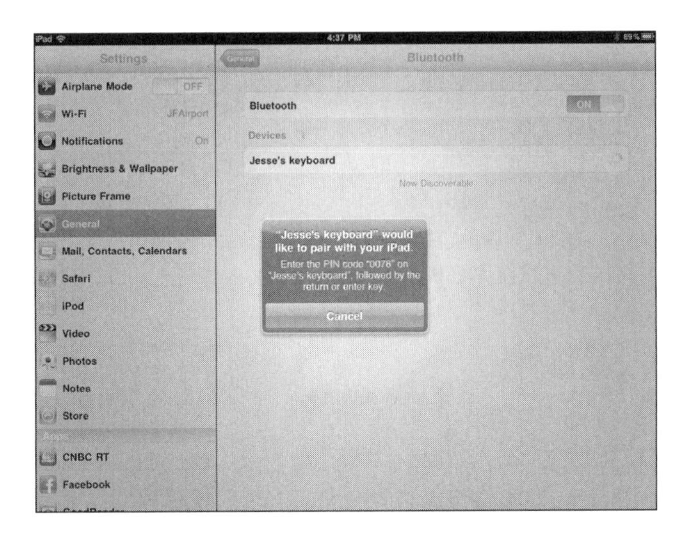

Figure A.3

Enter the keyboard PIN code.

Unpairing a Bluetooth Keyboard from a Computer

If you already have a keyboard paired with another device such as a computer, you need to unpair it.

First, make certain that the keyboard is turned on.

Open Bluetooth preferences from System Preferences or from Open Bluetooth Preferences in the menu bar if Bluetooth is displayed there (it is an option). The dialog shown previously in Figure A.1 opens. The device should be shown in the list at the left. Select it and click – to delete it from the list. This unpairs it. The connection is automatically broken because you cannot have a connection to an unpaired device. Do not simply disconnect the device.

Now you are ready to pair your keyboard with your iPad.

Index

A

B

C

J - K

M

N

O

P

Q - R

S

T

U

V

W

X - Y - Z